AQA Biology

2nd Edition

A LEVEL
YEAR 2

Glenn Toole
Susan Toole

OXFORD
UNIVERSITY PRESS

OXFORD
UNIVERSITY PRESS

Great Clarendon Street, Oxford, OX2 6DP, United Kingdom

Oxford University Press is a department of the University of Oxford. It furthers the University's objective of excellence in research, scholarship, and education by publishing worldwide. Oxford is a registered trade mark of Oxford University Press in the UK and in certain other countries

British Library Cataloguing in Publication Data
Data available

978-0-19-835770-4

10 9 8 7 6 5 4 3 2 1

MIX
Paper from
responsible sources
FSC
www.fsc.org   FSC® C007785

Paper used in the production of this book is a natural, recyclable product made from wood grown in sustainable forests. The manufacturing process conforms to the environmental regulations of the country of origin.

Printed in Great Britain by Bell and Bain Ltd, Glasgow

Message from AQA

This textbook has been approved by AQA for use with our qualification. This means that we have checked that it broadly covers the specification and we are satisfied with the overall quality. Full details of our approval process can be found on our website.

We approve textbooks because we know how important it is for teachers and students to have the right resources to support their teaching and learning. However, the publisher is ultimately responsible for the editorial control and quality of this book.

Please note that when teaching the *AQA AS or A-Level Biology* course, you must refer to AQA's specification as your definitive source of information. While this book has been written to match the specification, it cannot provide complete coverage of every aspect of the course.

A wide range of other useful resources can be found on the relevant subject pages of our website: www.aqa.org.uk.

AS/A Level course structure

This book has been written to support students studying for AQA A Level Biology. It covers the A Level Year 2 only content from the specification. The sections covered are shown in the contents list, which shows you the page numbers for the main topics within each section. There is also an index at the back to help you find what you are looking for. If you are studying for AS Biology, you will only need to know the content in the blue box for the AS exams.

AS exam

A level exam

Year 1 content

1 Biological molecules
2 Cells
3 Organisms exchange substances with their environment
4 Genetic information, variation, and relationships between organisms

Year 2 content

5 Energy transfers in and between organisms
6 Organisms respond to changes in their internal and external environment
7 Genetics, populations, evolution, and ecosystems
8 The control of gene expression

A Level exams will cover content from Year 1 and Year 2 and will be at a higher demand than the AS exams. You will also carry out practical activities throughout your course. There are **twelve** required practicals: six from the AS and six A-Level.

Contents

How to use this book

This book contains many different features. This book contains many different features. Each feature is designed to foster and stimulate your interest in Biology, as well as supporting and developing the skills you will need for your examination.

Terms that you will need to be able to define and understand are shown in **bold type** within the text.

Where terms are not explained within the same topic, they are highlighted in **bold orange text**. You can look these words up in the glossary.

Application features

These features contain important and interesting applications of biology in order to emphasise how scientists and engineers have used their scientific knowledge and understanding to develop new applications and technologies. There are also application features to develop your maths skills, with the icon √x, and to develop your practical skills, with the icon 🧪.

Extension features

These features contain material that is beyond the specification designed to stretch and provide you with a broader knowledge and understanding and lead the way into the types of thinking and areas you might study in further education. As such, neither the detail nor the depth of questioning will be required for the examinations. But this book is about more than getting through the examinations.

1 Extension and application features have questions that link the material with concepts that are covered in the specification. Answers can be found in the answers section at the back of the book.

Summary questions

1 These are short questions that test your understanding of the topic and allow you to apply the knowledge and skills you have aquired. The questions are ramped in order of difficulty.

2 √x Questions that will test and develop your mathematical and practical skills are labelled with the mathematical symbol (√x) and the practical symbol (🧪).

Introduction at the opening of each section summarises what you need to know.

A checklist to help you assess your knowledge from KS4, before starting work on the section.

Visual summaries of each section show how some of the key concepts of that section interlink with other sections.

A synoptic extension task to bring everything in the section together and start leading you towards higher study at university.

Summaries of the key practical and math skills of the section.

Section 5

[...] between organisms

[...] re a constant input of energy to maintain their highly [...] res and systems. Life depends on energy, usually [...] ing transferred continuously between organisms. [...] s live isolated lives but form part of interdependent [...] ach community interacts with other communities and with its non-living environment within ecosystems. While ecosystems as a whole remain relatively stable, their biotic and abiotic components are constantly changing. Ecosystems are maintained by light energy from the Sun that photosynthesising organisms absorb with chlorophyll to produce ATP and carbohydrates. These carbohydrates, and other substrates, are broken down by all organisms to produce the ATP needed for survival.

ATP production in both photosynthesis and respiration is formed when protons (hydrogen ions) diffuse down an electrochemical gradient through molecules of ATP synthase. This enzyme is embedded in the membranes of chloroplasts and mitochondria. As respiration is common to all organisms, and photosynthesis to all photoautotrophic ones, they provide indirect evidence for evolution.

The Sun constantly provides energy, which flows through ecosystems. In communities, molecules produced by photosynthetic organisms (producers) are consumed by other organisms such as bacteria, fungi, and animals (consumers). While the supply of energy from the Sun will be constant for the foreseeable future, other nutrients are finite and are recycled.

13 Energy and ecosystems

Working scientifically

The study of energy transfer in and between organisms provides many opportunities to carry out practical work and to develop practical skills. Required practical activities are:

- The use of chromatography to investigate the pigments isolated from leaves of different plants, e.g., leaves from shade-tolerant and shade-intolerant plants or leaves of different colours.
- Investigation into the effect of a named factor on the rate of dehydrogenase activity in extracts of chloroplasts.
- Investigation into the effect of a named variable on the rate of respiration of cultures of single-celled organisms.

In carrying out these activities you could develop practical skills such as:

- using appropriate apparatus to record a range of quantitative measurements
- using appropriate instrumentation to record quantitative measurements
- separating biological compounds using thin layer/paper chromatography
- using microbiological aseptic techniques.

You will require a range of mathematical skills. In particular the ability to recognise and make use of appropriate units in calculations, use fractions and percentages, and solve algebraic equations.

What you already know

The material in this unit is intended to be self-explanatory, but there is certain information from GCSE that will prove helpful to the understanding of this section. This information includes:

- ☐ Photosynthesis uses light energy to combine carbon dioxide and water to form glucose and oxygen.
- ☐ Light energy is absorbed by chlorophyll, which is found in chloroplasts in some plant cells and algae.
- ☐ The rate of photosynthesis may be limited by a shortage of light, low temperature or a shortage of carbon dioxide.
- ☐ Aerobic respiration takes place continuously in plants and animals during which oxygen is used to release energy from glucose.
- ☐ Aerobic respiration is summarised as: glucose + oxygen → carbon dioxide + water (+ energy)
- ☐ Energy released during respiration is used by the organism to build larger molecules from smaller ones.
- ☐ Radiation from the Sun is the source of energy for most communities of living organisms.
- ☐ The mass of living material (biomass) at each stage in a food chain is less than it was at the previous stage.
- ☐ The amounts of material and energy contained in the biomass of organisms is reduced at each successive stage in a food chain.
- ☐ Living things remove materials from the environment for growth and other processes and these materials are returned to the environment either in waste materials or when living things die and decay.
- ☐ Materials decay because they are broken down by microorganisms. The decay process releases substances that plants need to grow.
- ☐ Anaerobic respiration is the incomplete breakdown of glucose and produces lactic acid and it releases much less energy than during aerobic respiration.

2

Section 6 Organisms respond to changes in their environments

Practical skills

In this section you have met the following practical skills:

- How to carry out experiments to determine how plant growth factors such as auxins like IAA have their effects on cell growth and elongation.
- How to carry out an experiment to investigate the effects on blood sugar levels of consuming a glucose drink by diabetics and non-diabetics.

Maths skills

In this section you have met the following maths skills:

- Calculating percentage change in transmission speeds.
- Solving algebraic equations to determine the number of ATP molecules needed to contract a muscle fibre a specified distance.
- Translating information between graphical and numerical forms in calculating the number of action potentials in a given time.

Extension task

Using only the technique of a person catching a 30 cm ruler between his/her thumb and forefinger when the ruler is dropped, design and carry out a series of experiments to determine the distance travelled by the ruler before it is caught using:

1. the stimulus of sight only,
2. the stimulus of sound only,
3. the stimulus of touch only.

Using textbooks or the internet, research how to convert the distance travelled by the ruler before it is caught to the time taken for it to fall that distance. Calculate the average reaction time for each type of stimulus.

Suggest reasons for any differences you found between the reaction times for the three different stimuli.

You could also devise and carry out an experiment using the same technique to compare the reaction times between a person's dominant and non-dominant hand.

146

Mathematical section to support and develop your mathematical skills required for your course. Remember, at least 10% of your exam will involve mathematical skills.

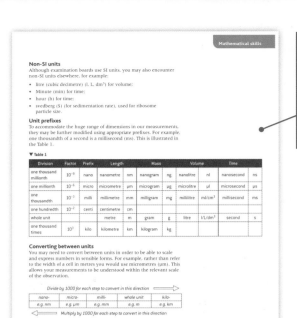

Mathematical skills

Non-SI units
Although examination boards use SI units, you may also encounter non-SI units elsewhere, for example:
- litre (cubic decimetre) (l, L, dm³) for volume;
- Minute (min) for time;
- hour (h) for time;
- svedberg (S) (for sedimentation rate), used for ribosome particle size.

Unit prefixes
To accommodate the huge range of dimensions in our measurements, they may be further modified using appropriate prefixes. For example, one thousandth of a second is a millisecond (ms). This is illustrated in the Table 1.

▼ Table 1

Division	Factor	Prefix	Length		Mass		Volume		Time	
one thousand millionth	10^{-9}	nano	nanometre	nm	nanogram	ng	nanolitre	nl	nanosecond	ns
one millionth	10^{-6}	micro	micrometre	μm	microgram	μg	microlitre	μl	microsecond	μs
one thousandth	10^{-3}	milli	millimetre	mm	milligram	mg	millilitre	ml/cm³	millisecond	ms
one hundredth	10^{-2}	centi	centimetre	cm						
whole unit			metre	m	gram	g	litre	l/L/dm³	second	s
one thousand times	10^3	kilo	kilometre	km	kilogram	kg				

Converting between units
You may need to convert between units in order to be able to scale and express numbers in sensible forms. For example, rather than refer to the width of a cell in metres you would use micrometres (μm). This allows your measurements to be understood within the relevant scale of the observation.

Divide by 1000 for each step to convert in this direction ⟶

nano- e.g. nm	micro- e.g. μm	milli- e.g. m	whole unit e.g. m	kilo- e.g. km

⟵ Multiply by 1000 for each step to convert in this direction

▲ Figure 1

Examples:
Convert 1 m to mm: 1 × 1000 = 1000 mm
Convert 1 m to μm: 1 × 1000 = 1000 mm, then 1000 × 1000 = 1 000 000 μm
Convert 11 to cm³: 1 × 1000 = 1000 cm³
Convert 20 000 μm to mm: 20 000 ÷ 1000 = 20 mm

565

Practical skills section with questions for each suggested practical on the specification. Remember, at least 15% of your exam will be based on practical skills.

Chapter 23 Practical skills

Practical skills are at the heart of Biology, a good foundation to take your skills to a higher level. Biology is a dynamic subject that constantly changes, largely as a result of developments in the specification there is a separate practical endorsement which 12 practicals across the two years of the A level course. The practicals will be reported separately on your A level certificate. Your A level as part of your A level qualification. It is assessed by your understanding and knowledge of practicals to the written account for 15% of the total assessment – the majority of

By undertaking the set practical activities in this course, manipulative skills with specific apparatus and technique a deeper understanding into the processes of scientific investigation, planning, implementing by making and processing measurements safely, analysing, and evaluating results will be reinforced and enhanced.

It is advantageous for you to answer practical questions when you have completed the practical – any questions on practical skills will have been written with the expectation that you will have carried out the practical activities. Having undertaken the practical, this helps with the teaching and learning of concepts in the specification. A richer practical experience will be gained if you do more practicals than the following twelve set practical activities in Table 1. Table 1 shows the 12 practicals which will be assessed in exams. For the practical endorsement the 12 practicals can consist of the required practicals or teacher devised practicals. For each activity, Table 1 references the relevant topic(s) in this book, the first six you will have already covered in your AS year of study.

▼ Table 1 A level required practical activities

	Practical	Topic
1	Investigation into the effect of a named variable on the rate of an enzyme-controlled reaction	1.8 Factors affecting enzyme action
2	Preparation of prepared squashes of cells from plant root tips; set-up and use of an optical microscope to identify the stages of mitosis in these stained squashes and calculation of a mitotic index	3.1 Methods of studying cells 3.7 Mitosis
3	Production of a dilution series of a solute to produce a calibration curve with which to identify the water potential of plant tissue	4.3 Osmosis
4	Investigation into the effect of a named variable on the permeability of cell-surface membranes	4.1 Structure of the cell-surface membrane
5	Dissection of animal or plant gas exchange systems, a mass transport system or of an organ within such a system	6.2 Gas exchange in insects 6.3 Gas exchange in fish 6.4 Gas exchange in the leaf of a plant 6.6 Mammalian lungs
6	Use of aseptic technique to investigate the effect of antimicrobial substances on microbial growth	9 Genetic diversity and adaptation
7	Use of chromatography to investigate the pigments isolated from leaves of different plants, e.g. leaves from shade-tolerant and shade intolerant plants or leaves of different colours	3.5.1 Photosynthesis
8	Investigation into the effect of a named factor on the rate of dehydrogenase activity in extracts of chloroplasts	3.5.1 Photosynthesis

581

Practice questions at the end of each chapter and each section, including questions that cover practical and maths skills. There are also additional practice questions at the end of the book.

Practice questions: Chapter 15

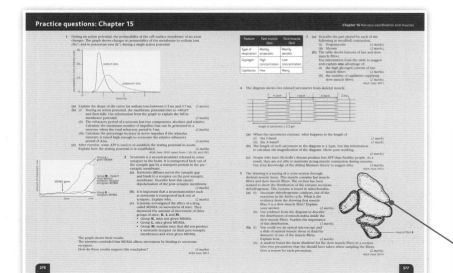

376

377

Kerboodle

This book is supported by next generation Kerboodle, offering unrivalled digital support for independent study, differentiation, assessment, and the new practical endorsement.

If your school subscribes to Kerboodle, you will also find a wealth of additional resources to help you with your studies and with revision:

- Study guides
- Maths skills boosters and calculation worksheets
- On your marks activities to help you achieve your best
- Practicals and follow up activities to support the practical endorsement
- Interactive objective tests that give question-by-question feedback
- Animations and revision podcasts
- Self-assessment checklists.

Revise with ease using the study guides to guide you through each chapter and direct you towards the resources you need.

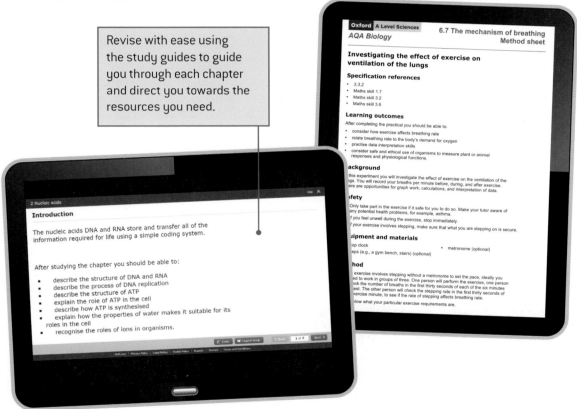

If you are a teacher reading this, Kerboodle also has plenty of further assessment resources, answers to the questions in the book, and a digital markbook along with full teacher support for practicals and the worksheets, which include suggestions on how to support and stretch your students. All of the resources that you need are pulled together into teacher guides that suggest a route through each chapter.

Section 5
Energy transfer in and between organisms

Introduction

Organisms require a constant input of energy to maintain their highly ordered structures and systems. Life depends on energy, usually from the Sun, being transferred continuously between organisms. Living organisms live isolated lives but form part of interdependent communities. Each community interacts with other communities and with its non-living environment within ecosystems. While ecosystems as a whole remain relatively stable, their biotic and abiotic components are constantly changing. Ecosystems are maintained by light energy from the Sun that photosynthesising organisms absorb with chlorophyll to produce ATP and carbohydrates. These carbohydrates, and other substrates, are broken down by all organisms to produce the ATP needed for survival.

ATP production in both photosynthesis and respiration is formed when protons (hydrogen ions) diffuse down an electrochemical gradient through molecules of ATP synthase. This enzyme is embedded in the membranes of chloroplasts and mitochondria. As respiration is common to all organisms, and photosynthesis to all photoautotrophic ones, they provide indirect evidence for evolution.

The Sun constantly provides energy, which flows through ecosystems. In communities, molecules produced by photosynthetic organisms (producers) are consumed by other organisms such as bacteria, fungi, and animals (consumers). While the supply of energy from the Sun will be constant for the foreseeable future, other nutrients are finite and are recycled.

Working scientifically

The study of energy transfer in and between organisms provides many opportunities to carry out practical work and to develop practical skills. Required practical activities are:

- The use of chromatography to investigate the pigments isolated from leaves of different plants, e.g., leaves from shade-tolerant and shade-intolerant plants or leaves of different colours.

- Investigation into the effect of a named factor on the rate of dehydrogenase activity in extracts of chloroplasts.

- Investigation into the effect of a named variable on the rate of respiration of cultures of single-celled organisms.

In carrying out these activities you could develop practical skills such as:

- using appropriate apparatus to record a range of quantitative measurements
- using appropriate instrumentation to record quantitative measurements
- separating biological compounds using thin layer/paper chromatography
- using microbiological aseptic techniques.

You will require a range of mathematical skills. In particular the ability to recognise and make use of appropriate units in calculations, use fractions and percentages, and solve algebraic equations.

What you already know

The material in this unit is intended to be self-explanatory, but there is certain information from GCSE that will prove helpful to the understanding of this section. This information includes:

○ Photosynthesis uses light energy to combine carbon dioxide and water to form glucose and oxygen.

○ Light energy is absorbed by chlorophyll, which is found in chloroplasts in some plant cells and algae.

○ The rate of photosynthesis may be limited by a shortage of light, low temperature or a shortage of carbon dioxide.

○ Aerobic respiration takes place continuously in plants and animals during which oxygen is used to release energy from glucose.

○ Aerobic respiration is summarised as: glucose + oxygen → carbon dioxide + water (+ energy)

○ Energy released during respiration is used by the organism to build larger molecules from smaller ones.

○ Radiation from the Sun is the source of energy for most communities of living organisms.

○ The mass of living material (biomass) at each stage in a food chain is less than it was at the previous stage.

○ The amounts of material and energy contained in the biomass of organisms is reduced at each successive stage in a food chain.

○ Living things remove materials from the environment for growth and other processes and these materials are returned to the environment either in waste materials or when living things die and decay.

○ Materials decay because they are broken down by microorganisms. The decay process releases substances that plants need to grow.

○ Anaerobic respiration is the incomplete breakdown of glucose and produces lactic acid and it releases much less energy than during aerobic respiration.

Learning objectives

→ Explain how the plant leaf is adapted to carry out photosynthesis.

→ Describe the main stages of photosynthesis.

Specification reference: 3.5.1

Synoptic link

The structure of ATP and of the chloroplast were considered in Topics 2.3, Energy and ATP, and 3.4, Eukaryotic cell structure, respectively. A review of these topics will help you to follow how both are linked in the process of photosynthesis as described here.

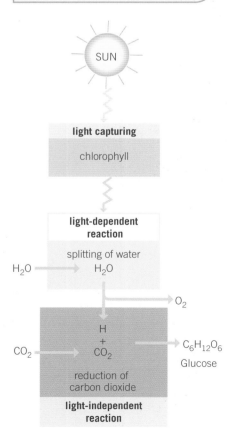

▲ **Figure 1** *Overview of photosynthesis*

Humans, along with almost every other living organism, owe their continued existence to photosynthesis. The energy we rely on, whether it comes from food when we respire or from the wood, coal, oil or gas that we burn in our homes, has been captured by photosynthesis from sunlight. Photosynthesis likewise produces the oxygen we breathe by releasing it from water molecules.

Life depends on continuous transfers of energy. How this energy enters an organism depends on its type of nutrition. In plants, energy in light is absorbed by chlorophyll and then transferred into the chemical energy of the molecules formed during photosynthesis. These molecules are used by the plant to produce **ATP** during respiration. Non-photosynthetic organisms feed on the molecules produced by plants and then also use them to make ATP during respiration.

Site of photosynthesis

The leaf is the main photosynthetic structure in eukaryotic plants. Chloroplasts are the cellular organelles within the leaf where photosynthesis takes place.

Structure of the leaf

Photosynthesis takes place largely in the leaf, the structure of which is shown in Figure 2. Leaves are adapted to bring together the three raw materials of photosynthesis (water, carbon dioxide, and light) and remove its products (oxygen and glucose). These adaptations include:

- a large surface area that absorbs as much sunlight as possible
- an arrangement of leaves on the plant that minimises overlapping and so avoids the shadowing of one leaf by another
- thin, as most light is absorbed in the first few micrometres of the leaf and the diffusion distance for gases is kept short
- a transparent **cuticle** and epidermis that let light through to the photosynthetic mesophyll cells beneath
- long, narrow upper mesophyll cells packed with chloroplasts that collect sunlight
- numerous stomata for gaseous exchange so that all mesophyll cells are only a short diffusion pathway from one
- stomata that open and close in response to changes in light intensity
- many air spaces in the lower mesophyll layer to allow rapid diffusion in the gas phase of carbon dioxide and oxygen
- a network of xylem that brings water to the leaf cells, and phloem that carries away the sugars produced during photosynthesis.

An outline of photosynthesis

The overall equation for photosynthesis is:

$$6CO_2 + 6H_2O \xrightarrow{\text{light}} C_6H_{12}O_6 + 6O_2$$

$$\text{carbon dioxide} \quad \text{water} \quad \text{glucose} \quad \text{oxygen}$$

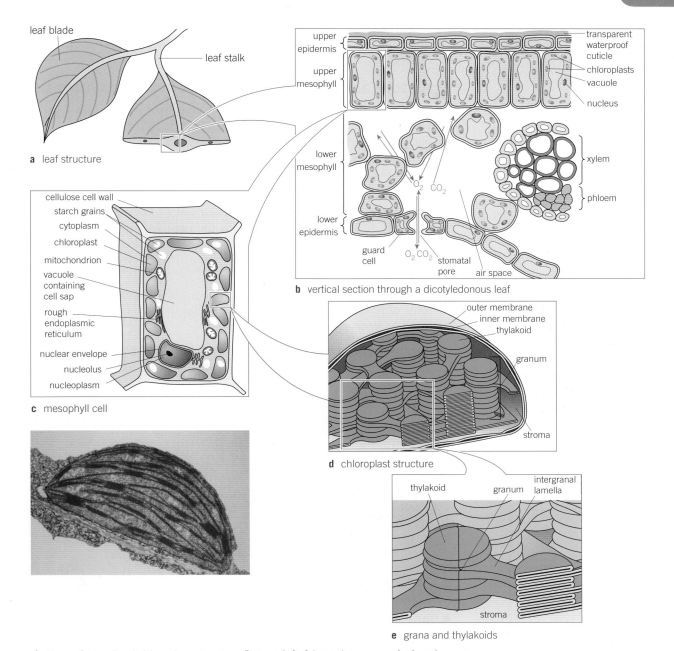

▲ **Figure 2** *Leaf and chloroplast structure. Bottom left, false colour transmission electron micrograph (TEM) of a chloroplast*

The equation shown is highly simplified. Photosynthesis is a complex metabolic pathway involving many intermediate reactions. It is a process of energy transferral in which some of the energy in light is conserved in the form of chemical bonds. There are three main stages to photosynthesis, see Figure 1:

1   **capturing of light energy** by chloroplast pigments such as chlorophyll
2   **the light-dependent reaction**, in which some of the light energy absorbed is conserved in chemical bonds. During the process an electron flow is created by the effect of light on chlorophyll, causing water to split (**photolysis**) into protons, electrons, and oxygen. The products are reduced NADP, ATP, and oxygen.

Practical link 🧪

Required practical 7. Use of chromatography to investigate pigments isolated from leaves of different plants, for example, leaves from shade-tolerant and shade-intolerant plants or leaves of different colours.

3   **the light-independent reaction**, in which these protons (hydrogen ions) are used to produce sugars and other organic molecules.

Structure and role of chloroplasts in photosynthesis

In eukaryotic plants, photosynthesis takes place within cell organelles called chloroplasts, the structure of which is shown in Figure 2d. These vary in shape and size but are typically disc-shaped, 2–10 µm long, and 1 µm in diameter. They are surrounded by a double membrane. Inside the chloroplast membranes are two distinct regions:

*   **The grana** are stacks of up to 100 disc-like structures called **thylakoids** where the light-dependent stage of photosynthesis takes place. Within the thylakoids is the photosynthetic pigment called chlorophyll. Some thylakoids have tubular extensions that join up with thylakoids in adjacent grana. These are called inter-granal lamellae.

*   **The stroma** is a fluid-filled matrix where the light-independent stage of photosynthesis takes place. Within the stroma are a number of other structures such as starch grains.

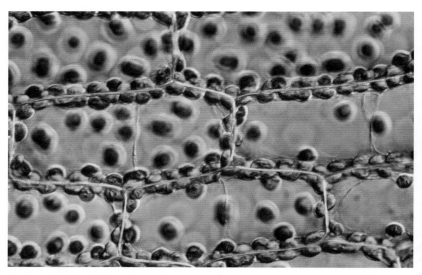

▲ **Figure 3** *Photomicrograph of a moss leaf showing cells that contain chloroplasts (green) around their margins*

Summary questions

1   List **two** molecules that are the raw materials of photosynthesis.

2   List **two** molecules that are the products of photosynthesis.

3   State in which parts of the chloroplast each of the following occur:

   a   the light-dependent reaction

   b   the light-independent reaction.

4   Name the products of each of the following:

   a   the light-dependent reaction

   b   the light-independent reaction.

The light-dependent reaction of photosynthesis involves the capture of light whose energy is used for two purposes:

- to add an inorganic phosphate (P_i) molecule to ADP, thereby making **ATP**
- to split water into H^+ ions (protons) and OH^- ions. As the splitting is caused by light, it is known as **photolysis**.

Oxidation and reduction

Before we look at what happens in the light-dependent reaction, it is necessary to understand what oxidation and reduction are.

When a substance gains oxygen or loses hydrogen the process is called **oxidation**. The substance to which oxygen has been added or hydrogen has been lost is said to be oxidised. When a substance loses oxygen, or gains hydrogen, the process is called **reduction**. In practice, when a substance is oxidised it loses electrons and when it is reduced it gains electrons. This is the more usual way to define oxidation and reduction. Oxidation results in energy being given out, whereas reduction results in it being taken in. Oxidation and reduction always take place together.

The making of ATP

When a chlorophyll molecule absorbs light energy, it boosts the energy of a pair of electrons within this chlorophyll molecule, raising them to a higher energy level. These electrons are said to be in an excited state. In fact the electrons become so energetic that they leave the chlorophyll molecule altogether. As result the chlorophyll molecule becomes ionised and so the process is called **photoionisation**. The electrons that leave the chlorophyll are taken up by a molecule called an **electron carrier**. Having lost a pair of electrons, the chlorophyll molecule has been oxidised. The electron carrier, which has gained electrons, has been reduced.

The electrons are now passed along a number of electron carriers in a series of oxidation-reduction reactions. These electron carriers form a transfer chain that is located in the membranes of the thylakoids. Each new carrier is at a slightly lower energy level than the previous one in the chain, and so the electrons lose energy at each stage. Some of this energy is used to combine an inorganic phosphate molecule with an ADP molecule in order to make ATP.

The precise mechanism by which ATP is produced can be explained by the **chemiosmotic theory**. This is described here and illustrated in Figure 1.

- Each thylakoid is an enclosed chamber into which protons (H^+) are pumped from the stroma using protein carriers in the thylakoid membrane called proton pumps.
- The energy to drive this process comes from electrons released when water molecules are split by light – photolysis of water (see later).

Learning objectives

→ Explain the processes of oxidation and reduction.

→ Explain how ATP is made during the light-dependent reaction.

→ Describe the role of photolysis in the light-dependent reaction.

→ Explain how chloroplasts are adapted to carry out the light-dependent reaction.

Specification reference: 3.5.1

Hint

Oxidation and reduction can each be described in three ways:

Oxidation – loss of electrons or loss of hydrogen or gain of oxygen.

Reduction – gain of electrons or gain of hydrogen or loss of oxygen.

Study tip

It will be useful here to revise the three ways in which something can be oxidised or reduced.

Synoptic link

An understanding of the light-independent reaction depends on knowledge of membrane structure, diffusion, and active transport. Now would be a good time to revise Topics 4.1, 4.2, and 4.4.

- The photolysis of water also produces protons which further increases their concentration inside the thylakoid space.

- Overall this creates and maintains a concentration gradient of protons across the thylakoid membrane with a high concentration inside the thylakoid space and a low concentration in the stroma.

- The protons can only cross the thylakoid membrane through ATP synthase channel proteins – the rest of the membrane is impermeable to protons. These channels form small granules on the membrane surface and so are also known as stalked granules.

- As the protons pass through these ATP synthase channels they cause changes to the structure of the enzyme which then catalyses the combination of ADP with inorganic phosphate to form ATP.

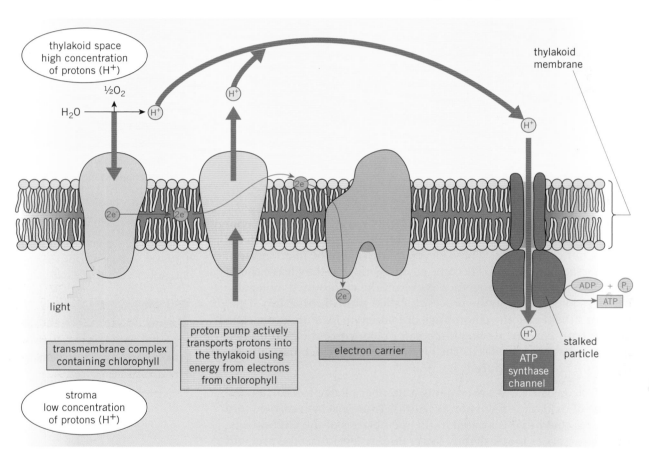

▲ **Figure 1** *Chemiosmosis in a thylakoid*

Photolysis of water

The loss of electrons when light strikes a chlorophyll molecule leaves it short of electrons. If the chlorophyll molecule is to continue absorbing light energy, these electrons must be replaced. The replacement electrons are provided from water molecules that are split using light energy. This photolysis of water also yields protons. The equation for this process is:

$$2H_2O \longrightarrow 4H^+ + 4e^- + O_2$$

water $\qquad$ protons $\quad$ electrons $\qquad$ oxygen

These protons pass out of the thylakoid space through the ATP synthase channels and are taken up by an electron carrier called NADP. On taking up the protons the NADP becomes reduced. The reduced NADP is the main product of the light-dependent stage and it enters the light-independent reaction (Topic 11.3) taking with it the electrons from the chlorophyll molecules. The reduced NADP is important because it is a further potential source of chemical energy to the plant. The oxygen by-product from the photolysis of water is either used in respiration or diffuses out of the leaf as a waste product of photosynthesis. Figure 2 summarises how ATP and reduced NADP are produced during the light-dependent stage of photosynthesis.

Hint

Reduced NADP is the most important product of the light-dependent reaction.

Hint

To picture how thylakoids are arranged in the grana, think of a thylakoid as a coin and the grana as a stack of many such coins, one on top of the other.

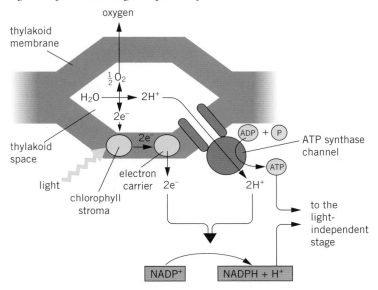

▲ **Figure 2** *Summary of light-dependent stage of photosynthesis*

Practical link

Required practical 8. Investigation into the effect of a named factor on the rate of dehydrogenase activity in extracts of chloroplasts.

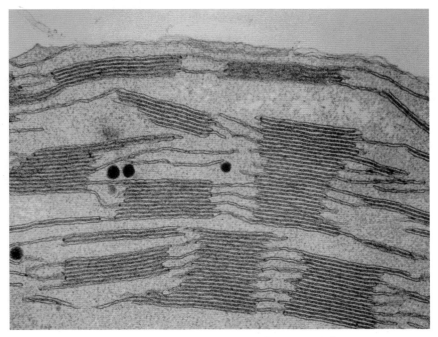

▲ **Figure 3** *False-colour TEM of grana in a chloroplast from a leaf of maize. The grana are made up of disc-like thylakoids where the light-dependent reaction of photosynthesis takes place*

Site of the light-dependent reaction

As we have seen, the light-dependent reaction of photosynthesis takes place in the thylakoids of chloroplasts. The thylakoids are disc-like structures that are stacked together in groups called grana.

Chloroplasts are structurally adapted to their function of capturing sunlight and carrying out the light-dependent reaction of photosynthesis in the following ways:

- The thylakoid membranes provide a large surface area for the attachment of chlorophyll, electron carriers and enzymes that carry out the light-dependent reaction.
- A network of proteins in the grana hold the chlorophyll in a very precise manner that allows maximum absorption of light.
- The granal membranes have ATP synthase channels within them, which catalyse the production of ATP. They are also selectively permeable which allows establishment of a proton gradient.
- Chloroplasts contain both DNA and ribosomes so they can quickly and easily manufacture some of the proteins involved in the light-dependent reaction.

Summary questions

1 State precisely where within a plant cell the electron carriers involved in the light-dependent reaction are found.

2 Describe what happens in the photolysis of water.

3 In each of the following, state whether the process involves oxidation or reduction of the molecule named.

   a An unsaturated fat molecule gains a hydrogen atom.

   b Oxygen is lost from a carbon dioxide molecule.

   c Light causes an electron to leave a chlorophyll molecule.

Chloroplasts and the light-dependent reaction $\sqrt{x}$

Figure 4 shows the structure of a chloroplast.

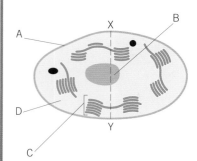

▲ Figure 4

1 Name the parts labelled A, C and D.
2 State in which of these labelled parts the light-dependent reaction takes place?
3 Structure B is used for storage. Suggest the name of the substance likely to be stored in B.
4 ATP is produced in the light-dependent reaction of photosynthesis. Suggest **two** reasons why plants cannot use this as their only source of ATP.
5 $\sqrt{x}$ The actual length of X—Y in this chloroplast is 2 μm. Calculate the magnification used in Figure 4. Show your working.

The products of the light-dependent reaction of photosynthesis, namely **ATP** and reduced **NADP**, are used to reduce glycerate 3-phosphate in the second stage of photosynthesis. Unlike the first stage, this stage does not require light directly and, in theory, occurs whether or not light is available. It is therefore called the light-independent reaction. In practice, it requires the products of the light-dependent stage and so rapidly ceases when light is absent. The light-independent reaction takes place in the stroma of the chloroplasts. The details of this stage were worked out by Melvin Calvin and his co-workers and so it is often referred to as the Calvin cycle.

The Calvin cycle

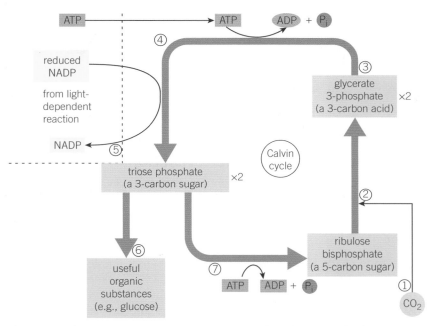

▲ **Figure 1** Summary of the light-independent reaction of photosynthesis (or Calvin cycle)

In the following account of the Calvin cycle, the numbered stages are illustrated in Figure 1.

1 Carbon dioxide from the atmosphere diffuses into the leaf through **stomata** and dissolves in water around the walls of the mesophyll cells. It then diffuses through the cell-surface membrane, cytoplasm and chloroplast membranes into the **stroma** of the chloroplast.
2 In the stroma, the carbon dioxide reacts with the 5-carbon compound **ribulose bisphosphate** (**RuBP**) a reaction catalysed by an enzyme called ribulose bisphosphate carboxylase, otherwise known as **rubisco**.
3 The reaction between carbon dioxide and RuBP produces two molecules of the 3-carbon **glycerate 3-phosphate** (**GP**).
4 Reduced NADP from the light-dependent reaction is used to reduce glycerate 3-phosphate to **triose phosphate** (**TP**) using energy supplied by ATP.

Learning objectives

→ Explain how carbon dioxide absorbed by plants is incorporated into organic molecules.

→ Describe the roles of ATP and reduced NADP in the light-independent reaction.

→ Describe the events in the Calvin cycle.

Specification reference: 3.5.1

5   The NADP is re-formed and goes back to the light-dependent reaction to be reduced again by accepting more protons.

6   Some triose phosphate molecules are converted to organic substances that the plant requires such as starch, cellulose, lipids, glucose, amino acids, and nucleotides..

7   Most triose phosphate molecules are used to regenerate ribulose bisphosphate using ATP from the light-dependent reaction.

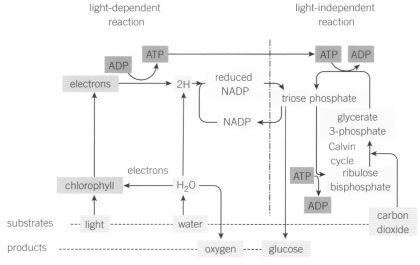

▲ **Figure 2** *Summary of photosynthesis*

Site of the light-independent reaction

The light-independent reaction of photosynthesis takes place in the stroma of the chloroplasts.

The chloroplast is adapted to carrying out the light-independent reaction of photosynthesis in the following ways:

- The fluid of the stroma contains all the enzymes needed to carry out the light-independent reaction. Stromal fluid is membrane-bound in the chloroplast which means a chemical environment which has a high concentration of enzymes and substrates can be maintained within it – as distinct from the environment of the cytoplasm.

- The stroma fluid surrounds the grana and so the products of the light-dependent reaction in the grana can readily diffuse into the stroma.

- It contains both DNA and ribosomes so it can quickly and easily manufacture some of the proteins involved in the light-independent reaction.

Summary questions

1   Describe the role of ribulose bisphosphate (RuBP) in the Calvin cycle.

2   State how the reduced NADP from the light-dependent reaction is used in the light-independent reaction.

3   Apart from reduced NADP, which other product of the light-dependent reaction is used in the light-independent reaction?

4   State precisely where in a plant cell the enzymes involved in the Calvin cycle are found.

5   Light is not required for the Calvin cycle to take place. Explain therefore why the Calvin cycle cannot take place for long in the absence of light.

 Factors affecting photosynthesis

In any complex process such as photosynthesis, the factors that affect its rate all operate together. However, the rate of the process at any given moment is not affected by all the factors, but rather by the one whose level is at the least favourable value. This factor is called a **limiting factor** because it limits the rate at which the whole process can take place. Changing only the levels of the other factors will not alter the rate of the process.

The law of limiting factors can therefore be expressed as:

  At any given moment, the rate of a physiological process is limited by the factor that is at its least favourable value.

When light is the limiting factor, the rate of photosynthesis is directly proportional to light intensity. As light intensity is increased, the volume of oxygen produced and carbon dioxide absorbed due to photosynthesis will increase to a point at which it is exactly balanced by the oxygen absorbed and the carbon dioxide produced by cellular respiration. At this point there will be no net exchange of gases into or out of the plant. This is known as the light **compensation point**. Further increases in light intensity will cause a proportional increase in the rate of photosynthesis and increasing volumes of oxygen will be given off and carbon dioxide taken up. A point will be reached at which further increases in light intensity will have no effect on photosynthesis. At this point some other factor, such as carbon dioxide concentration or temperature, is limiting the reaction. These events are illustrated in Figure 3.

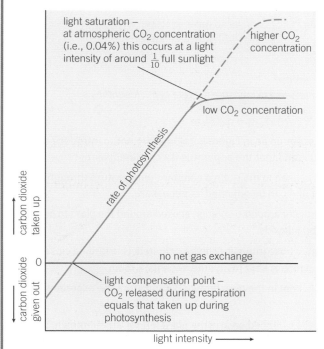

▲ **Figure 3** *Graph showing the effect of light intensity on the rate of photosynthesis as measured by the amount of CO_2 exchange*

Carbon dioxide is present in the atmosphere at a concentration of around 0.04% and is often the factor that limits the rate of photosynthesis under normal conditions. The optimum concentration of carbon dioxide for a consistently high rate of photosynthesis is 0.1% and growers of some greenhouse crops, such as tomatoes, enrich the air in the greenhouses with more carbon dioxide to provide higher yields. Carbon dioxide concentration affects enzyme activity, in particular the enzyme that catalyses the combination of ribulose bisphosphate with carbon dioxide in the light-independent reaction.

Provided that other factors are not limiting, the rate of photosynthesis increases in direct proportion to the temperature. Between the temperatures of 0 °C and 25 °C the rate of photosynthesis is approximately doubled for each 10 °C rise in temperature.

Figure 4 illustrates the influence of light intensity, carbon dioxide and temperature on the rate of photosynthesis.

1 0.1% carbon dioxide at 25 °C
2 0.04% carbon dioxide at 35 °C
3 0.04% carbon dioxide at 25 °C
4 0.04% carbon dioxide at 15 °C

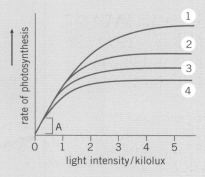

▲ **Figure 4**

1 State **one** measurement that could be taken to determine the rate of photosynthesis in this experiment.
2 Name the factor that is limiting the rate of photosynthesis over the region marked A on the graph. Explain your answer.
3 In the spring a commercial grower of tomatoes keeps his greenhouses at 25 °C and at a carbon dioxide concentration of 0.04%. The light intensity is 4 kilolux at this time of year. Using the graph, predict whether the tomato plants would grow more if the carbon dioxide level was raised to 0.1% or if the temperature was increased to 35 °C. Explain your answer.
4 Explain why there is no advantage in the grower heating his greenhouses on a dull day.
5 Using your knowledge of the light-independent reaction, explain why, at 25 °C, raising the level of carbon dioxide from 0.04% to 0.1% increases the amount of glucose produced.

▲ **Figure 5** *Student using a photosynthometer to measure the rate of photosynthesis*

 Measuring photosynthesis

The rate of photosynthesis in an aquatic plant such as Canadian pondweed (*Elodea*) can be found by measuring the volume of oxygen produced by using the apparatus (called a photosynthometer) illustrated in Figures 5 and 6.

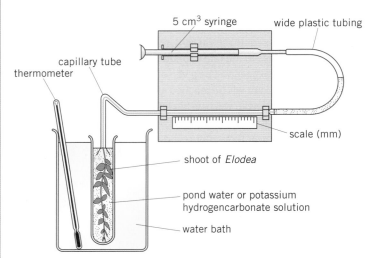

▲ **Figure 6** *Apparatus used to measure the rate of photosynthesis under various conditions*

- The apparatus is set up as in Figure 6, taking care not to introduce any air bubbles into it and that the apparatus is completely air-tight.
- The water bath is used to maintain a constant temperature throughout the experiment and can be adjusted as necessary.
- Potassium hydrogencarbonate solution is used around the plant to provide a source of carbon dioxide.
- A source of light, whose intensity can be adjusted, is arranged close to the apparatus, which is kept in an otherwise dark room.
- The apparatus is kept in the dark for two hours before the experiment begins.
- The light source is switched on and the plant left for 30 minutes to allow the air spaces in the leaves to fill with oxygen.
- Oxygen released by the plant during photosynthesis collects in the funnel end of the capillary tube above the plant.
- After 30 minutes this oxygen is drawn up the capillary tube by gently withdrawing the syringe until its volume can be measured on the scale, which is calibrated in mm^3.
- The gas is drawn up into the syringe, which is then depressed again before the process is repeated at the same light intensity four or five times, and the mean volume of oxygen produced per hour is calculated.
- The apparatus is left in the dark for 2 hours before the procedure is repeated with the light source set at a different light intensity.

1 Explain why the apparatus needs to be airtight.
2 Explain why the temperature of the water bath needs to be kept constant.
3 Suggest an advantage of providing an additional source of carbon dioxide.
4 Suggest a reason for carrying out the experiment in a room that is dark except for the light source.
5 Suggest why the plant is kept in the dark before the experiment begins.
6 Suggest why measuring the volume of gas produced by the plant in this experiment may not be an accurate measure of photosynthesis.

Using a lollipop to work out the light-independent reaction 🧪 √x̄

The details of the light-independent reaction were worked out by Melvin Calvin and his co-workers using his 'lollipop' experiment. It was so called because the apparatus, shown in Figure 7, resembled a lollipop.

In the experiment, single-celled algae are grown in the light in a thin transparent 'lollipop'. Radioactive hydrogencarbonate is injected into the 'lollipop'. This supplies radioactive carbon dioxide to the algae. At 5-second intervals, samples of the photosynthesising algae are dropped into hot methanol to stop chemical reactions instantly. The compounds in the algae are then separated out and those that are radioactive are identified. The results are given in Table 1.

▼ **Table 1**

Time / s	Substances found to be radioactive
0	carbon dioxide
5	glycerate 3-phosphate
10	glycerate 3-phosphate + triose phosphate
15	glycerate 3-phosphate + triose phosphate + glucose
20	glycerate 3-phosphate + triose phosphate + glucose + ribulose bisphosphate

- Algae are grown under light in the thin transparent lollipop.

- Radioactive ^{14}C in the form of hydrogencarbonate is injected.

- At intervals (seconds to minutes) samples of the photosynthesising algae are dropped into the hot methanol to stop chemical reactions instantly.

- The compounds in the algae are separated by two-way chromatography.

- The radioactive compounds are identified.

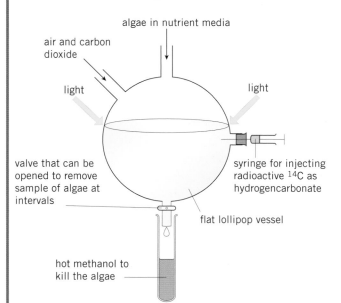

▲ **Figure 7** The 'lollipop' apparatus used by Melvin Calvin

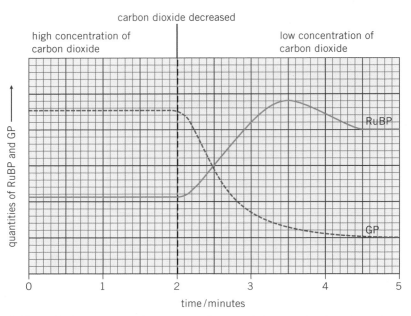

high concentration of carbon dioxide

carbon dioxide decreased

low concentration of carbon dioxide

▲ **Figure 8** *Apparatus used to measure the rate of photosynthesis under various conditions*

1 Suggest why the carbon dioxide supplied to the algae was radioactively labelled.

2 Explain how information in Table 1 provides evidence that glycerate 3-phosphate is converted into triose phosphate.

3 Suggest an explanation of how the hot methanol might stop further chemical reactions taking place.

In a further experiment, samples of algae were collected at 1-minute intervals over a period of five minutes. The quantities of glycerate 3-phosphate (GP) and ribulose bisphosphate (RuBP) were measured. At the beginning of the experiment, the concentration of carbon dioxide supplied was high. After two minutes the concentration of carbon dioxide was reduced. The graph in Figure 8 shows the results of this experiment.

4 Describe the effects on the quantities of GP and RuBP of the decrease in carbon dioxide after two minutes.

5 Suggest explanations for these changes to the levels of GP and RuBP.

1   Scientists investigated the effect of iron deficiency on the production of triose phosphate in sugar beet plants. They grew the plants under the same conditions with their roots in a liquid growth medium containing all the necessary nutrients. Ten days before the experiments, they transferred half the plants to a liquid growth medium containing no iron. The scientists measured the concentration of triose phosphate produced in these plants and in the control plants:

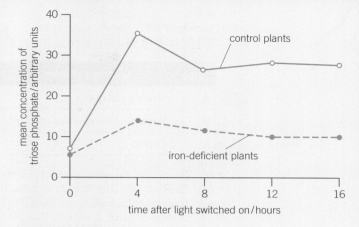

*   at the end of 6 hours in the dark
*   then for 16 hours in the light.

Their results are shown in the graph.

(a) (i)   The experiments were carried out at a high carbon dioxide concentration. Explain why. *(1 mark)*

(ii) Explain why it was important to grow the plants under the same conditions up to ten days before the experiment. *(1 mark)*

(iii) The plants were left in the dark for 6 hours before the experiment. Explain why. *(1 mark)*

(b) Iron deficiency reduces electron transport. Use this information and your knowledge of photosynthesis to explain the decrease in production of triose phosphate in the iron-deficient plants. *(4 marks)*

(c) Iron deficiency results in a decrease in the uptake of carbon dioxide. Explain why. *(2 marks)*

AQA June 2013

2   During photosynthesis, carbon dioxide reacts with ribulose bisphosphate (RuBP) to form two molecules of glycerate 3-phosphate (GP). This reaction is catalysed by the enzyme Rubisco. Rubisco can also catalyse a reaction between RuBP and oxygen to form one molecule of GP and one molecule of phosphoglycolate. Both the reactions catalysed by Rubisco are shown in **Figure 1**.

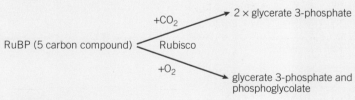

▲ **Figure 1**

(a) (i)   Where exactly in a cell is the enzyme Rubisco found? *(1 mark)*

(ii) Use the information provided to give the number of carbon atoms in **one** molecule of phosphoglycolate. *(1 mark)*

(b) Scientists investigated the effect of different concentrations of oxygen on the rate of absorption of carbon dioxide by leaves of soya bean plants. Their results are shown in **Figure 2**.
Use **Figure 1** to explain the results obtained in **Figure 2**. *(2 marks)*

(c) Use the information provided and your knowledge of the light-independent reaction to explain why the yield from soya bean plants is decreased at higher concentrations of oxygen. Phosphoglycolate is not used in the light-independent reaction. *(3 marks)*

AQA Jan 2013   ▲ **Figure 2**

3 A scientist investigated the uptake of radioactively labelled carbon dioxide in chloroplasts. She used three tubes, each containing different components of chloroplasts. She measured the uptake of carbon dioxide in each of these tubes. Her results are shown in the table.

Tube	Contents of tube	Uptake of radioactively labelled CO_2 / counts per minute
A	Stroma and grana	96 000
B	Stroma, ATP and reduced NADP	97 000
C	Stroma	4 000

(a) Name the substance which combines with carbon dioxide in a chloroplast. *(1 mark)*
(b) Explain why the results in tube **B** are similar to those in tube **A**. *(1 mark)*
(c) Use the information in the table to predict the uptake of radioactively labelled carbon dioxide if tube A was placed in the dark. Explain your answer. *(2 marks)*
(d) Use your knowledge of the light-independent reaction to explain why the uptake of carbon dioxide in tube **C** was less than the uptake in tube **B**. *(2 marks)*
(e) DCMU is used as a weed killer. It inhibits electron transfer during photosynthesis. The addition of DCMU to tube A decreased the uptake of carbon dioxide. Explain why. *(2 marks)*
AQA June 2012

4 (a) The concentrations of carbon dioxide in the air at different heights above ground in a forest changes over a period of 24 hours. Use your knowledge of photosynthesis to describe these changes and explain why they occur. *(5 marks)*
(b) In the light-independent reaction of photosynthesis, the carbon in carbon dioxide becomes carbon in triose phosphate. Describe how. *(5 marks)*
(c) Microorganisms make the carbon in polymers in a dead worm available to cells in a leaf. Describe how. *(5 marks)*
AQA June 2010

5 Scientists investigated the effects of temperature and light intensity on the rate of photosynthesis in creeping azalea. They investigated the effect of temperature on the net rate of photosynthesis at three different light intensities. They also investigated the effect of temperature on the rate of respiration. The graph shows the results.

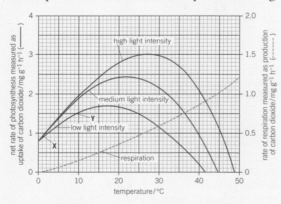

(a) (i) Name the factors that limited the rate of photosynthesis between **X** and **Y**. *(1 mark)*
(ii) Use information from the graph to explain your answer. *(2 marks)*
(b) Use information from the graph to find the gross rate of photosynthesis at 20°C and medium light intensity.

(1 mark)

(c) Creeping azalea is a plant which grows on mountains. Scientists predict that in the area where this plant grows the mean summer temperature is likely to rise from 20°C to 23°C. It is also likely to become much cloudier. Describe and explain how these changes are likely to affect the growth of creeping azalea. *(3 marks)*
AQA Jan 2011

12 Respiration
12.1 Glycolysis

We have seen in Chapter 11 that photosynthesis transfers energy in the form of sunlight into the chemical energy of carbohydrates such as glucose. We also saw, in Topic 2.3, that this glucose cannot be used directly by cells as a source of energy. Instead, cells use ATP as their immediate energy source. The formation of ATP from the break down of glucose takes place during the process of cellular respiration. There are two different forms of cellular respiration depending on whether oxygen is involved or not:

- **Aerobic respiration** requires oxygen and produces carbon dioxide, water and much ATP.
- **Anaerobic respiration** takes place in the absence of oxygen and produces lactate (in animals) or ethanol and carbon dioxide (in plants and fungi) but only a little ATP in both cases.

Aerobic respiration can be divided into four stages:

1 **glycolysis** – the splitting of the 6-carbon glucose molecule into two 3-carbon pyruvate molecules
2 **link reaction** – the 3-carbon pyruvate molecules enter into a series of reactions which lead to the formation of acetylcoenzyme A, a 2-carbon molecule.
3 **Krebs cycle** – the introduction of acetylcoenzyme A into a cycle of oxidation-reduction reactions that yield some ATP and a large quantity of reduced NAD and FAD (Topic 12.2)
4 **oxidative phosphorylation** – the use of the electrons, associated with reduced NAD and FAD, released from the Krebs cycle to synthesise ATP with water produced as a by-product.

The main respiratory pathways are summarised in Figure 1.

Glycolysis is the initial stage of both aerobic and anaerobic respiration. It occurs in the cytoplasm of all living cells and is the process by which

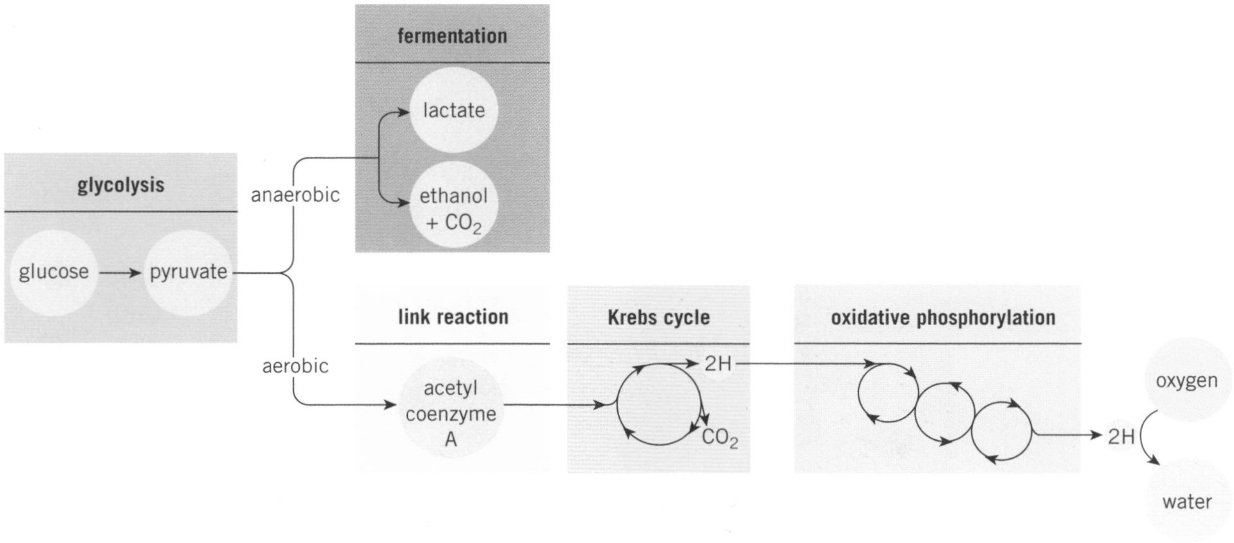

Learning objectives
→ Outline where glycolysis fits into the overall process of respiration.
→ Describe the main stages of glycolysis and its products.

Specification reference: 3.5.2

▲ **Figure 1** *Summary of respiratory pathways*

a hexose (6-carbon) sugar, usually glucose, is split into two molecules of the 3-carbon molecule, pyruvate. Although there are a number of smaller enzyme-controlled reactions in glycolysis, these can be conveniently grouped into four stages:

1 **phosphorylation of glucose to glucose phosphate**. Before it can be split into two, glucose must first be made more reactive by the addition of two phosphate molecules (phosphorylation). The phosphate molecules come from the **hydrolysis** of two ATP molecules to ADP. This provides the energy to activate glucose and lowers the **activation energy** for the enzyme-controlled reactions that follow (Topic 1.7).

2 **splitting of the phosphorylated glucose**. Each glucose molecule is split into two 3-carbon molecules known as triose phosphate.

3 **oxidation of triose phosphate**. Hydrogen is removed from each of the two triose phosphate molecules and transferred to a hydrogen-carrier molecule known as NAD to form reduced NAD.

4 **the production of ATP**. Enzyme-controlled reactions convert each triose phosphate into another 3-carbon molecule called pyruvate. In the process, two molecules of ATP are regenerated from ADP.

The events of glycolysis are summarised in Figure 2.

Hint

It must be remembered that for each molecule of glucose at the start of the process, two molecules of triose phosphate are produced. Therefore the yields must be doubled, that is, four molecules of ATP and two molecules of reduced NAD.

Hint

Gluc**ose**, a sugar, is oxidised to pyruv**ate**, an acid.

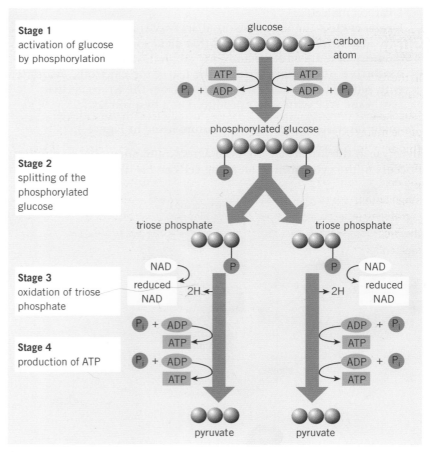

▲ **Figure 2** *Summary of glycolysis*

Energy yields from glycolysis

The overall yield from one glucose molecule undergoing glycolysis is therefore:

- two molecules of ATP (four molecules of ATP are produced, but two were used up in the initial phosphorylation of glucose and so the net increase is two molecules)
- two molecules of reduced NAD (these have the potential to provide energy to produce more ATP as we shall see in Topic 12.3)
- two molecules of pyruvate.

Glycolysis is a universal feature of every living organism and therefore provides indirect evidence for evolution. The enzymes for the glycolytic pathway are found in the cytoplasm of cells and so glycolysis does not require any organelle or membrane for it to take place. It does not require oxygen and therefore it can take place whether or not it is present. In the absence of oxygen the pyruvate produced by glycolysis can be converted into either lactate or ethanol during anaerobic respiration. This is necessary in order to re-oxidise NAD so that glycolysis can continue. This is explained, along with details of the reactions, in Topic 12.4. Anaerobic respiration, however, yields only a small fraction of the potential energy stored in the pyruvate molecule. In order to release the remainder of this energy, most organisms use oxygen to break down pyruvate further.

Summary questions

In the following passage, state the most suitable word to replace each of the numbers 1–10.

Glycolysis takes place in the (1) of cells and begins with the activation of the main respiratory substrate, namely the hexose sugar called (2). This activation involves the addition of two (3) molecules provided by two molecules of (4). The resultant activated molecule is known as (5) and in the next stage of glycolysis it is split into two molecules called (6). The third stage entails the oxidation of these molecules by the removal of (7), which is transferred to a carrier called (8). The final stage is the production of the 3-carbon molecule (9), which also results in the formation of two molecules of (10).

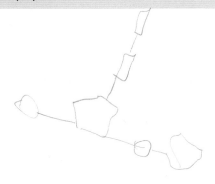

The pyruvate molecules produced during glycolysis possess potential energy that can only be released in a process called the Krebs cycle. Before they can enter the Krebs cycle, these pyruvate molecules must first be oxidised in a procedure known as the **link reaction**. In eukaryotic cells both the Krebs cycle and the link reaction take place exclusively inside mitochondria.

The link reaction

The pyruvate molecules produced in the cytoplasm during glycolysis are actively transported into the matrix of mitochondria. Here pyruvate undergoes a series of reactions during which the following changes take place:

- The pyruvate is oxidised to acetate. In this reaction, the 3-carbon pyruvate loses a carbon dioxide molecule and two hydrogens. These hydrogens are accepted by NAD to form reduced NAD, which is later used to produce ATP (Topic 12.4).

- The 2-carbon acetate combines with a molecule called coenzyme A (CoA) to produce a compound called **acetylcoenzyme A**.

The overall equation can be summarised as:

$$\text{pyruvate} + \text{NAD} + \text{CoA} \rightarrow \text{acetyl CoA} + \text{reduced NAD} + CO_2$$

The Krebs cycle

The Krebs cycle was named after the British biochemist, Hans Krebs, who worked out its sequence. The Krebs cycle involves a series of oxidation-reduction reactions that take place in the matrix of mitochondria. Its events are illustrated in Figure 1 and can be summarised as follows:

- The 2-carbon acetylcoenzyme A from the link reaction combines with a 4-carbon molecule to produce a 6-carbon molecule.

- In a series of reactions this 6-carbon molecule loses carbon dioxide and hydrogen to give a 4-carbon molecule and a single molecule of ATP produced as a result of substrate-level phosphorylation (Topic 2.3).

- The 4-carbon molecule can now combine with a new molecule of acetylcoenzyme A to begin the cycle again.

For each molecule of pyruvate, the link reaction and the Krebs cycle therefore produce:

- **reduced coenzymes** such as NAD and FAD. These have the potential to provide energy to produce ATP molecules by oxidative phosphorylation (Topic 12.3) and are therefore the important products of Krebs cycle

- one molecule of ATP

- three molecules of carbon dioxide.

As two pyruvate molecules are produced for each original glucose molecule, the yield from a single glucose molecule is double the quantities above.

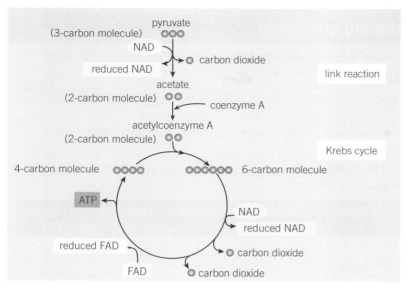

▲ **Figure 1** *Summary of the link reaction and the Krebs cycle*

Hint

Only a small amount of ATP is formed directly by the Krebs cycle. The vast majority of potential energy is carried away from the Krebs cycle by reduced NAD and reduced FAD and only later converted to ATP.

A summary of the link reaction and the Krebs cycle is shown in Figure 1.

Coenzymes

Despite their name, coenzymes are not enzymes. They are molecules that some enzymes require in order to function. Coenzymes play a major role in photosynthesis and respiration where they carry hydrogen atoms from one molecule to another. Examples include:

- **NAD**, which is important throughout respiration
- **FAD**, which is important in the Krebs cycle
- **NADP**, which is important in photosynthesis (Topic 11.2).

In respiration, NAD is the most important carrier. It works with dehydrogenase enzymes that catalyse the removal of hydrogen atoms from substrates and transfer them to other molecules involved in oxidative phosphorylation (Topic 12.3).

The significance of the Krebs cycle

The Krebs cycle performs an important role in the cells of organisms for four reasons:

- It breaks down macromolecules into smaller ones – pyruvate is broken down into carbon dioxide.
- It produces hydrogen atoms that are carried by NAD to the electron transfer chain and provide energy for oxidative phosphorylation. This leads to the production of ATP that provides metabolic energy for the cell.
- It regenerates the 4-carbon molecule that combines with acetylcoenzyme A, which would otherwise accumulate.
- It is a source of intermediate compounds used by cells in the manufacture of other important substances such as fatty acids, amino acids and chlorophyll.

Hint

The breakdown products of lipids and amino acids can enter the Krebs cycle as respiratory substrates. See Topic 12.3.

Summary questions

1 State how many carbon molecules there are in a single molecule of pyruvate.

2 Name the 2-carbon molecule that pyruvate is converted to during the link reaction.

3 State precisely in which part of the cell the Krebs cycle takes place.

4 Table 1 lists statements about some biochemical processes in a plant cell. State whether each of the letters a–r represents true or false.

▼ Table 1

Statement	Glycolysis	Krebs cycle	Light-dependent reaction of photosynthesis
ATP is produced	a	b	c
ATP is needed	d	e	f
NAD is reduced	g	h	i
NADP is reduced	j	k	l
CO_2 is produced	m	n	o
CO_2 is needed	p	q	r

Coenzymes in respiration

Coenzymes such as NAD are important in respiration. They help enzymes to function by carrying hydrogen atoms from one molecule to another. Scientists can model the way coenzymes work in cells using a blue dye called methylene blue. It can accept hydrogen atoms and so become reduced. Reduced methylene blue is colourless.

methylene blue + hydrogen ⟶ reduced methylene
(blue colour)   blue (colourless)

In an investigation into respiration in yeast, three test tubes were set up as follows:

Tube A	Tube B	Tube C
2 cm^3 yeast suspension	2 cm^3 distilled water	2 cm^3 yeast suspension
2 cm^3 glucose solution	2 cm^3 glucose solution	2 cm^3 distilled water
1 cm^3 methylene blue	1 cm^3 methylene blue	1 cm^3 methylene blue

All three tubes were incubated at a temperature of 30 °C. The colour of each tube was recorded at the start of the experiment and after 5 and 15 minutes. The results are shown in the table below:

Time / min	Colour of tube contents		
	Tube A	Tube B	Tube C
0	blue	blue	blue
5	colourless	blue	blue
15	colourless	blue	pale blue

1 Tube B acts as a control. Explain why this control was necessary in this investigation.

2 Using your knowledge of respiration, suggest an explanation for the colour change after 15 minutes in:
   a tube A   b tube C.

3 How might the results in tube A after 15 minutes have been different if the experiment had been carried out at 70 °C? Explain your answer.

4 After 20 minutes the contents of tube A were mixed with air by shaking it vigorously, turning the methylene blue back to a blue colour. Suggest a reason for this colour change.

5 Suggest why conclusions made only on the basis of the results of this experiment may not be reliable.

12.3 Oxidative phosphorylation

So far in the process of **aerobic** respiration, we have seen how hexose sugars such as glucose are split (glycolysis) and how the 3-carbon pyruvate that results is fed into the Krebs cycle to yield carbon dioxide and hydrogen atoms. The carbon dioxide is a waste product and is removed during the process of gaseous exchange. The hydrogen atoms (or more particularly the electrons they possess) are valuable as a potential source of energy. These hydrogen atoms are carried by the coenzymes NAD and FAD into the next stage of the process, **oxidative phosphorylation**. This is the mechanism by which some of the energy of the **electrons** within the hydrogen atoms is conserved in the formation of **adenosine triphosphate (ATP)**.

Oxidative phosphorylation and mitochondria

Mitochondria are organelles that are found in **eukaryotic cells**. Each mitochondrion is bounded by a smooth outer membrane and an inner one that is folded into extensions called cristae. The inner space, or matrix, of the mitochondrion contains proteins lipids, and traces of DNA.

Mitochondria are the site of oxidative phosphorylation. Within the inner folded membrane (cristae) are the enzymes and other proteins involved in oxidative phosphorylation and hence ATP synthesis.

As mitochondria play a vital role in respiration it is hardly surprising that they occur in greater numbers in metabolically active cells, such as those of the muscles, liver and epithelial cells, which carry out active transport. The mitochondria in these cells also have more densely packed cristae which provide a greater surface area of membrane incorporating enzymes and other proteins involved in oxidative phosphorylation.

The electron transfer chain and the synthesis of ATP

The synthesis of ATP by oxidative phosphorylation involves the transfer of electrons down a series of electron carrier molecules which together form the **electron transfer chain**. The process takes place as follows:

- The hydrogen atoms produced during glycolysis and the Krebs cycle combine with the coenzymes NAD and FAD.
- The reduced NAD and FAD donate the electrons of the hydrogen atoms they are carrying to the first molecule in the electron transfer chain.
- The electrons pass along a chain of electron transfer carrier molecules in a series of **oxidation-reduction** reactions. As the electrons flow along the chain, the energy they release causes the active transport of protons across the inner mitochondrial membrane and into inter-membranal space.
- The protons accumulate in the inter-membranal space before they diffuse back into the mitochondrial matrix through ATP synthase channels embedded in the inner mitochondrial membrane.

Learning objectives
→ Describe where oxidative phosphorylation takes place.
→ Explain how ATP is synthesised during oxidative phosphorylation.
→ Explain the role of oxygen in aerobic respiration.

Specification reference: 3.5.2

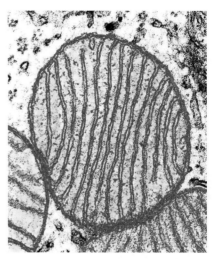

▲ **Figure 1** *Coloured TEM of a sectioned mitochondrion (red and yellow). It has two membranes: an outer surrounding membrane and an inner membrane that forms folds called cristae, seen here as red lines. The cristae are the sites of oxidative phosphorylation*

Synoptic link

Re-reading about mitochondria in Topic 3.4 and membranes in Topic 4.1 will help you follow the processes of the electron transport chain.

Hint

Remember that a single hydrogen atom is made up of one proton (H^+) and one electron (e^-).

- At the end of the chain the electrons combine with these protons and oxygen to form water. Oxygen is therefore the final acceptor of electrons in the electron transfer chain.

The process described above is the chemiosmotic theory of oxidative phosphorylation and is summarised in Figure 2. You will notice that it involves the same types of processes as those used to explain photophosphorylation in the light-dependent stage of photosynthesis (Topic 11.2).

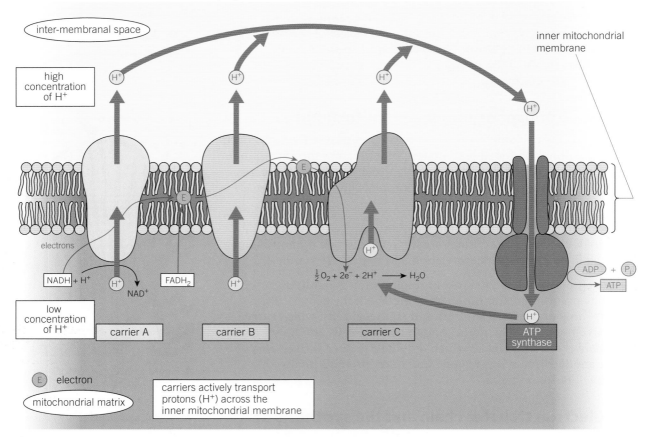

▲ **Figure 2** *Summary of the chemiosmotic theory of oxidative phosphorylation*

The importance of oxygen in respiration is to act as the final acceptor of the hydrogen atoms produced in glycolysis and the Krebs cycle. Without its role in removing hydrogen atoms at the end of the chain, the hydrogen ions (protons) and electrons would 'back up' along the chain and the process of respiration would come to a halt.

Releasing energy in stages

In general, the greater the energy that is released in a single step, the more of it is released as heat and the less there is available for more useful purposes. When energy is released a little at a time, more of it can be harvested for the benefit of the organism. For this reason, the electrons carried by NAD and FAD are not transferred in one explosive step. Instead they are passed along a series of electron transfer carrier molecules, each of which is at a slightly lower energy level (Figure 3). The electrons therefore move down an energy gradient. The transfer of electrons down this gradient allows their energy to be released gradually and therefore more usefully.

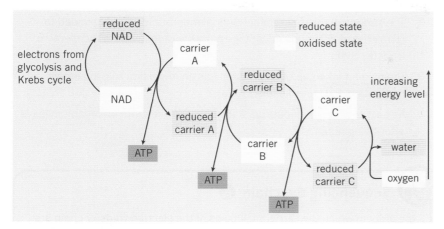

▲ **Figure 3** *Summary of electron transfer chain*

Alternative respiratory substrates

Sugars are not the only substances which can be oxidised by cells to release energy. Both lipids and protein may, in certain circumstances, be used as respiratory substrates, without first being converted to carbohydrate.

Respiration of lipids

Before being respired, lipids are first hydrolysed to glycerol and fatty acids. The glycerol is then phosphorylated and converted to triose phosphate which enters the glycolysis pathway and subsequently the Krebs cycle. The fatty acid component is broken down into 2-carbon fragments which are converted to acetyl coenzyme A. This then enters the Krebs cycle.

The oxidation of lipids produces 2-carbon fragments of carbohydrate and many hydrogen atoms. The hydrogen atoms are used to produce ATP during oxidative phosphorylation. For this reason lipids release more than double the energy of the same mass of carbohydrate.

Respiration of protein

Protein is another potential source of energy. It is first hydrolysed to its constituent amino acids. These have their amino group removed (deamination) before entering the respiratory pathway at different points depending on the number of carbon atoms they contain. 3-carbon compounds are converted to pyruvate, while 4- and 5-carbon compounds are converted to intermediates in the Krebs cycle.

Practical link

Required practical 9. Investigation into the effect of a named variable on the rate of respiration of cultures of single-celled organisms.

Study tip

Oxygen is used as the final acceptor of hydrogen atoms at the end of the electron transfer chain. It is therefore used to form water and not carbon dioxide, as you may think.

Summary questions

1 The processes that occur in the electron transfer chain are also known as oxidative phosphorylation. Suggest why this term is used.

2 The surface of the inner mitochondrial membrane is highly folded to form cristae. State **one** advantage of this arrangement to the electron transfer chain.

3 The oxygen taken up by organisms has an important role in aerobic respiration. Explain this role.

4 As part of which molecule does the oxygen taken into an organism leave after being respired?

 Sequencing the chain

The order in which the carrier molecules of the electron transfer chain are arranged can be determined experimentally. The experiments rely on the fact that each transfer of electrons between one molecule and the next is catalysed by a specific enzyme. In a series of experiments, three different inhibitors, 1, 2, and 3, are added to four electron transfer molecules, A, B, C and D. Table 1 shows whether the molecules A–D are oxidised or reduced after the inhibitor is added.

▼ Table 1

Inhibitor added	Electron transfer molecules			
	A	B	C	D
1	reduced	oxidised	reduced	oxidised
2	oxidised	oxidised	reduced	oxidised
3	reduced	oxidised	reduced	reduced

1 Using the information in the table, state the order of the electron transfer molecules in this chain. Explain your answer.

12.4 Anaerobic respiration

We saw in Topic 12.3, that oxygen is needed if the hydrogen atoms produced in **glycolysis** and the **Krebs cycle** are to be used in the production of ATP. What happens if oxygen is temporarily or permanently unavailable to a tissue or a whole organism?

In the absence of oxygen, neither the Krebs cycle nor the electron transfer chain can continue because soon all the FAD and NAD will be reduced. No FAD or NAD will be available to take up the H^+ produced during the Krebs cycle and so the enzymes stop working. This leaves only the anaerobic process of glycolysis as a potential source of ATP. For glycolysis to continue, its products of pyruvate and hydrogen must be constantly removed. In particular, the hydrogen must be released from the reduced NAD in order to regenerate **NAD**. Without this, the already tiny supply of NAD in cells will be entirely converted to reduced NAD, leaving no NAD to take up the hydrogen newly produced from glycolysis. Glycolysis will then grind to a halt. The replenishment of NAD is achieved by the pyruvate molecule from glycolysis accepting the hydrogen from reduced NAD. The oxidised NAD produced can then be used in further glycolysis.

In eukaryotic cells, only two types of anaerobic respiration occur with any regularity:

- In plants, and in microorganisms such as yeast, the pyruvate is converted to ethanol and carbon dioxide.
- In animals, the pyruvate is converted to lactate.

Production of ethanol in plants and some microorganisms

Anaerobic respiration leading to the production of ethanol occurs in organisms such as certain bacteria and fungi (e.g., yeast) as well as in some cells of higher plants, for example, root cells under waterlogged conditions.

The pyruvate molecule formed at the end of glycolysis loses a molecule of carbon dioxide and accepts hydrogen from reduced NAD to produce ethanol. The summary equation for this is:

$$\text{pyruvate} + \text{reduced NAD} \rightarrow \text{ethanol} + \text{carbon dioxide} + \text{oxidised NAD}$$

This form of anaerobic respiration in yeast has been exploited by humans for thousands of years in the brewing industry. In brewing, ethanol is the important product. Yeast is grown in anaerobic conditions in which it ferments natural carbohydrates in plant products, such as grapes (wine production) or barley seeds (beer production) into ethanol.

Production of lactate in animals

Anaerobic respiration leading to the production of lactate occurs in animals as a means of overcoming a temporary shortage of oxygen. Clearly, such a mechanism has considerable survival value, for example, in a baby mammal in the period immediately after birth, and in an animal living in water where the amount of oxygen may sometimes be very low.

Learning objectives

→ Explain how energy is released by respiration in the absence of oxygen.

→ Explain how ethanol is produced by anaerobic respiration.

→ Explain how lactate is produced by anaerobic respiration.

Specification reference: 3.5.2

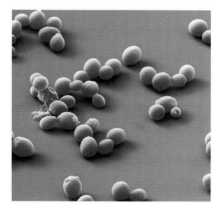

▲ **Figure 1** *Coloured SEM of yeast cells. Yeast produces ethanol and carbon dioxide during anaerobic respiration making it useful in brewing*

Hint

During strenuous exercise, muscles carry out aerobic respiration. If this cannot supply ATP fast enough, they also carry out some anaerobic respiration as well. It is a case not of one or the other but of both together.

▲ **Figure 2** *During strenuous exercise, muscles may temporarily respire anaerobically*

Summary questions

The diagram below shows the relationship between some respiratory pathways.

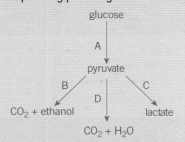

1 State which of the pathways, A, B, C or D, apply to each of the following statements. There may be more than one answer in each case.

 a Only occurs in the presence of oxygen.

 b Takes place in animals.

 c Produces ATP.

 d Is carried out by yeast in the absence of oxygen.

 e Produces reduced NAD.

 f Regenerates NAD from reduced NAD.

 g Is known as glycolysis.

However, lactate production occurs most commonly in muscles as a result of strenuous exercise. In these conditions oxygen may be used up more rapidly than it can be supplied and therefore an oxygen debt occurs. It is often essential, however, that the muscles continue to work despite the shortage of oxygen, for example, if the organism is fleeing from a predator. When oxygen is in short supply, NAD from glycolysis can accumulate and must be removed. To achieve this, each pyruvate molecule produced takes up the two hydrogen atoms from the reduced NAD produced in glycolysis to form lactate as shown below:

pyruvate + reduced NAD → lactate + oxidised NAD

At some point the lactate produced is oxidised back to pyruvate. This can then be either further oxidised to release energy or converted into glycogen. This happens when oxygen is once again available. In any case, lactate will cause cramp and muscle fatigue if it is allowed to accumulate in the muscle tissue. As lactate is an acid it also causes pH changes which affects enzymes. Although muscle has a certain tolerance to lactate, it is nevertheless important that it is removed by the blood and taken to the liver to be converted to glycogen. Figure 3 shows how the NAD needed for glycolysis to continue is regenerated in both common forms of anaerobic respiration.

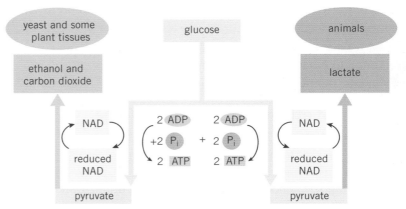

▲ **Figure 3** *How the NAD needed for glycolysis is regenerated in various organisms*

Energy yields from anaerobic and aerobic respiration

Energy from cellular respiration is derived in two ways:

- substrate-level phosphorylation in glycolysis and the Krebs cycle. This is the direct transfer of phosphate from a respiratory intermediate to ADP to produce ATP.

- oxidative phosphorylation in the electron transfer chain. This is the indirect linking of energy from phosphate to ADP to produce ATP involving energy from the hydrogen atoms that are carried on NAD and FAD. Cells produce most of their ATP in this way.

In anaerobic respiration, pyruvate is converted to either ethanol or lactate. Consequently it is not available for the Krebs cycle. Therefore in anaerobic respiration neither the Krebs cycle nor the electron transfer chain can take place. The only ATP that can be produced by anaerobic respiration is therefore that formed by glycolysis.

Investigating where certain respiratory pathways take place in cells

Most people are aware that cyanide is a very potent poison that causes death rapidly. It is lethal because it is a non-competitive inhibitor of the final enzyme in the electron transport chain. This enzyme is called cytochrome oxidase and it catalyses the addition of the hydrogen ions and electrons to oxygen to form water. The inhibition of cytochrome oxidase causes hydrogen ions and electrons to accumulate on their carrier molecules, bringing the electron transport chain and Krebs cycle to a halt.

To determine where in the cell some of the respiratory pathways take place, scientists carried out the following experiment involving cyanide.

- Mammalian liver cells were broken up (homogenised) and the resulting homogenate was centrifuged.
- Portions containing only nuclei, ribosomes, mitochondria, and the remaining cytoplasm were separated out.

> **Synoptic link**
>
> To help you follow the experiment described in the extension and to answer the questions, it is necessary to understand cell fractionation and enzyme inhibition. It is therefore advisable to review Topics 3.1 and 1.9.

- Samples of each portion, and of the complete homogenate, were incubated as follows:

  – with glucose   – with glucose and cyanide
  – with pyruvate and cyanide   – with pyruvate

After incubation the presence or absence of carbon dioxide and lactate in each sample was recorded. The results are shown in Table 1, in which ✓ = present and ✗ = absent.

▼ **Table 1** *Presence (✓) or absence (✗) of CO₂ and lactate in sample shown*

Incubated with	Complete homogenate		Nuclei only		Ribosomes only		Mitochondria only		Remaining cytoplasm only	
	Carbon dioxide	Lactate	Carbon dioxide	Lactate	Carbon dioxide	Lactate	Carbon dioxide	Lactate	Carbon dioxide	Lactate
Glucose	✓	✓	✗	✗	✗	✗	✗	✗	✗	✓
Pyruvate	✓	✓	✗	✗	✗	✗	✓	✗	✗	✓
Glucose + cyanide	✗	✓	✗	✗	✗	✗	✗	✗	✗	✓
Pyruvate + cyanide	✗	✓	✗	✗	✗	✗	✗	✗	✗	✓

1 Briefly describe how the different portions of the homogenate may have been separated out by centrifuging.

2 From the results of this experiment, name two organelles that appear not to be involved in respiration. Explain your answer.

3 In which cell organelle would you expect to find the enzymes of the Krebs cycle?
Explain how the results in the table support your answer.

4 Suggest which portion of the homogenate contains the enzymes that convert pyruvate into lactate.

5 Explain why lactate is produced in the presence of cyanide but carbon dioxide is not.

6 Explain why carbon dioxide can be produced by the complete homogenate when none of the separate portions can do so.

7 Suggest which two products might be formed if glucose was incubated with cytoplasm from yeast cells.

8 Giving your reason in each case, assess the relative number of mitochondria in the following: xylem vessel, liver cell, red blood cell, epithelial cell of intestine, myofibril (muscle fibre).

9 Mature red blood cells do not possess mitochondria. Suggest two advantages of this to the functioning of these cells.

1 (a) The table contains statements about three biological processes. Copy and complete the table with a tick if the statement in the first column is true, for each process.

	Photosynthesis	Anaerobic respiration	Aerobic respiration
ATP produced			
Occurs in organelles			
Electron transport chain involved			

(3 marks)

(b) Write a simple equation to show how ATP is synthesised from ADP.   *(1 mark)*

(c) Give two ways in which the properties of ATP make it a suitable source of energy in biological processes.   *(2 marks)*

(d) Humans synthesise more than their body mass of ATP each day. Explain why it is necessary for them to synthesise such a large amount of ATP.   *(2 marks)*

AQA June 2011

2 (a) The table contains statements about three stages of respiration. Copy and complete the table with a tick if the statement in the first column is true for each stage of respiration in an animal.

	Glycolysis	Link reaction	Krebs cycle
Occurs in mitochondria			
Carbon dioxide produced			
NAD is reduced			

(3 marks)

(b) The following reaction occurs in the Krebs cycle.

Succinate $\xrightarrow{\text{Enzyme}}$ Fumarate

A scientist investigated the effect of the enzyme inhibitor malonate on this reaction. The structure of malonate is very similar to the structure of succinate. The scientist added malonate and the respiratory substrate, pyruvate, to a suspension of isolated mitochondria. She also bubbled oxygen through the suspension.

(i) Explain why the scientist did not use glucose as the respiratory substrate for these isolated mitochondria.   *(2 marks)*

(ii) Explain how malonate inhibits the formation of fumarate from succinate.   *(2 marks)*

(iii) The scientist measured the uptake of oxygen by the mitochondria during the investigation. The uptake of oxygen decreased when malonate was added. Explain why.   *(2 marks)*

AQA June 2013

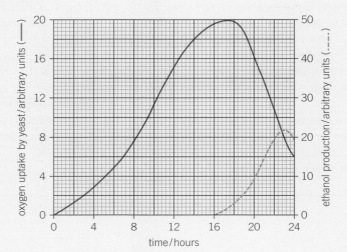

3   Yeast is a single-celled organism. A student investigated respiration in a population of
    yeast growing in a sealed container. His results are shown in the graph.
    (a) Calculate the rate of oxygen uptake in arbitrary units per hour between 2 and
        4 hours.   (*1 mark*)
    (b) (i) Use the information provided to explain the changes in oxygen uptake during
             this investigation.   (*3 marks*)
        (ii) Use the information provided to explain the changes in production of ethanol
             during this investigation.   (*2 marks*)
    (c) Sodium azide is a substance that inhibits the electron transport chain in respiration.
        The student repeated the investigation but added sodium azide after 4 hours.
        Suggest and explain how the addition of sodium azide would affect oxygen uptake
        and the production of ethanol.   (*3 marks*)
                                                                            AQA Jan 2013

4   A student investigated the rate of gas exchange in aerobically
    respiring seeds using the apparatus shown in the diagram. She
    carried out two experiments.
        • In Experiment **1**, she put potassium hydroxide solution
          in the beaker. Potassium hydroxide solution absorbs
          carbon dioxide.
        • In Experiment **2**, she put water in the beaker.
    (a) Both experiments were carried out at the same
        temperature. Explain why.   (*2 marks*)
    (b) (i) The level of coloured liquid in the right-hand
             side of the manometer tube went down during
             Experiment 1. Explain why.
                                                  (*3 marks*)

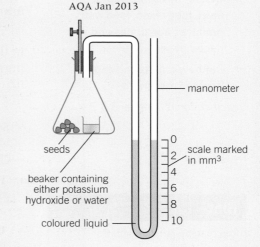

The results from both experiments are shown in the table.

Experiment	Solution in the beaker	Distance moved by the meniscus in the right hand side of the manometer (mm)
1	Potassium hydroxide	6.4
2	Water	1.2

    (ii) The diameter of the manometer tube was 1mm.
         Use these results to calculate the volume of carbon dioxide produced
         during experiment 1.   (*3 marks*)
    (c) The student repeated Experiment **1** using seeds which were respiring anaerobically.
        What would happen to the level of coloured liquid in the right-hand side of
        the manometer tube? Explain your answer.   (*2 marks*)
                                                    AQA June 2012 (apart from 4 (b) (ii))

Learning objectives

→ Explain how energy enters an ecosystem.

→ Explain how energy is transferred between the organisms in the ecosystem.

→ Define the terms: trophic level, food chain, food web, producer, consumer, and decomposer.

→ Define biomass and explain how it is measured.

Specification reference: 3.5.3

Hint

You may be familiar with the following terms:

- Herbivore – an animal that eats plants (producers) and is therefore a primary consumer.

- Carnivore – an animal that eats animals and may therefore be a secondary or a tertiary consumer.

- Omnivore – an animal that eats both plants and animals and is therefore a primary consumer and also a secondary or a tertiary consumer.

Study tip

Make sure you learn simple definitions of ecological terms and use them appropriately.

The organisms found in any ecosystem rely on a source of energy to carry out all their activities. The ultimate source of this energy for almost all organisms is sunlight, which is conserved as chemical energy by plants. Most plants use sunlight in making organic compounds from carbon dioxide in the air or water that surrounds them. These organic compounds include sugars, most of which are used by the plants as respiratory substrates. The remainder are used to make other groups of biological molecules. These biological molecules form the **biomass** of plants that is the means by which energy is passed between other organisms.

In this chapter we shall look at how this energy is transferred, how nutrients are cycled and how we use artificial fertilisers to supplement natural nutrients in order to improve productivity. Before we do, let us first recap some of the basic terminology of ecology.

Organisms can be divided into three groups according to how they obtain their energy and nutrients. These three groups are – producers, consumers, and saprobionts.

- **Producers** are photosynthetic organisms that manufacture organic substances using light energy, water, carbon dioxide, and mineral ions.

- **Consumers** are organisms that obtain their energy by feeding on (consuming) other organisms rather than using the energy of sunlight directly. Animals are consumers. Those that directly eat producers (green plants) are called **primary consumers** because they are the first in the chain of consumers. Those animals eating primary consumers are called **secondary consumers** and those eating secondary consumers are called **tertiary consumers**. Secondary and tertiary consumers are usually predators but they may also be scavengers or parasites.

- **Saprobionts** (decomposers) are a group of organisms that break down the complex materials in dead organisms into simple ones. In doing so, they release valuable minerals and elements in a form that can be absorbed by plants and so contribute to recycling. The majority of this work is carried out by fungi and bacteria.

- A **food chain** describes a feeding relationship in which the producers are eaten by primary consumers. These in turn are eaten by secondary consumers, which are then eaten by tertiary consumers. In a long food chain the tertiary consumers may in turn be eaten by further consumers called quaternary consumers. Each stage in this chain is referred to as a **trophic level**. The arrows on food chain diagrams represent the direction of energy flow.

- **Food webs** – in reality, most animals do not rely on a single food source and within a single habitat many food chains will be linked together to form a food web. The problem with food webs is their complexity. In practice, it is likely that all organisms within a habitat, even within an ecosystem, will be linked to others in the food web.

Biomass

Biomass is the total mass of living material in a specific area at a given time. The fresh mass is quite easy to assess, but the presence of varying amounts of water makes it unreliable. Measuring the mass of carbon or dry mass overcomes this problem but, because the organisms must be killed, it is usually only made on a small sample, and this sample may not be representative. Biomass is measured using dry mass per given area, in a given time. More specifically it is measured in grams per square metre ($g\,m^{-2}$) where an area is being sampled, for example, on grassland or a seashore. Where a volume is being sampled, for example, in a pond or an ocean, it is measured in grams per cubic metre ($g\,m^{-3}$).

The chemical energy store in dry mass can be estimated using **calorimetry**. In bomb calorimetry, a sample of dry material is weighed and is then burnt in pure oxygen within a sealed chamber called a bomb. The bomb is surrounded by a water bath and the heat of combustion causes a small temperature rise in this water. As we know how much heat (energy) is required to raise the temperature of 1g of water by 1°C, if we know the volume of water and the temperature rise, we can calculate the energy released from the mass of burnt biomass in units such as $kJ\,kg^{-1}$.

▲ **Figure 1** *The snake (tertiary consumer) is swallowing an insect-eating frog (secondary consumer) on a plant leaf (producer)*

Summary questions

The diagram below shows a simplified food web within an aquatic ecosystem.

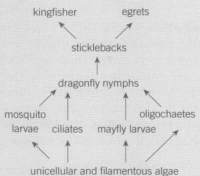

1 State which organisms are secondary consumers.

2 State which organisms carry out photosynthesis.

3 State which organisms are at the fourth trophic level.

4 Explain what the arrows in the diagram show.

5 When the organisms in this web die they will be broken down by bacteria and fungi. Name the general term used to describe these bacteria and fungi.

Learning objectives

→ Calculate the percentage of energy that is transferred from one trophic level to the next.

→ Explain how energy is lost along a food chain.

→ Explain what is meant by gross primary productivity and net primary productivity.

Specification reference: 3.5.3

The Sun is the source of energy for **ecosystems**. However, as little as one % of this light energy may be captured by green plants and so made available to organisms in the food chain. These organisms in turn pass on only a small fraction of the energy that they receive to each successive stage in the chain. How then is so much energy lost?

Plants normally convert between one % and three % of the Sun's energy available to them into organic matter. Most of the Sun's energy is not converted to organic matter by photosynthesis because:

- over 90% of the Sun's energy is reflected back into space by clouds and dust or absorbed by the atmosphere
- not all wavelengths of light can be absorbed and used for photosynthesis
- light may not fall on a chlorophyll molecule
- a factor, such as low carbon dioxide levels, may limit the rate of photosynthesis.

The total quantity of the chemical energy store in plant biomass, in a given area or volume, in a given time is called the **gross primary production** (**GPP**). However, plants use 20–50% of this energy in respiration. The chemical energy store which is left when these losses to respiration have been taken into account, is called **net primary productivity** (**NPP**).

net primary production   =   gross primary production   −   respiratory losses
$$NPP \qquad\qquad\qquad GPP \qquad\qquad\qquad R$$

The net primary production is available for plant growth and reproduction. It is also available to other trophic levels in the ecosystem, such as consumers and decomposers. Usually less than 10% of this net primary production in plants can be used by primary consumers for growth. Secondary and tertiary consumers are slightly more efficient, transferring up to about 20% of the energy available from their prey into their own bodies. The low percentage of energy transferred at each stage is the result of the following:

- Some of the organism is not consumed.
- Some parts are consumed but cannot be digested and are therefore lost in faeces.
- Some of the energy is lost in excretory materials, such as urine.
- Some energy losses occur as heat from respiration and lost to the environment. These losses are high in mammals and birds because of their high body temperature. Much energy is needed to maintain their body temperature when heat is constantly being lost to the environment.

The net production of consumers can therefore be calculated as:

$$N = I - (F + R)$$

where:

N represents the net production
I represents the chemical energy store of ingested food
F represents the energy lost in faeces and urine
R represents the energy lost in respiration.

Energy flow along food chains, showing the percentage transferred at each trophic level, is summarised in Figure 1.

It is the relative inefficiency of energy transfer between trophic levels that explains why:

* most food chains have only four or five trophic levels because insufficient energy is available to support a large enough breeding population at trophic levels higher than these
* the total mass of organisms in a particular place (biomass) is less at higher trophic levels
* the total amount of energy available is less at each level as one moves up a food chain.

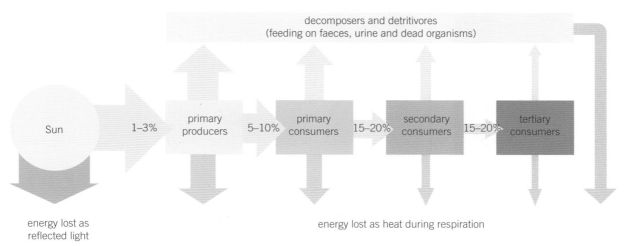

▲ **Figure 1** *Energy flow through different trophic levels of a food chain. The arrows are not to scale and give only an idea of the proportion of energy transferred at each stage. Likewise, the figures for % energy transfer between trophic levels are only a rough average as they vary considerably between different plants, animals, and habitats*

Summary questions √x̄

1 State **three** reasons for the small percentage of energy transferred at each trophic level.

2 Explain why most food chains rarely have more than four trophic levels.

3 √x̄ An area of vegetation 5 m by 5 m produces 4×10^4 kJ of potential energy in a year. Calculate the gross primary production of this area.

Hint

When making calculations involving energy transfer, always remember that energy cannot be created or destroyed. In this type of question, this means that the total amount of energy entering a box must equal the amount of energy in the box plus the amount leaving the box.

Hint

If you were ever in any doubt about the considerable loss of energy from organisms, just think about how much food you have eaten in your whole life – and all there is to show for it is what you are now.

Maths link $\sqrt{x}$

MS 2.4, see Chapter 22.

▲ **Figure 2** *Only about 10% of the energy in the plant being eaten by this swallowtail butterfly larva will be used for its growth*

Maths link $\sqrt{x}$

MS 0.1 and 2.4, see Chapter 22.

 ## Calculating the efficiency of energy transfers

Data are often presented showing the amount of energy available at each trophic level of a food chain. The energy available is usually measured in kilojoules per square metre per year ($kJ\ m^{-2}\ year^{-1}$). It is often useful to calculate the efficiency of the energy transfer between each trophic level of these food chains. This is calculated as follows:

$$\text{percentage efficiency} = \frac{\text{energy available after the transfer}}{\text{energy available before the transfer}} \times 100$$

Let us take an example. Look at Figure 3, which shows the amount of energy available at different trophic levels in a lake in the USA. Suppose we wanted to calculate the percentage efficiency of the transfer of energy from trout to humans. We would make the calculation as follows.

Energy available after the transfer = $50\ kJ\ m^{-3}\ year^{-1}$
(i.e., energy available to humans)

Energy available before the transfer = $250\ kJ\ m^{-3}\ year^{-1}$
(i.e., energy available to trout)

$$\text{percentage efficiency} = \frac{50}{250} \times 100 = \frac{5000}{250} = 20\%$$

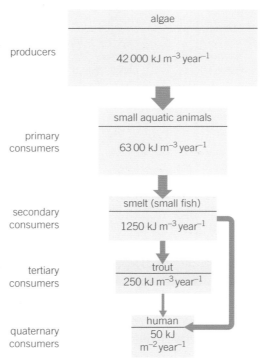

	algae
producers	$42\,000\ kJ\ m^{-3}\ year^{-1}$

	small aquatic animals
primary consumers	$63\,00\ kJ\ m^{-3}\ year^{-1}$

	smelt (small fish)
secondary consumers	$1250\ kJ\ m^{-3}\ year^{-1}$

	trout
tertiary consumers	$250\ kJ\ m^{-3}\ year^{-1}$

	human
quaternary consumers	$50\ kJ$ $m^{-2}\ year^{-1}$

▲ **Figure 3** *Food chain in Cayuga Lake, New York State. Figures illustrate the relative amount of energy available at each trophic level in the food chain*

1 $\sqrt{x}$ Using Figure 3, calculate the percentage efficiency of energy transfer between:
   a primary consumers and secondary consumers
   b algae and humans.
Show your working in both cases.

Adding up the totals

Figure 4 shows the flow of energy through a terrestrial ecosystem each year. All the values are in $kJ\,m^{-2}\,year^{-1}$.

Maths link

MS 2.4 and 3.1, see Chapter 22.

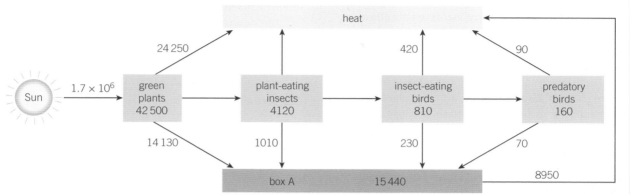

▲ **Figure 4**

1   Give the name of the group of organisms represented by Box A.
2   Which group of organisms are secondary consumers?
3   Calculate the percentage efficiency with which light energy is transferred to energy in green plants. Show your working.
4   State **three** reasons why so little of the solar energy is transferred to energy in green plants.
5   Calculate the amount of energy that is lost as heat from plant-eating insects. Show your working.

 ## Productivity and farming practices

Many farming practices are employed as methods of increasing yields by increasing the efficiency of energy transfer along the food chains which produce our food. As energy passes along a food chain only a small percentage passes from one organism in the chain to the next. This is because much of the energy is lost as heat during respiration. Any practice that reduces the respiratory losses in a human food chain will therefore reduce energy loss and increase the yield.

One farming practice that achieves this is the intensive rearing of domestic livestock. This is about converting the smallest possible quantity of food energy into the greatest quantity of animal mass. One way to achieve this is to minimise the energy losses from domestic animals during their lifetime. This means that more of the food energy taken in by the animals will be converted into body mass, ready to be passed on to the next link in the food chain, namely us. Energy conversion can be made more efficient by ensuring that as much energy from

Maths link

MS 0.3 and 3.1, see Chapter 22.

respiration as possible goes into growth rather than other activities or other organisms. This is achieved by keeping animals in confined spaces, such as small enclosures, barns or cages, a practice often called factory farming. This increases the energy-conversion rate because:

*   movement is restricted and so less energy is used in muscle contraction
*   the environment can be kept warm in order to reduce heat loss from the body (most intensively reared species are warm-blooded)
*   feeding can be controlled so that the animals receive the optimum amount and type of food for maximum growth with no wastage
*   predators are excluded so that there is no loss to other organisms in the food web.

1 Suggest a reason why keeping animals in the dark for longer periods might improve the energy conversion rate.

Another farming practice that increases the efficiency of energy transfer is to reduce losses to non-human food chains by simplifying food webs. In other words, to reduce or eliminate organisms that are part of a food web and which compete with the plant or animal that is being farmed. Weeds compete with crop plants for water, mineral ions, carbon dioxide, space, and light. As these resources are often in limited supply, any amount taken by the pest means less is available for the crop plant. Insect pests may damage the leaves of crops, limiting their ability to photosynthesise and thus reducing their productivity. Alternatively, they may be in direct competition with humans, eating the crop itself. Many crops are now grown in **monoculture**, and this enables insect and fungal pests to spread rapidly. Pests of domesticated animals may cause disease. The animals may not grow as rapidly, be unfit for human consumption or die – all of which lead to reduced productivity.

2 Pesticides are used to increase productivity. Suggest how their use might sometimes reduce productivity.

The aim of pest control is to simplify the food web and so limit the effect of pests on productivity to a commercially acceptable level. In other words, to balance the cost of pest control with the benefits it brings. The problem is that at least two different interests are involved: the farmer who has to satisfy our demand for cheap food while still making a living, and the **conservation** of natural resources, which will enable us to continue to

have food in the future. The trick is to balance these two, often conflicting, interests.

The graph in Figure 5 shows the effects of weeds on the productivity of two crops: wheat and soya bean.

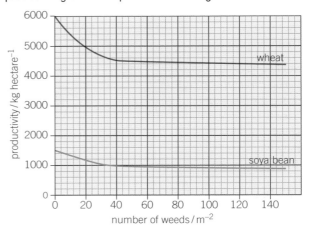

▲ Figure 5

3 Describe the effects of weeds on the productivity of wheat.
4 A herbicide that reduces the number of weeds from 40 m^{-2} to 0 m^{-2} is applied to both crops. Which crop would show the greatest percentage change in productivity?
5 $\sqrt{x}$ It will cost a farmer £100 to treat each hectare of his wheat crop with a herbicide. The herbicide will reduce the number of weeds from 40 m^{-2} to 20 m^{-2}. He can sell his wheat at £150 a tonne (one tonne = 1000 kg). Is it economically worthwhile for the farmer to apply the herbicide to his crop? Use calculations to support your answer.

A mighty problem

The two-spotted spider mite, *Tetranychus urticae*, is an important pest of crops, especially those in greenhouses. Control is mostly achieved using chemicals. However, the spider mite has increasingly developed resistance to these chemicals and they are therefore less effective in controlling its populations.

Studies have been carried out to investigate the use of biological control to combat spider mites. In one such study the predatory mite *Phytoseiulus persimilis* was used to test its effectiveness against the two-spotted spider mite. This predatory mite feeds on the spider mite. Mites were introduced into two separate groups of 100 bean plants as follows:

1 Describe and explain the differences between the spider mite populations in experiment 1 and experiment 2.
2 Comment on the effectiveness of predatory mites in controlling populations of spider mites.
3 Predict what the levels of the two populations in experiment 2 might be over a period of 150–300 days if the experiment was continued. Explain the reasons for the levels you suggest.

• Experiment 1 – spider mites only

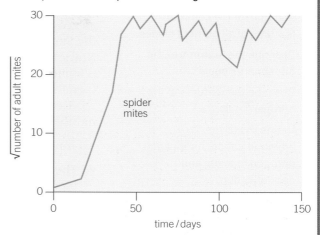

• Experiment 2 – spider mites and predatory mites

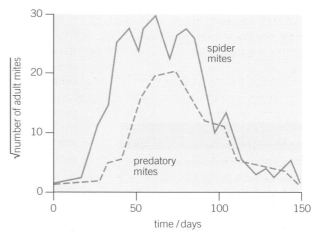

▲ **Figure 6**

▲ **Figure 7** *Biological pest control: orange predatory mite* (Phytoseiulus persimilis) *attacking the red spider mite* (Tetranychus urticae)

Learning objectives

→ Summarise the common features of all nutrient cycles.

→ Describe the features of the phosphorus cycle.

→ Describe the features of the nitrogen cycle.

→ Define the terms ammonification, nitrification, nitrogen fixation, and denitrification.

→ Explain the roles of saprobiotic organisms in nutrient recycling.

Specification reference: 3.5.4

Hint

Although the weathering of rocks releases inorganic ions, the rate is inadequate to sustain most communities. Recycling of inorganic ions is therefore essential.

We saw in Topic 13.1, that energy enters an ecosystem as sunlight and is lost as heat. This heat cannot be recycled. The flow of energy through an ecosystem is therefore in one direction, that is, it is linear. Provided the Sun continues to supply energy to Earth, this is not a problem. Nutrients, by contrast, do not have an extraterrestrial source. There is limited availability of nutrient ions in a usable form. It is important therefore that elements such as carbon, nitrogen, and phosphorus are recycled. The flow of nutrients within an ecosystem is not linear, but mostly cyclic.

All nutrient cycles have one simple sequence at their heart.

- The nutrient is taken up by producers (plants) as simple, inorganic molecules.

- The producer incorporates the nutrient into complex organic molecules.

- When the producer is eaten, the nutrient passes into consumers (animals).

- It then passes along the food chain when these animals are eaten by other consumers.

- When the producers and consumers die, their complex molecules are broken down by saprobiontic microorganisms (decomposers) that release the nutrient in its original simple form. The cycle is then complete. The role of these saprobionts in nutrient cycles cannot be overestimated. They are in many ways the driving forces that ensure that nutrients are released for reuse. Without them nutrients would remain locked up as part of complex molecules that cannot be taken up and used again by plants.

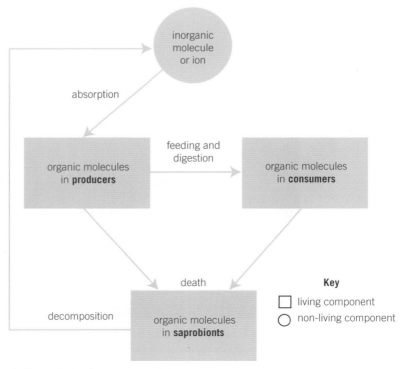

▲ **Figure 1** *Basic sequence of all nutrient cycles*

Although other processes and non-living sources are also involved, it is this sequence, illustrated in Figure 1, that forms the basis of all nutrient cycles.

The nitrogen cycle

Living organisms require a source of nitrogen from which to manufacture proteins, nucleic acids and other nitrogen-containing compounds. Although 78% of the atmosphere is nitrogen, there are very few organisms that can use nitrogen gas directly. Plants take up most of the nitrogen they require in the form of nitrate ions (NO_3^-), from the soil. These ions are absorbed, using **active transport**, by the roots. This is where nitrogen enters the living component of the ecosystem. Animals obtain nitrogen-containing compounds by eating and digesting plants.

Nitrate ions are very soluble and easily leach (wash) through the soil, beyond the reach of plant roots. In natural ecosystems, the nitrate concentrations are restored largely by the recycling of nitrogen-containing compounds. In agricultural ecosystems, the concentration of soil nitrate can be further increased by the addition of fertilisers. When plants and animals die, the process of decomposition begins, in a series of steps by which microorganisms replenish the nitrate concentrations in the soil. This release of nitrate ions by decomposition is most important because, in natural ecosystems, there are very few nitrate ions available from other sources.

There are four main stages in the nitrogen cycle (Figure 2), **ammonification**, **nitrification**, **nitrogen fixation** and **denitrification**, each of which involves saprobiontic microorganisms.

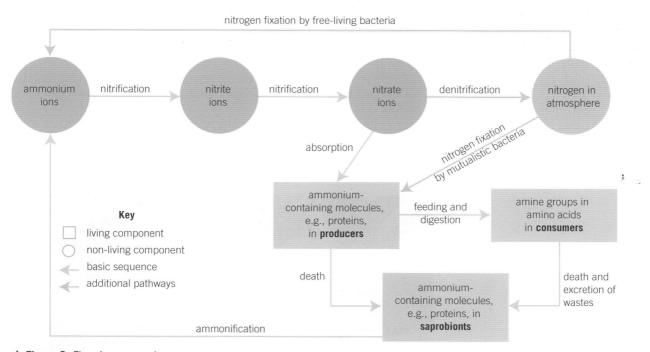

▲ **Figure 2** *The nitrogen cycle*

Ammonification

Ammonification is the production of ammonia from organic nitrogen-containing compounds. In nature, these compounds include urea (from the breakdown of excess amino acids) and proteins, nucleic acids and vitamins (found in faeces and dead organisms). Saprobiontic microorganisms, mainly fungi and bacteria, feed on faeces and dead organisms materials, releasing ammonia, which then forms ammonium ions in the soil. This is where nitrogen returns to the non-living component of the ecosystem.

Nitrification

Plants use light energy to produce organic compounds. Some bacteria, however, obtain their energy from chemical reactions involving inorganic ions. One such reaction is the conversion of ammonium ions to nitrate ions. This is an **oxidation** reaction and so releases energy. It is carried out by free-living soil microorganisms called nitrifying bacteria. This conversion occurs in two stages:

1   oxidation of ammonium ions to nitrite ions (NO_2^-)

2   oxidation of nitrite ions to nitrate ions (NO_3^-).

Nitrifying bacteria require oxygen to carry out these conversions and so they require a soil that has many air spaces. To raise productivity, it is important for farmers to keep soil structure light and well aerated by ploughing. Good drainage also prevents the air spaces from being filled with water and so prevents air being forced out of the soil.

Nitrogen fixation

This is a process by which nitrogen gas is converted into nitrogen-containing compounds. It can be carried out industrially and also occurs naturally when lightning passes through the atmosphere. By far the most important form of nitrogen fixation is carried out by microorganisms, of which there are two main types:

- **free-living nitrogen-fixing bacteria**. These bacteria reduce gaseous nitrogen to ammonia, which they then use to manufacture amino acids. Nitrogen-rich compounds are released from them when they die and decay.
- **mutualistic nitrogen-fixing bacteria**. These bacteria live in nodules on the roots of plants such as peas and beans (Figure 3). They obtain carbohydrates from the plant and the plant acquires amino acids from the bacteria.

▲ **Figure 3** *Nitrogen-fixing nodules on the roots of a pea plant allow the plant to use free nitrogen in the atmosphere and soil. Mutualistic bacteria in the nodules fix the nitrogen, transforming it into a form usable by the plant*

In many ecosystems, the availability of nitrates is the factor that limits plant growth. As plants are the primary producers, this means that nitrate availability affects the whole ecosystem.

Study tip

The word nitrogen is often misused. Nitrogen is an element which forms a part of ions, such as nitrites and nitrates, as well as part of complex molecules, such as proteins and nucleic acids. Do not use the term nitrogen when referring to these substances. Instead, use nitrogen-containing ions or nitrogen-containing molecules.

Denitrification

When soils become waterlogged, and have a low oxygen concentration, the type of microorganism present changes. Fewer **aerobic** nitrifying and nitrogen-fixing bacteria are found, and there is an increase in **anaerobic** **denitrifying bacteria**. These convert soil nitrates into gaseous nitrogen. This reduces the availability of nitrogen-containing compounds for plants. For land to be productive, the soils on which crops grow must therefore be kept well aerated to prevent the build-up of denitrifying bacteria.

As with any nutrient cycle, the delicate balance can be easily upset by human activities. Some of the effects of these activities are considered in Topic 13.4, Use of natural and artificial fertilisers.

▲ **Figure 4** *Ploughing helps to aerate the soil and so prevents the build-up of denitrifying bacteria that can reduce the level of soil nitrates*

The phosphorus cycle

Phosphorus is an important biological element as it is a component of ATP, phospholipids and nucleic acids. Life therefore depends on it being constantly recycled.

In the carbon and nitrogen cycles the main reservoir of each element is in the atmosphere. In the phosphorus cycle however the main reservoir is in mineral form rather than in the atmosphere – in fact the phosphorus cycle lacks a gaseous phase altogether.

Phosphorus exists mostly as phosphate ions (PO_4^{3-}) in the form of sedimentary rock deposits. These have their origins in the seas but are brought to the surface by the geological uplifting of rocks. The weathering and erosion of these rocks helps phosphate ions to become dissolved and so available for absorption by plants which incorporate them into their biomass. The phosphate ions pass into animals which feed on the plants. Excess phosphate ions are excreted by animals and may accumulate in waste material such as guano formed from the excretory products of some sea birds.

On the death of plants and animals, decomposers such as certain bacteria and fungi break them down releasing phosphate ions into the water or soil. Some phosphate ions remain in parts of animals, such as bones or shells, that are very slow to breakdown. Phosphate ions in excreta, released by decomposition and dissolved out of rocks, are transported by streams and rivers into lakes and oceans where they form sedimentary rocks thus completing the cycle.

The phosphorus cycle is illustrated in Figure 5.

> **Synoptic link**
>
> To revise the part phosphorus plays in the structure of biological molecules consult Topic 1.5, Lipids, Topic 2.1, Structure of RNA and DNA, and 2.3, Energy and ATP.

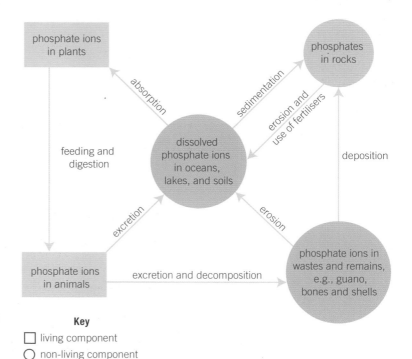

Key
☐ living component
○ non-living component

▲ **Figure 5** *The phosphorus cycle*

The role of mycorrhizae in nutrient cycles

Mycorrhizae are associations between certain types of fungi and the roots of the vast majority of plants. The fungi act like extensions of the plant's root system and vastly increase the total surface area for the absorption of water and minerals. The mycorrhiza acts like a sponge and so holds water and minerals in the neighbourhood of the roots. This enables the plant to better resist drought and to take up inorganic ions more readily. The mycorrhiza plays a part in nutrient cycles by improving the uptake of relatively scarce ions such as phosphates ions.

The mycorrhizal relationship between plants and fungi is a mutualistic one. The plant benefits from improved water and inorganic ion uptake while the fungus receives organic compounds such as sugars and amino acids from the plant.

Summary questions

In the following passage, suggest the most appropriate word to replace each of the numbers in brackets.

A few organisms can convert nitrogen gas into compounds useful to other organisms in a process known as (**1**). These organisms can be free-living or live in a relationship with certain (**2**). Most plants obtain their nitrogen by absorbing (**3**) from the soil through their (**4**) by active transport. They then convert this to (**5**), which is passed to animals when they eat the plants. On death, (**6**) break down these organisms, releasing (**7**), which can then be oxidised to form nitrite ions by (**8**) bacteria. Further oxidation by the same type of bacteria forms (**9**) ions. These ions may be converted back to atmospheric nitrogen by the activities of (**10**) bacteria.

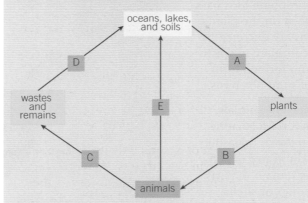

▲ **Figure 6**

11 Figure 6 is a simplified illustration of the phosphorus cycle. Each box represents a process. Name the process in each of the boxes A, B, C, D and E.

13.4 Use of natural and artificial fertilisers

Agricultural ecosystems increase the efficiency of energy transfer along human food chains. In doing so they improve productivity. One farming practice that contributes to this improved productivity is the use of fertilisers. Let us see how this is achieved.

The need for fertilisers

All plants need mineral ions, especially nitrates, from the soil. Much food production in the developed world is intensive, that is, it is concentrated on specific areas of land that are used repeatedly to achieve maximum yield from the crops and animals grown on them. Intensive food production makes large demands on the soil because mineral ions are continually taken up by the crops being grown on it. These crops are either used directly as food or as fodder for animals that are then eaten. Either way, the mineral ions that the crops have absorbed from the soil are removed.

In natural ecosystems the minerals that are removed from the soil by plants are returned when the plant is decomposed by microorganisms on its death. In agricultural systems the crop is harvested and then transported from its point of origin for consumption. The urine, faeces and dead remains of the consumer are rarely returned to the same area of land. Under these conditions the concentrations of the mineral ions in agricultural land will fall. It is therefore necessary to replenish these mineral ions because, otherwise, their reduced concentrations will become the main limiting factor to plant growth. Productivity will consequently be reduced. To offset this loss of mineral ions, fertilisers need to be added to the soil. These fertilisers are of two types:

- **natural (organic) fertilisers**, which consist of the dead and decaying remains of plants and animals as well as animal wastes such as manure, slurry and bone meal
- **artificial (inorganic) fertilisers**, which are mined from rocks and deposits and then converted into different forms and blended together to give the appropriate balance of minerals for a particular crop. Compounds containing the three elements, nitrogen, phosphorus, and potassium are almost always present.

Research suggests that a combination of natural and artificial fertilisers gives the greatest long-term increase in productivity. However, it is important that minerals are added in appropriate quantities as there is a point at which further increases in the quantity of fertiliser no longer results in increased productivity. This is illustrated in Figure 1.

How fertilisers increase productivity

Plants require minerals for their growth. Let us look at nitrogen as an example. Nitrogen is an essential component of amino acids, ATP, and nucleotides in DNA. Both are needed for plant growth. Where nitrate ions are readily available, plants are likely to develop earlier, grow taller and have a greater leaf area. This increases the rate of photosynthesis and improves crop productivity. There can be no doubt that nitrogen-containing fertilisers have been of considerable benefit in providing us with cheaper food. It is

Learning objectives

→ Explain why fertilisers are needed in agricultural ecosystems.

→ Distinguish between natural and artificial fertilisers.

→ Explain how fertilisers increase productivity.

Specification reference: 3.5.4

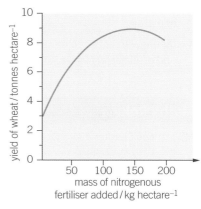

▲ **Figure 1** *The effect of different quantities of nitrogenous fertiliser on the yield of wheat*

▲ **Figure 2** *Cattle slurry, a natural fertiliser, being spread onto a crop of wheat*

Summary questions

1 Explain why fertilisers are needed in an agricultural ecosystem.

2 Using Figure 1, determine what concentration of fertiliser you would advise a farmer to apply to a field of wheat.

3 Suggest a reason why, after a certain point, the addition of more fertiliser no longer improves the productivity of a crop.

4 Distinguish between natural and artificial fertilisers.

Maths link

MS 3.1, see Chapter 22.

estimated that the use of fertilisers has increased agricultural food production in the UK by around 100% since 1955.

Different forms of nitrogen-containing fertiliser

Nitrogen-containing fertiliser can be applied to crops in a number of different forms. These include ammonium salts, animal manure, the ground-up bones of animals (bone meal), and urea (a waste product found in the urine of mammals). An investigation was carried out in which the same crop was grown on six separate plots of land each of the same area. No nitrogen-containing fertiliser was added to the first plot. To each of the remaining five plots, a different form of nitrogen-containing fertiliser was added at the rate of 140 kg total nitrogen per hectare. The graph in Figure 3 shows the results of the investigation.

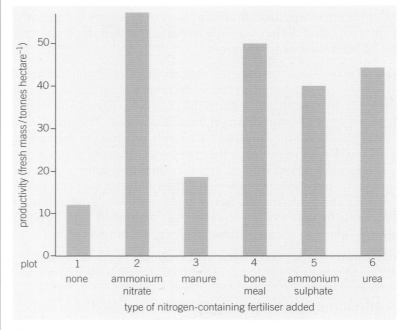

▲ Figure 3

1 State which forms of nitrogen used in the investigation are natural fertilisers.

2 Suggest why the investigation included a plot to which no nitrogen-containing fertiliser was added.

3 Suggest how the addition of nitrogen-containing fertiliser, in whatever form, increased productivity.

4 The mass of each fertiliser used was different in each case. Suggest why this was necessary.

5 It is sometimes claimed that nitrogen-containing fertilisers in the form of ammonium salts increase productivity of crops better than other forms of nitrogen-containing fertilisers. State, with your reasons, whether or not you think the results of this experiment support this view.

6 The increase in productivity when manure was applied was lower than for other forms of nitrogen-containing fertiliser. This is because the manure has to break down before its nitrogen is released and this process takes a few months. Suggest how a farmer who spreads manure on his/her crops, might use this information in order to improve productivity.

In natural ecosystems minerals such as nitrate ions, which are removed from the soil by plants, are mainly returned when the plant is decomposed. However, as we saw in Topic 13.4, in agricultural systems the crop is removed and so the nitrate is not returned and has to be replaced. This is done by the addition of natural or artificial fertilisers.

Effects of nitrogen-containing fertilisers

Nitrogen is an essential component of biological molecules such as proteins and is needed for growth and, therefore, an increase in the area of leaves. This increases the rate of photosynthesis and improves crop productivity. There can be no doubt that nitrogen-containing fertilisers have benefited us considerably by providing us with cheaper food. Most of this increase is due to additional nitrogen (Figure 1). The use of nitrogen-containing fertilisers has also had some detrimental effects. These include:

- **reduced species diversity**, because nitrogen-rich soils favour the growth of grasses, nettles and other rapidly growing species. These out-compete many other species, which die as a result. Species-rich hay meadows, such as the one in the photograph (Figure 3), only survive when soil nitrogen concentrations are low enough to allow other species to compete with the grasses
- **leaching**, which may lead to pollution of watercourses
- **eutrophication**, caused by leaching of fertiliser into watercourses.

Leaching

Leaching is the process by which nutrients are removed from the soil. Rainwater will dissolve any soluble nutrients, such as nitrate ions, and carry them deep into the soil, eventually beyond the reach of plant roots. The leached nitrate ions find their way into watercourses, such as streams and rivers, that in turn may drain into freshwater lakes. Here they may have a harmful effect on humans if the river or lake is a source of drinking water. Very high nitrate ion concentrations in drinking water can prevent efficient oxygen transport in babies and a link to stomach cancer in humans has been suggested. The leached nitrate ions are also harmful to the environment as they can cause eutrophication.

Eutrophication

Eutrophication is the process by which nutrient concentrations increase in bodies of water. It is a natural process that occurs mostly in freshwater lakes and the lower reaches of rivers. Eutrophication consists of the following sequence of events:

1 In most lakes and rivers there is naturally very low concentration of nitrate and so nitrate ions are a limiting factor for plant and algal growth.

2 As the nitrate ion concentration increases as a result of leaching, it ceases to be a limiting factor for the growth of plants and algae whose populations both grow.

3 As algae mostly grow at the surface, the upper layers of water become densely populated with algae. This is called an 'algal bloom'.

Learning objectives

→ Describe the main environmental effects of using nitrogen-containing fertilisers.

→ State the meanings of leaching and eutrophication.

→ Explain how leaching and eutrophication affect the environment.

Specification reference: 3.5.4

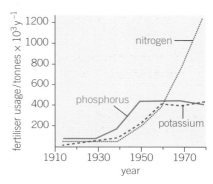

▲ **Figure 1** *Use of different types of fertilisers in the UK*

▲ **Figure 2** *Low species diversity in a field grown for silage that has had nitrogen-containing fertiliser added*

▲ **Figure 3** *High species diversity in a meadow grown for hay without the addition of nitrogen-containing fertiliser*

▲ **Figure 4** *Algal bloom in a canal as a result of eutrophication caused by nitrogen-containing fertiliser run-off*

4 This dense surface layer of algae absorbs light and prevents it from penetrating to lower depths.

5 Light then becomes the limiting factor for the growth of plants and algae at lower depths and so they eventually die.

6 The lack of dead plants and algae is no longer a limiting factor for the growth of saprobiontic bacteria and so these populations too grow, using the dead organisms as food.

7 The saprobiontic bacteria require oxygen for their respiration, creating an increased demand for oxygen.

8 The concentration of oxygen in the water is reduced and nitrates are released from the decaying organisms.

9 Oxygen then becomes the limiting factor for the population of **aerobic** organisms, such as fish. These organisms ultimately die as the oxygen is used up altogether.

10 Without the aerobic organisms, there is less competition for the **anaerobic** organisms, whose populations now rise.

11 The anaerobic organisms further decompose dead material, releasing more nitrates and some toxic wastes, such as hydrogen sulphide, which make the water putrid.

Organic manures, animal slurry, human sewage, ploughing old grassland and natural leaching can all contribute to eutrophication, but the leaching of artificial fertilisers is the main cause.

Summary questions

1 Explain what is meant by eutrophication.

2 Explain how an increase in algal growth at the surface can lead to the death of plants growing beneath them.

3 Explain how the death of these plants can result in the death of animals such as fish.

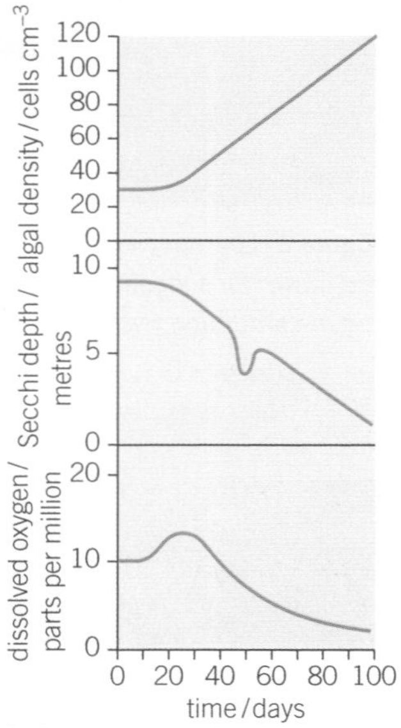

Maths link √x̄

MS 3.1, see Chapter 22.

Troubled waters √x̄

A farmer applied a large quantity of fertiliser to fields next to a small lake. A period of heavy rain followed. After 10 days, scientists monitoring the lake noticed changes to the algal population, the clarity of the water and the levels of dissolved oxygen. These changes are shown in the three graphs in Figure 5. Secchi depth is a measure of the clarity of water. Measurements are taken by lowering a black-and-white disc (called a Secchi disc) into the water and recording the depth at which it is no longer visible.

1 Suggest a reason why the changes in the lake do not occur until 10 days after the application of the fertiliser to the fields.

2 Explain a possible cause of the increase in the density of algae after 10 days.

3 Describe and explain the relationship between the density of algae and water clarity in the lake.

4 Describe and explain changes to the levels of dissolved oxygen over the 100-day period.

▲ **Figure 5**

1   Scientists constructed a mathematical model. They used this model to estimate the transfer of energy through consumers in a natural grassland ecosystem. The table shows their results.

	Energy transferred as percentage of energy in biomass of producers				
	Ingested Food (I)	Absorbed from gut (A)	Egested (E)	Net Production (N)	Respired (R)
Primary consumers					
Mammals	25.00	12.50	12.50	0.25	12.25
Insects	4.00	1.60	2.40	0.64	0.96
Secondary consumers					
Mammals	0.16	0.13	0.03	0.003	0.127
Insects	0.17	0.135	0.035	0.040	0.095

(a)   Copy and complete the equation to show how net production is calculated from the energy in ingested food.
$$P =$$ *(1 mark)*

(b)   Describe and explain how intensive rearing of domestic livestock would affect
   (i)   the figure for **A** in the first row of the table   *(1 mark)*
   (ii)   the figure for **R** in the first row of the table.   *(1 mark)*

(c)   (i)   Calculate the ratio of **R : A** for mammalian primary consumers.   *(1 mark)*
   (ii)   The **R : A** ratio is higher in mammalian primary consumers than in insect primary consumers. Suggest a reason for this higher value.   *(1 mark)*

(d)   The scientists tested their model by comparing the values it predicted with actual measured values. The graph shows their results.

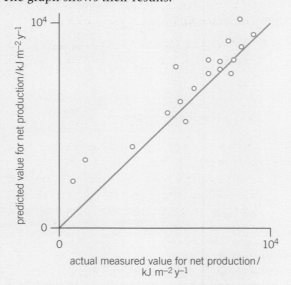

Are the values predicted by the model supported by the actual measured values?
Evaluate the evidence in the graph.
                                                                                          (3 marks)
                                                                      AQA June 2010

2　Scientists measured the mean temperature in a field each month between March and October. The table shows their results.

Month	Mean temperature / °C
March	9.0
April	11.0
May	14.0
June	17.0
July	20.0
August	18.0
September	16.0
October	14.0

(a)　The gross productivity of the plants in the field was highest in July. Use the data in the table to explain why. *(2 marks)*

(b)　(i)　Give the equation that links gross productivity and net productivity. *(1 mark)*

　　　(ii)　The net productivity of the plants in the field was higher in August than in July. Use the equation in part **(b)(i)** and your knowledge of photosynthesis and respiration to suggest why. *(2 marks)*

(c)　A horse was kept in the field from March to October. During the summer months, the horse was able to eat more than it needed to meet its minimum daily requirements. Suggest how the horse used the extra nutrients absorbed. *(1 mark)*

(d)　The horse's mean energy expenditure was higher in March than it was in August. Use information in the table to suggest why. *(2 marks)*

AQA June 2011

3　The diagram shows the nitrogen cycle.

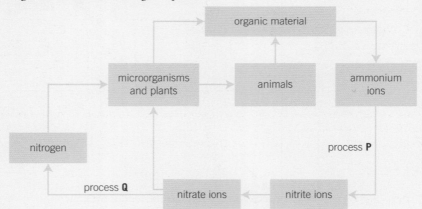

(a)　(i)　Name process **P**. *(1 mark)*

　　　(ii)　Name process **Q**. *(1 mark)*

(b)　Leguminous crop plants have nitrogen-fixing bacteria in nodules on their roots. On soils with a low concentration of nitrate ions, leguminous crops often grow better than other types of crop. Explain why. *(2 marks)*

(c)　Applying very high concentrations of fertiliser to the soil can reduce plant growth. Use your knowledge of water potential to explain why. *(2 marks)*

AQA Jan 2013

4   Scientists investigated the effect of a mycorrhizal fungus on the growth of pea plants with a nitrate fertiliser or an ammonium fertiliser. The fertilisers were identical, except for nitrate or ammonium.

The scientists took pea seeds and sterilised their surfaces. They planted the seeds in soil that had been heated to 85 °C for 2 days before use. The soil was sand that contained no mineral ions useful to the plants.

(a)   Explain why the scientists sterilised the surfaces of the seeds and grew them in soil that had been heated to 85 °C for 2 days.   *(2 marks)*

(b)   Explain why it was important that the soil contained no mineral ions useful to the plants.   *(1 mark)*

The pea plants were divided into four groups, A, B, C and D.
- Group A – heat-treated mycorrhizal fungus added, nitrate fertiliser
- Group B – mycorrhizal fungus added, nitrate fertiliser
- Group C – heat-treated mycorrhizal fungus added, ammonium fertiliser
- Group D – mycorrhizal fungus added, ammonium fertiliser

The heat-treated fungus had been heated to 120 °C for 1 hour.

(c)   Explain how groups A and C act as controls.   *(2 marks)*

After 6 weeks, the scientists removed the plants from the soil and cut the roots from the shoots. They dried the plant material in an oven at 90 °C for 3 days. They then determined the mean dry masses of the roots and shoots of each group of pea plants.

(d)   (i)   Suggest what the scientists should have done during the drying process to be sure that all of the water had been removed from the plant samples.   *(2 marks)*

The scientists' results are shown in **Table 3**.

▼ Table 3

Treatment	Mean dry mass / g per plant (± standard deviation)	
	Root	Shoot
A – heat-treated fungus and nitrate fertiliser	0.40 (±0.05)	1.01 (±0.12)
B – fungus and nitrate fertiliser	1.61 (±0.28)	9.81 (±0.33)
C – heat-treated fungus and ammonium fertiliser	0.34 (±0.03)	0.96 (±0.26)
D – fungus and ammonium fertiliser	0.96 (±0.18)	4.01 (± 0.47)

(ii)   the values of dry mass recorded for the shoots of the plants in group D were   3.4, 4.5, 4.2, 3.9, 4.1g.

calculate the mean dry mass and the standard deviation, s, of this mean.

Use the formula $s = \sqrt{\dfrac{\Sigma(x - \bar{x})^2}{n-1}}$ *(5 marks)*

(iii)   comment on the reliability of the means shown in the table   *(2 marks)*

(e)   What conclusions can be drawn from the data in **Table 3** about the following?

The effects of the fungus on growth of the pea plants.

The effects of nitrate fertiliser and ammonium fertiliser on growth of the pea plants.   *(4 marks)*

AQA Specimen 2014 (apart from 5 (d) (ii) and(iii))

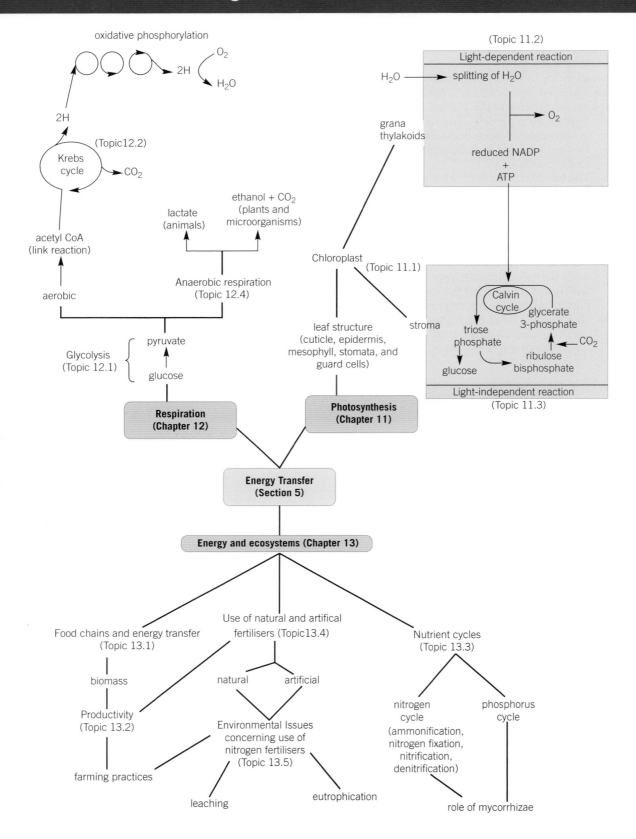

Practical skills

In this section you have met the following practical skills:

How to use appropriate apparatus to record a range of quantitative measurements in experiments such as:

- investigating the effect of named environmental variables on the rate of photosynthesis

- using a redox indicator such as methylene blue to investigate dehydrogenase activity.

Maths skills

In this section you have met the following maths skills:

- calculating gross primary productivity and deriving the appropriate units.

- calculating the efficiency of energy transfers within ecosystems.

- calculating percentage yields of crops.

- interpreting data from a variety of tables and graphs.

- translating information between graphical, numerical, and algebraic forms.

Extension

Design and carry out an experiment to determine whether there is a significant difference between the average surface area of leaves on nettle plants that are growing in a sunny position compared to leaves on nettle plants growing in a shaded position.

In planning and carrying out your experiment, consider how you will:

- select the sunny and shaded positions

- choose the individual nettle plants

- select the sample leaves from the chosen plants

- decide on the number of samples you will take

- measure the surface area of the leaves

- test whether any differences you find are statistically significant.

1 **(a)** A student measured the rate of aerobic respiration of a woodlouse using the apparatus shown in Figure 1.

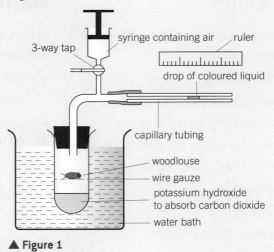

▲ **Figure 1**

(i) The student closed the tap. After thirty minutes the drop of coloured liquid had moved to the left. Explain why the drop of coloured liquid moved to the left. *(3 marks)*

(ii) What measurements should the student have taken to calculate the rate of aerobic respiration in mm^3 of oxygen $g^{-1} h^{-1}$? *(3 marks)*

(b) DNP inhibits respiration by preventing a proton gradient being maintained across membranes. When DNP was added to isolated mitochondria the following changes were observed
- less ATP was produced
- more heat was produced
- the uptake of oxygen remained constant

Explain how DNP caused these changes *(3 marks)*

AQA Jan 2011

2 **(a)** Ecologists investigated photosynthesis in two species of plant found in woodland. One of the species was adapted to growing in bright sunlight (sun plant) and the other was adapted to growing in the shade (shade plant). The ecologists' results are shown in Figure 2.

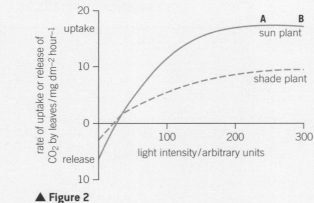

▲ **Figure 2**

(i) Give two factors which could be limiting the rate of photosynthesis in the sun plant between points A and B on Figure 2. *(1 mark)*

(ii) Explain why CO_2 uptake is a measure of net productivity. *(1 mark)*

(iii) Use the information in Figure 1 to explain how the shade plant is better adapted than the sun plant to growing at low light intensities. *(2 marks)*

AQA June 2014

3 A student investigated the rate of anaerobic respiration in yeast. She put 5g of yeast into a glucose solution and placed this mixture in the apparatus shown in Figure 3. She then recorded the total volume of gas collected every ten minutes for 1 hour.

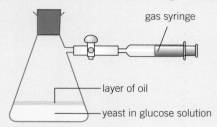

gas syringe

layer of oil

yeast in glucose solution

▲ **Figure 3**

(a) Explain why a layer of oil is required in this investigation. *(1 mark)*

(b) The student's results are shown in Table 1.

▼ **Table 1**

Time / minutes	Total volume of gas collected / cm^3
10	0.3
20	0.9
30	1.9
40	3.1
50	5.0
60	5.2

(c) (i) Calculate the rate of gas production in cm^3 g^{-1} min^{-1} during the first 40 minutes of this investigation. Show your working. *(2 marks)*

(ii) Suggest why the rate of gas production decreased between 50 and 60 minutes. *(1 mark)*

(iii) Yeast can also respire aerobically. The student repeated the investigation with a fresh sample of yeast in glucose solution, but without the oil. All other conditions remained the same.
Explain what would happen to the volume of gas in the syringe if the yeast were only respiring aerobically. *(2 marks)*

(d) Respiration produces more ATP per molecule of glucose in the presence of oxygen than it does when oxygen is absent. Explain why. *(2 marks)*

AQA June 2014

4 (a) Name the type of bacteria which convert:
(i) nitrogen in the air into ammonium compounds
(ii) nitrites into nitrates *(2 marks)*

(b) (i) Other than spreading fertilisers, describe and explain how **one** farming practice results in addition of nitrogen-containing compounds to a field. *(2 marks)*

(ii) Describe and explain how **one** farming practice results in the removal of nitrogen-containing compounds from a field. *(2 marks)*

AQA June 2006

5 Starfish feed on a variety of invertebrate animals that are attached to rocks on the seashore. The diagram shows part of a food web involving a species of starfish.

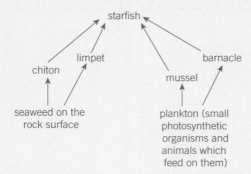

(a) Explain why a starfish can be described as both a secondary and a tertiary consumer. *(1 mark)*

(b) When starfish feed on mussels they leave behind the empty shell. Explain how quadrats could be used to determine the percentage of mussels that had been eaten by starfish on a rocky shore. *(3 marks)*

(c) Table 2 shows the composition of the diet of starfish.

▼ **Table 2**

	Prey species			
	chitons	limpets	mussels	barnacles
Percentage of total number of animals eaten	3	5	27	65
Energy provided by each species as a percentage of total energy intake	42	5	38	15

(i) The percentage of barnacles in the diet is much higher than the percentage of energy they provide. Suggest **one** explanation for this difference. *(1 mark)*

(ii) Table 2 shows that the amount of energy provided by chitons is greater than the amount of energy provided by limpets. Calculate the number of limpets a starfish would need to eat in order to obtain the same amount of energy as it would obtain from one chiton. *(1 mark)*

AQA June 2006

6 The progress of the light-dependent reaction was investigated using DCPIP indicator solution to compare the activity of chlorophyll extract with that of intact chloroplasts. The following results were obtained.

Time / s	% absorption of red light (extract)	% absorption of red light (intact chloroplasts)
60	98	96
90	92	94
120	86	88
150	74	84
180	63	76
210	55	70
240	41	64
270	35	58
300	28	55
330	26	53
360	24	52

(a) Plot a suitable graph to show the progress of the reaction in each sample. (*4 marks*)

(b) Use the graph to determine the percentage difference between the maximum rates of reaction of the two samples. (*3 marks*)

7 **(a)** The biochemical pathway of aerobic respiration involves a number of different steps. Name **one** step in which carbon dioxide is produced. (*1 marks*)

In an investigation, scientists transferred slices of apple from air to anaerobic conditions in pure nitrogen gas. They measured the rate of carbon dioxide production.

(b) The scientists kept the temperature constant throughout the investigation. Explain how a decrease in temperature would affect the rate of carbon dioxide production. (*2 marks*)

(c) When the apple slices were transferred to nitrogen, the following biochemical pathway took place.

pyruvic acid   ethanol

Use this pathway to explain the part played by reduced NAD when the apple slices were transferred to nitrogen. (*2 marks*)

(d) The rate of carbon dioxide production was higher when the apple slices were in nitrogen than when they were in the air. Explain why. (*3 marks*)

AQA Jan 2010

Section 6
Organisms respond to changes in their environment

Introduction

Multicellular organisms are able to respond to stimuli that originate both from outside and from within their bodies. By doing so, they can avoid harmful environments while at the same time maintaining an internal environment that provides the optimum conditions for their metabolism. These organisms control their activities through a combination of growth factors, hormones, and nerve impulses. A stimulus is a change in the internal or external environment. Stimuli are detected by receptors which are specific to one type of stimulus. A coordinator formulates a suitable response to a stimulus and an effector produces the response.

There are two forms of coordination in most multicellular animals – nervous and hormonal. The nervous system allows rapid communication between one part of an organism and another. It comprises nerve cells that pass electrical impulses along their length. The nerve impulse releases a chemical messenger on to its target cell that usually leads to a rapid, short-lived, and localised response. The hormonal system produces slower responses. Hormones are produced in endocrine glands and stimulate their target cells via the blood stream. Each hormone is specific to particular receptors that are only present on their target cells. Hormonal responses are usually slow, long-lasting, and widespread.

It is advantageous to maintain a relatively constant internal environment. Not only can chemical reactions take place at a predictable rate but also the organism has a greater degree of independence from the external environment. The maintenance of a constant internal environment is called homeostasis. Responding to changes in the internal and external environment is no less important to survival in plants. Plants lack contractile tissue and do not move from place to place. At the molecular level, some plant responses are rapid where they use hormone-like substances to control their responses to stimuli such as light and gravity.

Working scientifically

In studying how organisms respond to changes in their environment there will be opportunities to perform practical exercises and so develop practical skills. Required practical activities are:

- Investigating the effect of an environmental variable on the movement of an animal using either a choice chamber or a maze.

- Producing a dilution series of a glucose solution and using colorimetric techniques to produce a calibration curve with which to identify the concentration of glucose in an unknown 'urine' sample.

In performing these experiments you will have the chance to develop practical skills such as:

- using a colorimeter to record quantitative measurements
- using laboratory glassware apparatus to make up serial dilutions
- using a light microscope at high power and low power
- safely and ethically using organisms to measure animal responses.

You will be able to develop a range of mathematical skills. In particular the ability to use appropriate units when calculating the maximum frequency of impulse conduction.

What you already know:

The information in this unit is intended to be self-explanatory, but there is certain knowledge from GCSE that will prove beneficial to the understanding of this section. This information includes:

- ◯ Cells called receptors detect stimuli (changes in the environment). Information from receptors passes along cells (neurones) in nerves to the brain. The brain coordinates the response. Reflex actions are automatic and rapid.

- ◯ In a simple reflex action impulses from a receptor pass along a sensory neurone to the central nervous system. At a junction (synapse) between a sensory neurone and a relay neurone in the central nervous system, a chemical is released that causes an impulse to be sent along a relay neurone.

- ◯ Many processes within the body are coordinated by chemical substances called hormones. Hormones are secreted by glands and are usually transported to their target organs by the bloodstream.

- ◯ Plants are sensitive to light, moisture and gravity; their shoots grow towards light and against the force of gravity while their roots grow towards moisture and in the direction of the force of gravity.

- ◯ Plants produce hormones to coordinate and control growth. Auxin controls phototropism and gravitropism (geotropism).

- ◯ Waste products to be removed from the body include urea, produced in the liver by the breakdown of amino acids and removed by the kidneys in the urine.

- ◯ A healthy kidney produces urine by filtering the blood and then reabsorbing all the sugar, dissolved ions and water needed by the body and finally releasing urea, excess ions, and water as urine.

- ◯ The blood glucose concentration of the body is monitored and controlled by the pancreas through the production of the hormone insulin.

- ◯ A second hormone, glucagon, is produced in the pancreas when blood glucose levels fall. This causes glycogen to be converted into glucose.

- ◯ Type 1 diabetes is a disease in which a person's blood glucose concentration may rise to a high level because the pancreas does not produce enough of the hormone insulin.

Learning objectives

→ Define a stimulus and a response.

→ Examine the advantage to organisms of being able to respond to stimuli.

→ Describe taxes, kineses, and tropisms.

→ Explain how each type of response increases an organism's chances of survival.

Specification reference: 3.6.1.1

Practical link 🧪

Required practical 10. Investigation into the effect of an environmental variable on the movement of an animal using either a choice chamber or a maze.

▲ **Figure 1** *Woodlice exhibit a behaviour called kinesis, which ensures that they spend most of their time in the dark moist conditions that prevent them from drying out and hence aid their survival*

Stimulus and response

A **stimulus** is a detectable change in the internal or external environment of an organism that leads to a **response** in the organism. The ability to respond to stimuli is a characteristic of life and increases the chances of survival for an organism. For example, to be able to detect and move away from harmful stimuli, such as predators and extremes of temperature, or to detect and move towards a source of food clearly aid survival. Those organisms that survive have a greater chance of raising offspring and of passing their alleles to the next generation. There is always, therefore, a selection pressure favouring organisms with more appropriate responses.

Stimuli are detected by **receptors**. Receptors are specific to one type of stimulus. A coordinator formulates a suitable response to a stimulus. Coordination may be at the molecular level or involve a large organ such as the brain. A response is produced by an **effector**. This response may be at the molecular level or involve the behaviour of a whole organism. One means of communication in large, multicellular organisms occurs via chemicals called hormones, which is a relatively slow process found in both plants and animals (Topics 14.2 and 16.3).

In addition to hormonal communication, animals have another, more rapid, means of communication – the nervous system. Their nervous systems usually have many different receptors and control effectors. Each receptor and effector is linked to a central **coordinator** of some type. The coordinator acts like a switchboard, connecting information from each receptor with the appropriate effector. The sequence of events can therefore involve either chemical control or nerve cells and may be summarised as:

stimulus → receptor → coordinator → effector → response

Let us look first at the simplest forms of response to stimuli and how they can increase an organism's chances of survival.

Taxes

A **taxis** is a simple response whose direction is determined by the direction of the stimulus. As a result, a motile organism responds directly to environmental changes by moving its whole body either towards a favourable stimulus or away from an unfavourable one. Taxes are classified according to whether the movement is towards the stimulus (positive taxis) or away from the stimulus (negative taxis) and also by the nature of the stimulus. Some examples are given below:

• Single-celled algae will move towards light (positive phototaxis). This increases their chances of survival since, being photosynthetic, they require light to manufacture their food.

- Earthworms will move away from light (negative phototaxis). This increases their chances of survival because it takes them into the soil, where they are better able to conserve water, find food and avoid some predators.
- Some species of bacteria will move towards a region where glucose is more highly concentrated (positive chemotaxis). This increases their chances of survival because they use glucose as a source of food.

Kineses

A **kinesis** is a form of response in which the organism does not move towards or away from a stimulus. Instead, it changes the speed at which it moves and the rate at which it changes direction. If an organism crosses a sharp dividing line between a favourable and an unfavourable environment, its rate of turning increases. This raises its chances of a quick return to a favourable environment. However, if it moves a considerable distance into an unfavourable environment its rate of turning may slowly decrease so that it moves in long straight lines before it turns, often very sharply. This type of response tends to bring the organism into a new region with favourable conditions. It is important when a stimulus is less directional. Humidity and temperature, for example, do not always produce a clear gradient from one extreme to another.

An example of a kinesis occurs in woodlice. Woodlice lose water from their bodies in dry conditions. When they move from a damp area into a dry one, they move more rapidly and change direction more often. This increases their chance of moving back into the damp area. Once back in the damp area, they slow down and change direction less often. This means they are more likely to stay within the damp area. However, if after some time spent changing direction rapidly they are in the damp area, their behaviour changes. Instead they move rapidly in straight lines, which increases their chances of moving through the dry area and into a new damp one. In this way they spend more time in favourable damp conditions than in less favourable drier ones. This prevents them drying out and so increases their chances of survival.

Tropisms

A **tropism** is the growth of part of a plant in response to a directional stimulus. In almost all cases the plant part grows towards (positive response) or away from (negative response) the stimulus. Again, the type of response is named after the stimulus. Examples, and the survival value of the response, include the following:

- Plant shoots grow towards light (positive phototropism) and away from gravity (negative gravitropism) so that their leaves are in the most favourable position to capture light for photosynthesis.
- Plant roots grow away from light (negative phototropism) and towards gravity (positive gravitropism). In both cases the response increases the probability that roots will grow into the soil, where they are better able to absorb water and mineral ions.

We will learn more about tropisms in Topic 14.2.

▲ **Figure 2** *Cress seedlings exhibiting phototropism. The seedlings at the bottom have been grown with light directed at them from the left-hand side*

Summary questions

For each of the following statements, name the type of response described and the survival value of the response.

1 Some species of bacteria move away from the waste products that they produce.

2 The sperm cells of a moss plant are attracted towards a chemical produced by the female reproductive organ of another moss plant.

3 The young stems of seedlings grow away from gravity.

Unlike animals, plants have no nervous system. Nevertheless, in order to survive, plants respond to changes in both their external and internal environments. For example, plants respond to:

- **light**. Shoots grow towards light (i.e., are positively phototropic) because light is needed for photosynthesis.
- **gravity**. Plants need to be firmly anchored in the soil. Roots are sensitive to gravity and grow in the direction of its pull (i.e., they are positively gravitropic).
- **water**. Almost all plant roots grow towards water (i.e., are positively hydrotropic) in order to absorb it for use in photosynthesis and other metabolic processes, as well as for support.

Plant responses to external stimuli involve hormone-like substances or, more correctly, **plant growth factors**. The latter term is more descriptive because:

- they exert their influence by affecting growth and, they may be made by cells located throughout the plant rather than in particular organs
- unlike animal hormones, some plant growth factors affect the tissues that release them rather than acting on a distant target organ.

Plant growth factors are produced in small quantities. An example of a plant growth factor is **indoleacetic acid** (**IAA**), which belongs to a group of substances called auxins. Among other things, IAA controls plant cell elongation.

Control of tropisms by IAA

We learnt in Topic 14.1 that a tropism is the directional growth of a plant in response to a directional stimulus. In the case of light, we can observe that a young shoot will grow towards light that is directed at it from one side (unilateral light). This is known as **positive phototropism**.

Phototropism in flowering plants

The response of the shoots of flowering plants to unilateral light is due to the following sequence of events:

1 Cells in the tip of the shoot produce IAA, which is then transported down the shoot.
2 The IAA is initially transported evenly throughout all regions as it begins to move down the shoot.
3 Light causes the movement of IAA from the light side to the shaded side of the shoot.
4 A greater concentration of IAA builds up on the shaded side of the shoot than on the light side.
5 As IAA causes elongation of shoot cells and there is a greater concentration of IAA on the shaded side of the shoot, the cells on this side elongate more.
6 The shaded side of the shoot elongates faster than the light side, causing the shoot tip to bend towards the light.

IAA also controls the bending of roots in response to light. However, whereas a high concentration of IAA increases cell elongation in shoots, it inhibits cell elongation in roots. For example, an IAA concentration of 10 parts per million increases shoot cell elongation by 200% but decreases root cell elongation by 100% (Figure 1). As a result, in roots the elongation of cells is greater on the light side than on the shaded side and so roots bend away from light, that is, they are negatively phototropic.

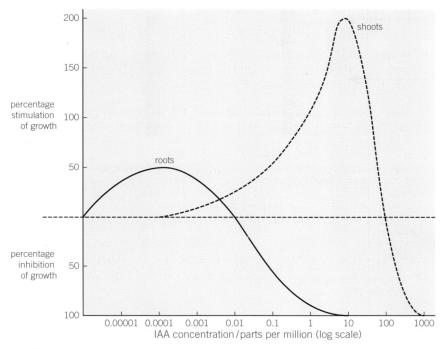

▲ **Figure 1** *Relationship between cell elongation and IAA concentration in shoots and roots*

Gravitropism in flowering plants

The response of a horizontally-growing root to gravity is as follows:

1 Cells in the tip of the root produce IAA, which is then transported along the root.
2 The IAA is initially transported to all sides of the root.
3 Gravity influences the movement of IAA from the upper side to the lower side of the root.
4 A greater concentration of IAA builds up on the lower side of the root than on the upper side.
5 As IAA inhibits the elongation of root cells and there is a greater concentration of IAA on the lower side, the cells on this side elongate less than those on the upper side.
6 The relatively greater elongation of cells on the upper side compared to the lower side causes the root to bend downwards towards the force of gravity.

In shoots, the greater concentration of IAA on the lower side increases cell elongation and causes this side to elongate more than the upper side. As a result the shoot grows upwards away from the force of gravity.

> **Hint**
>
> In many plants, gravity leads to a change in the distribution of IAA carrier proteins that export IAA from cells.

▲ **Figure 2** *Gravitropism is a plant response to the Earth's gravitational field. This bean plant shows a turn in its stem which occurred after the pot was tipped over. The response also occurs in the dark, showing that it is not phototropism*

These events are summarised in Figure 3.

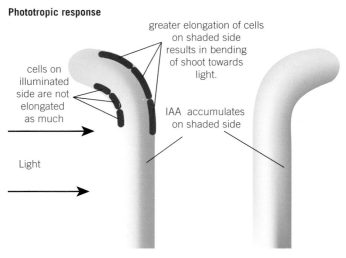

Phototropic response

cells on illuminated side are not elongated as much

greater elongation of cells on shaded side results in bending of shoot towards light.

IAA accumulates on shaded side

Light

shoot
high IAA concentration stimulates cell elongation causing a positive phototropic response

root
high IAA concentration inhibits cell elongation causing a negative phototropic response

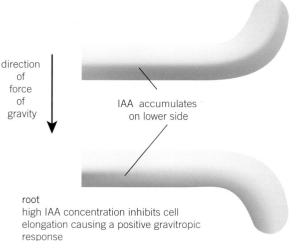

Gravitropism

shoot
high IAA concentration stimulates cell elongation causing a negative gravitropic response

direction of force of gravity

IAA accumulates on lower side

root
high IAA concentration inhibits cell elongation causing a positive gravitropic response

▶ **Figure 3** *Mechanism of IAA action in the phototropic and gravitropic responses of shoots and roots*

Role of IAA in elongation growth

The transport of IAA is in one direction, namely away from the tip of shoots and roots where it is produced. IAA has a number of effects on plant cells including increasing the plasticity (ability to stretch) of their cell walls. The response only occurs on young cell walls where cells are able to elongate. As the cells mature they develop greater rigiditiy – therefore older parts of the shoot/root will not be able to respond. The proposed explanation of how IAA increases the plasticity of cells is called the acid growth hypothesis. It involves the active transport of hydrogen ions from the cytoplasm into spaces in the cell wall causing the cell wall to become more plastic allowing the cell to elongate by expansion.

The elongation of cells on one side only of a stem or root can lead to them bending (Figure 3). This is the means by which plants respond relatively quickly to environmental stimuli like light and gravity. These responses can be explained in terms of the stimuli causing uneven distribution of IAA, as described earlier, as it moves away from the tip of the stem or root.

Summary questions

1 Explain how the movement of IAA in shoots helps a plant to survive.

2 Suggest **two** advantages to a plant of having roots that respond to gravity by growing in the direction of its force.

3 Consider the following facts about IAA:

   i They are easily made synthetically

   ii They are readily absorbed by plants

   iii They are not easily broken down

   iv They are lethal to some plants in low concentrations

   v Narrow-leaved plants are less easily killed than broad-leaved plants.

   Suggest ways in which these facts might be relevant to agricultural practice.

 ### Discovering the role of IAA in tropisms

No less a person than Charles Darwin was one of the earliest scientists to investigate the response of plant shoots to light. He observed that young grass shoots grow towards the window (i.e., they were positively phototropic). Being curious, he proposed the hypothesis that the stimulus of the light was detected by the tip of the shoot, which was therefore the source of the response. He tested his hypothesis in a series of experiments in which he removed the tips of shoots or covered the tips with lightproof covers.

These experiments and the results are summarised in Figure 4 (experiments 1–3).

> **1** Which of Darwin's three experiments acted as a control?

Expt no.	Method	Result	Explanation
1	Unilateral light → Plant shoot	Shoot bends towards light	The shoot is positively phototropic. Bending occurs behind the tip.
2	Shoot tip removed, Light → Tip discarded	No response	The tip must either detect the stimulus or produce the messenger (or both) as its removal prevents any response.
3	Lightproof cover is placed over intact tip of shoot, Light →	No response	The light stimulus must be detected by the tip.

▲ **Figure 4** *Darwin's experiments to show that it is the tips of shoots that are the source of the phototropic response*

Once it had been shown that the tip is the light-sensitive region of the shoot but that the response (bending) occurs lower down the shoot, some scientists proposed another hypothesis, that a chemical substance was being produced in the tip and transported down the stem, where it caused a response. Others disagreed and put forward an alternative theory, that it was an electrical signal that was passing from the tip and causing the response. One scientist, Peter Boysen-Jensen, carried out a further set of experiments

designed to prove the hypothesis that the 'messenger' was a chemical. In these experiments, he used mica, which conducts electricity but not chemicals, and gelatin, which conducts chemicals but not electricity. His experiments and the results are summarised in Figure 5 (experiments 4–6).

2 Suggest an explanation for the results in experiment 5.

EXPT NO.	METHOD	RESULT	EXPLANATION
4	Thin, impermeable barrier of mica. Light 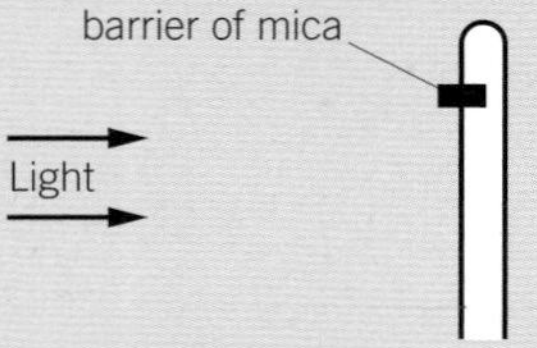	Movement of chemical down shaded side. Bends towards the light 	Mica on the illuminated side of the shoot allows the hormone to pass only down the shaded side where it increases growth and causes bending.
5	Mica inserted on shaded side. Light 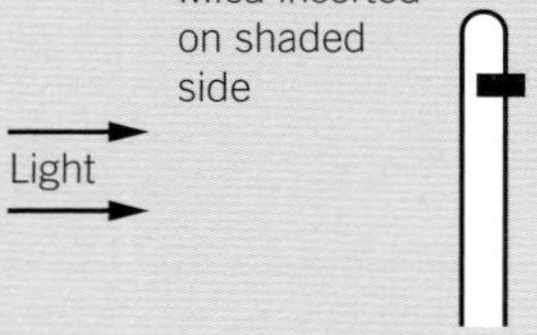	Movement of chemical down shaded side is prevented by mica. No response 	
6	Tip removed, gelatin block inserted and tip replaced. Light. Gelatin block 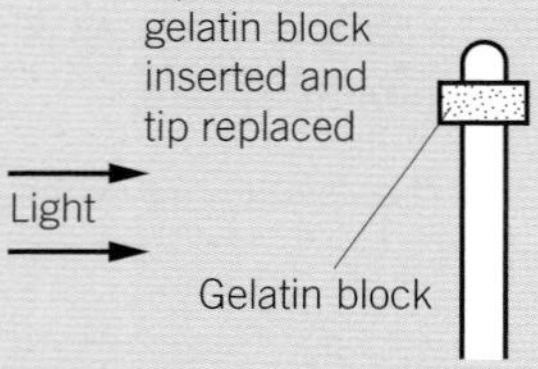	Movement of chemical down shaded side. Bends towards the light 	As gelatin allows chemicals to pass through it, but not electrical signals, the bending which occurs must be due to a chemical passing from the tip.

▲ **Figure 5** *Boysen-Jensen's experiments to show the nature of the 'messenger' in the phototropic response*

Boysen-Jenson's experiments stimulated another scientist, Arpad Paal, to investigate how the chemical messenger worked. He removed the tips of shoots and placed them on one side of the cut surface. He kept the shoots in total darkness throughout the experiment.

His experiment and its results are shown in Figure 6 (experiment 7).

3 Suggest an explanation for the results in experiment 7.

EXPT NO.	METHOD	RESULT
7	Darkness. 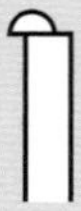 Tips removed and then replaced but displaced to one side 	 Shoots bend towards side where no tip is present 

▲ **Figure 6** *Paal's experiment on the action of the 'messenger'*

So far, it had been established that bending was due to a chemical which was produced in the tip and caused growth on the shaded side of the shoot. This chemical was later shown to be indoleacetic acid (IAA). The next question was how did light cause the uneven distribution of IAA? Different theories were put forward, including:

- Light inhibits IAA production in the tip and so it is only produced on the shaded side.

- Light destroys the IAA as it passes down the light side of the shoot.

- IAA is transported from the light side to the shaded side of the shoot.

This prompted Winslow Briggs and his associates to test these hypotheses. They set up experiments as shown in Figure 7 (experiments 8–10).

Study experiments 8, 9 and 10.

4 Suggest reasons for using a glass plate in experiments 9 and 10.

5 State which of the three theories the results tend to support. Give reasons for your answer.

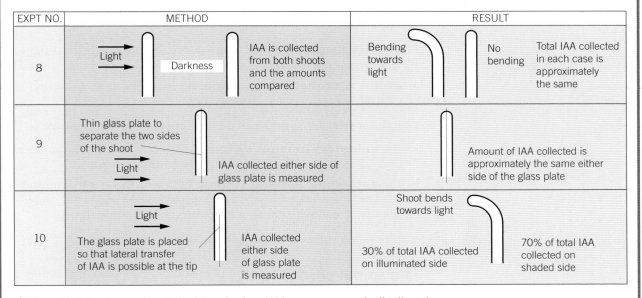

EXPT NO.	METHOD	RESULT
8	Light → Darkness — IAA is collected from both shoots and the amounts compared	Bending towards light — No bending — Total IAA collected in each case is approximately the same
9	Thin glass plate to separate the two sides of the shoot — Light → IAA collected either side of glass plate is measured	Amount of IAA collected is approximately the same either side of the glass plate
10	Light → The glass plate is placed so that lateral transfer of IAA is possible at the tip — IAA collected either side of glass plate is measured	Shoot bends towards light — 30% of total IAA collected on illuminated side — 70% of total IAA collected on shaded side

▲ **Figure 7** *Briggs's experiments to determine how IAA becomes unevenly distributed*

The simplest type of nervous response to a stimulus is a reflex arc. Before considering how a reflex arc works, it is helpful to understand how the millions of **neurones** in a mammalian body are organised.

Nervous organisation

The nervous system has two major divisions:

* the **central nervous system (CNS)**, which is made up of the brain and spinal cord
* the **peripheral nervous system (PNS)**, which is made up of pairs of nerves that originate from either the brain or the spinal cord.

The peripheral nervous system is divided into:

* **sensory neurones**, which carry nerve impulses (electrical signals) from receptors towards the central nervous system
* **motor neurones**, which carry nerve impulses away from the central nervous system to **effectors**.

The motor nervous system can be further subdivided as follows:

* the **voluntary nervous system**, which carries nerve impulses to body muscles and is under voluntary (conscious) control
* the **autonomic nervous system**, which carries nerve impulses to glands, **smooth muscle** and cardiac muscle and is not under voluntary control, that is, it is involuntary (subconscious).

A summary of nervous organisation is given in Figure 1.

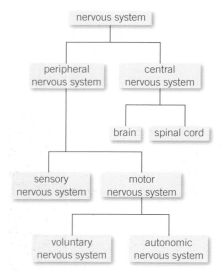

▲ **Figure 1** *Nervous organisation*

The spinal cord

The spinal cord is a column of nervous tissue that runs along the back and lies inside the vertebral column for protection. Emerging at intervals along the spinal cord are pairs of nerves as shown in Figure 2.

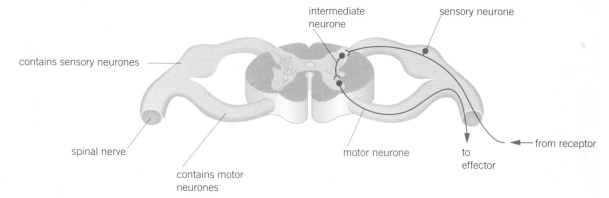

▲ **Figure 2** *Section through spinal cord showing the neurones of a reflex arc*

A reflex arc

You will have noticed that you immediately withdraw your hand if you place it on a hot or sharp object. You do not stop to consider any alternative actions. The response is rapid, short-lived, localised and totally involuntary. Indeed, by the time the brain has received nerve impulses from the receptors in the hand, the muscles in the arm have already pulled the hand clear of the danger. This type of involuntary response to a sensory stimulus is called a **reflex**. The pathway of neurones involved in a reflex is known as a **reflex arc**.

Reflex arcs, such as the withdrawal reflex described above, involve just three neurones. One of the neurones is in the spinal cord and so this type of reflex is also called a spinal reflex. The main stages of a spinal reflex arc, such as withdrawing the hand from a hot object, are described below (the numbers relate to the stages shown in Figure 3).

1 the **stimulus** – heat from the hot object
2 a **receptor** – temperature receptors in the skin on the back of the hand, which generates nerve impulses in the sensory neurone
3 a **sensory neurone** – passes nerve impulses to the spinal cord
4 a **coordinator** (intermediate neurone) – links the sensory neurone to the motor neurone in the spinal cord
5 a **motor neurone** – carries nerve impulses from the spinal cord to a muscle in the upper arm
6 an **effector** – the muscle in the upper arm, which is stimulated to contract
7 the **response** – pulling the hand away from the hot object.

> **Hint**
>
> Remember the sequence: stimulus, receptor, sensory neurone, intermediate neurone, motor neurone, effector, response.

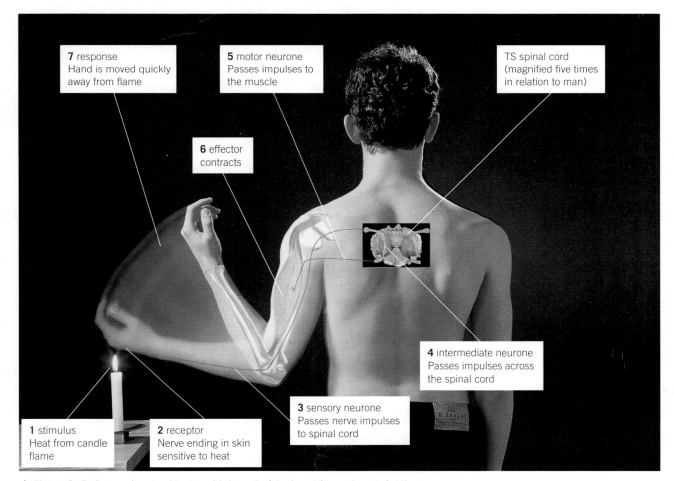

7 response
Hand is moved quickly away from flame

5 motor neurone
Passes impulses to the muscle

TS spinal cord
(magnified five times in relation to man)

6 effector
contracts

4 intermediate neurone
Passes impulses across the spinal cord

3 sensory neurone
Passes nerve impulses to spinal cord

1 stimulus
Heat from candle flame

2 receptor
Nerve ending in skin sensitive to heat

▲ **Figure 3** *Reflex arc involved in the withdrawal of the hand from a heat stimulus*

Importance of reflex arcs

Any action that makes survival more likely is clearly of value. Reflexes are involuntary – the actions they control do not need to be considered, because there is only one obvious course of action, that is, to remove the hand from the hot object. Reflex actions are important for the following reasons:

- They are involuntary and therefore do not require the decision-making powers of the brain, thus leaving it free to carry out more complex responses. In this way, the brain is not overloaded with situations in which the response is always the same. Some impulses are nevertheless sent to the brain, so that it is informed of what is happening and can sometimes override the reflex if necessary.

- They protect the body from harm. They are effective from birth and do not have to be learnt.

- They are fast, because the neurone pathway is short with very few, typically one or two, **synapses** where neurones communicate with each other (synapses are the slowest link in a neurone pathway). This is important in withdrawal reflexes.

- The absence of any decision-making process also means the action is rapid.

Summary questions

In the following passage give the word that best replaces the number in brackets.

The nervous system has two main divisions, the central nervous system (CNS), comprising the (1) and (2), and the peripheral nervous system (PNS). The peripheral nervous system is made up of the (3) nerves that carry impulses away from the CNS and (4) nerves that carry impulses towards the CNS. A spinal reflex is an (5) response that involves the spinal cord. An example is the withdrawal reflex, for example, the withdrawing of the hand from a hot object. The sequence of events begins with the heat from the hot object, which acts as the (6). This is detected by a (7) in the skin on the back of the hand, which creates nerve impulses that pass along a (8) neurone into the spinal cord. The impulse then passes to an (9) neurone, in the central region of the spinal cord. The impulse leaves the spinal cord via a (10) neurone. This neurone stimulates a muscle of the upper arm to contract and withdraw the hand from the object. Structures such as these that bring about a response to a stimulus are called (11).

The central nervous system receives sensory information from its internal and external environment through a variety of receptors, each type responding to a different and specific type of stimulus. Sensory reception is the function of these receptors, whereas sensory perception involves making sense of the information from the receptors. This is largely a function of the brain. The concepts of stimulus and response were covered in Topic 14.1. We shall now look in detail at one receptor – the **Pacinian corpuscle**.

Features of sensory reception as illustrated by the Pacinian corpuscle

Pacinian corpuscles respond to changes in mechanical pressure. As with all sensory receptors, a Pacinian corpuscle:

- **is specific to a single type of stimulus**. In this case, it responds only to mechanical pressure. It will not respond to other stimuli, such as heat, light, or sound.

- **produces a generator potential by acting as a transducer**. All stimuli involve a change in some form of energy. It is the role of the transducer to convert the change in form of energy by the stimulus into a form, namely nerve impulses, that can be understood by the body. The stimulus always involves a change in some form of energy, for example, heat, light, sound, or mechanical energy. The nerve impulse is also a form of energy. Receptors therefore convert, or transduce, one form of energy into another. Receptors in the nervous system convert the energy of the stimulus into a nervous impulse known as a **generator potential**. For example, the Pacinian corpuscle, whose action is described below, transduces the mechanical energy of the stimulus into a generator potential.

Structure and function of a Pacinian corpuscle

Pacinian corpuscles respond to mechanical stimuli such as pressure. They occur deep in the skin and are most abundant on the fingers, the soles of the feet and the external genitalia. They also occur in joints, ligaments and tendons, where they enable the organism to know which joints are changing direction. The single sensory neurone of a Pacinian corpuscle is at the centre of layers of tissue, each separated by a gel. This gives it the appearance of an onion when cut vertically (Figure 2). How does this structure transduce the mechanical energy of the stimulus into a generator potential?

In Topic 4.1 we saw that plasma membranes contain channel proteins that span them. These proteins have channels along which ions can be transported. Each channel is specific. Sodium channels, for example, carry only sodium ions.

The sensory neurone ending at the centre of the Pacinian corpuscle has a special type of sodium channel in its plasma membrane. This is called a **stretch-mediated sodium channel**. These channels are so-called because their permeability to sodium changes when they are deformed, for example, by stretching. The Pacinian corpuscle functions as follows:

- In its normal (resting) state, the stretch-mediated sodium channels of the membrane around the neurone of a Pacinian corpuscle are

Learning objectives

→ Describe the main features of sensory reception.

→ Describe the structure of a Pacinian corpuscle and explain how it works.

→ Explain how receptors work together in the eye.

Specification reference: 3.6.1.2

▲ **Figure 1** *Pacinian corpuscle*

Synoptic link

Details of carrier proteins in membranes and how they function to transport ions through them is covered in Topics 4.4 and 4.5. Revision of this material will help you to fully understand what follows here and in the next chapter.

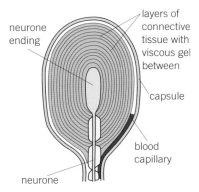

▲ **Figure 2** *Structure of a Pacinian corpuscle*

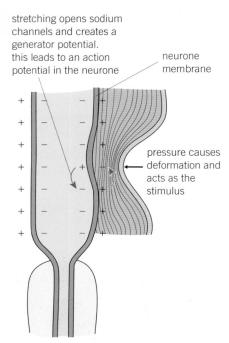

stretching opens sodium channels and creates a generator potential. this leads to an action potential in the neurone

neurone membrane

pressure causes deformation and acts as the stimulus

▲ **Figure 3** *Creation of a generator potential in a Pacinian corpuscle*

▼ **Table 1** *Differences between rod and cone cells*

Rod cells	Cone cells
Rod-shaped	Cone-shaped
Greater numbers than cone cells	Fewer numbers than rod cells
Distribution – more at the periphery of the retina, absent at the fovea	Fewer at the periphery of the retina, concentrated at the fovea
Give poor visual acuity	Give good visual acuity
Sensitive to low-intensity light	Not sensitive to low-intensity light
One type only	Three types each responding to different wavelengths of light

too narrow to allow sodium ions to pass along them. In this state, the neurone of the Pacinian corpuscle has a resting potential (Topic 15.2).

- When pressure is applied to the Pacinian corpuscle, it is deformed and the membrane around its neurone becomes stretched (Figure 3).
- This stretching widens the sodium channels in the membrane and sodium ions diffuse into the neurone.
- The influx of sodium ions changes the potential of the membrane (i.e., it becomes **depolarised**), thereby producing a generator potential.
- The generator potential in turn creates an action potential (nerve impulse) (Topic 15.2) that passes along the neurone and then, via other neurones, to the central nervous system.

These events are illustrated in Figure 3.

Receptors working together in the eye

The light receptor cells of the mammalian eye are found on its innermost layer, the retina. The millions of light receptor cells found in the retina are of two main types: rod cells and cone cells. Both rod and cone cells act as **transducers** by conserving light energy into the electrical energy of a nerve impulse.

Rod cells

Rod cells cannot distinguish different wavelengths of light and therefore lead to images being seen only in black and white. Rod cells are more numerous than cone cells – there are around 120 million in each eye.

Many rod cells are connected to a single sensory neurone in the optic nerve (Figure 4). Rod cells are used to detect light of very low intensity. A certain threshold value has to be exceeded before a **generator potential** is created in the bipolar cells to which they are connected. As a number of rod cells are connected to a single bipolar cell (= retinal convergence), there is a much greater chance that the threshold value will be exceeded than if only a single rod cell were connected to each bipolar cell. This is due to summation which is explained in Topic 15.5. As a result, rod cells allow us to see in low light intensity (i.e., at night), although only in black and white.

In order to create a generator potential, the pigment in the rod cells (rhodopsin) must be broken down. There is enough energy from low-intensity light to cause this breakdown. This explains why rod cells respond to low-intensity light.

A consequence of many rod cells linking to a single bipolar cell is that light received by rod cells sharing the same neurone will only generate a single impulse travelling to the brain regardless of how many of the neurones are stimulated. This means that, in perception, the brain cannot distinguish between the separate sources of light that stimulated them. Two dots close together cannot be resolved and so will appear as a single blob. Rod cells therefore give low **visual acuity**.

Cone cells

Cone cells are of three different types, each responding to a different range of wavelengths of light. Depending upon the proportion of each type that is stimulated, we can perceive images in full colour.

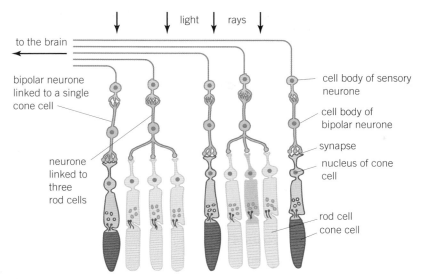

▲ **Figure 4** *Microscopic structure of the retina*

In each human eye, there are around 6 million cone cells, often with their own separate bipolar cell connected to a sensory neurone in the optic nerve (see Figure 4). This means that the stimulation of a number of cone cells cannot be combined to help exceed the threshold value and so create a generator potential. As a result, cone cells only respond to high light intensity and not to low light intensity.

In addition, cone cells contain different types of pigment from that found in rod cells. The pigment in cone cells (iodopsin) requires a higher light intensity for its breakdown. Only light of high intensity will therefore provide enough energy to break it down and create a generator potential. There are three different types of cone cell, each containing a specific type of iodopsin. As a result, each cone cell is sensitive to a different specific range of wavelengths.

Each cone cell has its own connection to a single bipolar cell, which means that, if two adjacent cone cells are stimulated, the brain receives two separate impulses. The brain can therefore distinguish between the two separate sources of light that stimulated the two cone cells. This means that two dots close together can be resolved and will appear as two dots. Therefore cone cells give very accurate vision, that is, they have good visual acuity.

The distribution of rod and cone cells on the retina is uneven. Light is focused by the lens on the part of the retina opposite the pupil. This point is known as the fovea. The fovea therefore receives the highest intensity of light. Therefore cone cells, but not rod cells, are found at the fovea. The concentration of cone cells diminishes further away from the fovea. At the peripheries of the retina, where light intensity is at its lowest, only rod cells are found.

Table 1 summarises the differences between rod and cone cells.

All this shows how the distribution of rod and cone cells, and the connections they make in the optic nerve, can explain the differences in sensitivity and visual acuity in mammals. By having different types of light receptor, each responding to different stimuli, mammals can benefit from good all-round vision both day and night.

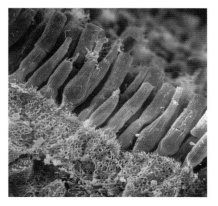

▲ **Figure 5** *False colour SEM of rod and cone cells in the retina of the eye. Rod cells (brown) are long nerve cells responding to dim light while cone cells (green) detect colour*

Summary questions

1   Describe a stretch-mediated sodium channel.

2   Describe the sequence of events by which pressure on a Pacinian corpuscle results in the creation of a generator potential.

3   Explain why brightly coloured objects often appear grey in dim light.

4   At night, it is often easier to see a star in the sky by looking slightly to the side of it rather than directly at it. Suggest why this is so.

Learning objectives

→ Describe the autonomic nervous system.

→ Explain how the autonomic nervous system controls heart rate.

→ Explain the role chemical and pressure receptors play in the processes controlling the heart rate.

Specification reference: 3.6.1.3

Although we are not aware of it, much of the sensory information reaching our central nervous system comes from receptors within our bodies responding to internal stimuli. All the internal systems of our body need to operate efficiently and be ready to adapt to meet the changing demands made upon them. This requires the coordination of a vast amount of information. This information comes from the monitoring of all our internal environment – a process that takes place continuously. Before investigating one example, how heart rate is controlled, let us first look at the part of the nervous system responsible for this type of control – the autonomic nervous system.

The autonomic nervous system

Autonomic means self-governing. The autonomic nervous system controls the involuntary (subconscious) activities of internal muscles and glands. It has two divisions:

- **the sympathetic nervous system**. In general, this stimulates effectors and so speeds up any activity. It acts rather like an emergency controller. It controls effectors when we exercise strenuously or experience powerful emotions. In other words, it helps us to cope with stressful situations by heightening our awareness and preparing us for activity (the fight or flight response).

- **the parasympathetic nervous system**. In general, this inhibits effectors and so slows down any activity. It controls activities under normal resting conditions. It is concerned with conserving energy and replenishing the body's reserves.

The actions of the sympathetic and parasympathetic nervous systems normally oppose one another. In other words they are **antagonistic**. If one system contracts a muscle, then the other relaxes it. The activities of internal glands and muscles are therefore regulated by a balance of the two systems. Let us look at one such example, the control of heart rate.

Control of heart rate

The muscle of the heart is known as cardiac muscle. It is myogenic, that is, its contraction is initiated from within the muscle itself, rather than by nervous impulses from outside (neurogenic), as is the case with other muscles. Within the wall of the right atrium of the heart is a distinct group of cells known as the **sinoatrial node (SAN)**. It is from here that the initial stimulus for contraction originates. The sinoatrial node has a basic rhythm of stimulation that determines the beat of the heart. For this reason it is often referred to as the pacemaker. The sequence of events that controls the basic heart rate is:

- A wave of electrical excitation spreads out from the sinoatrial node across both atria, causing them to contract.

- A layer of non-conductive tissue (atrioventricular septum) prevents the wave crossing to the ventricles.

Synoptic link

The cardiac cycle is covered in Topic 7.5, and the structure of the heart in Topic 7.4. This includes information on how the sinoatrial node controls the heart beat – information that is relevant here.

- The wave of excitation enters a second group of cells called the **atrioventricular node (AVN)**, which lies between the atria.
- The atrioventricular node, after a short delay, conveys a wave of electrical excitation between the ventricles along a series of specialised muscle fibres called **Purkyne tissue** which collectively make up a structure called the **bundle of His**.
- The bundle of His conducts the wave through the atrioventricular septum to the base of the ventricles, where the bundle branches into smaller fibres of Purkyne tissue.
- The wave of excitation is released from the Purkyne tissue, causing the ventricles to contract quickly at the same time, from the bottom of the heart upwards.

These events are summarised in Figure 1.

Modifying the resting heart rate

The resting heart rate of a typical adult human is around 70 beats per minute. However, it is essential that this rate can be altered to meet varying demands for oxygen. During exercise, for example, the resting heart rate may need to more than double.

Changes to the heart rate are controlled by a region of the brain called the **medulla oblongata**. This has two centres concerned with heart rate:

- a centre that **increases heart rate**, which is linked to the sinoatrial node by the sympathetic nervous system
- a centre that **decreases heart rate**, which is linked to the sinoatrial node by the parasympathetic nervous system.

Which of these centres is stimulated depends upon the nerve impulses they receive from two types of receptor, which respond to stimuli of either chemical or pressure changes in the blood.

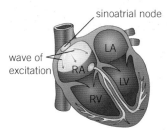

a wave of electrical activity spreads out from the sinoatrial node

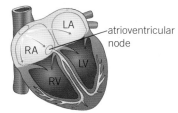

b wave spreads across both atria causing them to contract and reaches the atrioventricular node

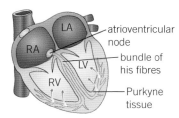

c atrioventricular node conveys wave of electrical activity between the ventricles along the bundle of His and releases it at the apex, causing the ventricles to contract.

▲ **Figure 1** *Control of the heart rate*

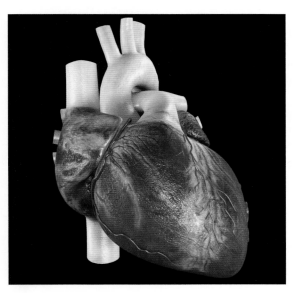

▲ **Figure 2** *Structure of the human heart.*

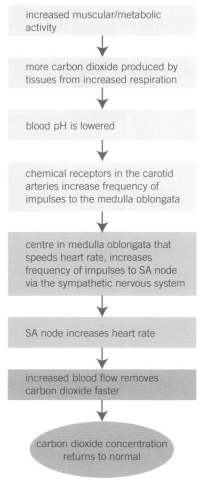

increased muscular/metabolic activity

⬇

more carbon dioxide produced by tissues from increased respiration

⬇

blood pH is lowered

⬇

chemical receptors in the carotid arteries increase frequency of impulses to the medulla oblongata

⬇

centre in medulla oblongata that speeds heart rate, increases frequency of impulses to SA node via the sympathetic nervous system

⬇

SA node increases heart rate

⬇

increased blood flow removes carbon dioxide faster

⬇

carbon dioxide concentration returns to normal

▲ **Figure 3** *Effects of exercise on cardiac output (SA node = sinoatrial node)*

Control by chemoreceptors

Chemoreceptors are found in the wall of the carotid arteries (the arteries that serve the brain). They are sensitive to changes in the pH of the blood that result from changes in carbon dioxide concentration. In solution, carbon dioxide forms an acid and therefore lowers pH. The process of control works as follows:

- When the blood has a higher than normal concentration of carbon dioxide, its pH is lowered.
- The chemoreceptors in the wall of the carotid arteries and the aorta detect this and increase the frequency of nervous impulses to the centre in the medulla oblongata that increases heart rate.
- This centre increases the frequency of impulses via the sympathetic nervous system to the sinoatrial node. This, in turn, increases the rate of production of electrical waves by the sinoatrial node and therefore increases the heart rate.
- The increased blood flow that this causes leads to more carbon dioxide being removed by the lungs and so the carbon dioxide concentration of the blood returns to normal.
- As a consequence the pH of the blood rises to normal and the chemoreceptors in the wall of the carotid arteries and aorta reduce the frequency of nerve impulses to the medulla oblongata.
- The medulla oblongata reduces the frequency of impulses to the sinoatrial node, which therefore leads to a reduction in the heart rate.

This process is summarised in Figure 3, which shows the sequence of events that follows changes in activity levels.

Control by pressure receptors

Pressure receptors occur within the walls of the carotid arteries and the aorta. They operate as follows:

- **When blood pressure is higher than normal**, pressure receptors transmit more nervous impulses to the centre in the medulla oblongata that decreases heart rate. This centre sends impulses via the parasympathetic nervous system to the sinoatrial node of the heart, which leads to a decrease in the rate at which the heart beats.
- **When blood pressure is lower than normal**, pressure receptors transmit more nervous impulses to the centre in the medulla oblongata that increases heart rate. This centre sends impulses via the sympathetic nervous system to the sinoatrial node, which increases the rate at which the heart beats.

Figure 4 summarises the control of heart rate

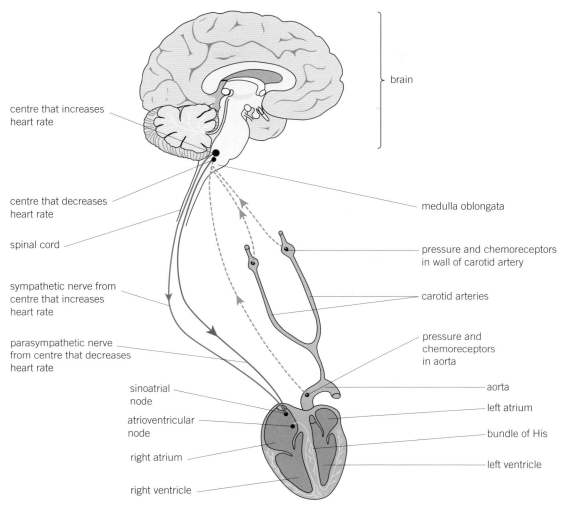

▲ **Figure 4** *Control of heart rate*

Summary questions

1 Describe the function of the autonomic nervous system.

2 Distinguish between the functions of the sympathetic and parasympathetic nervous systems.

3 Suppose the parasympathetic nerve connections from the medulla oblongata to the sinoatrial node were cut. Suggest what might happen if a person's blood pressure increases above normal.

4 The nerve connecting the carotid artery to the medulla oblongata of a person is cut. This person then undertakes some strenuous exercise. Suggest what might happen to the person's:

   a heart rate

   b blood carbon dioxide concentration.

   Explain your answers.

1   IAA is a specific growth factor
    (a) Name the process by which IAA moves from the growing regions of a plant
         shoot to other tissues.   (1 mark)
    (b) (i)   When a young shoot is illuminated from one side, IAA stimulates growth
              on the shaded side. Explain why growth on the shaded side helps to
              maintain the leaves in a favourable environment.   (2 marks)

Temperature °C	0	5	10	15	20	25
Rate of uptake of NAA by lower surface of leaf / arbitrary units	0	4	10	15	26	36
Rate of uptake of NAA by upper surface (arbitrary units)	0	2	4	6	9	12

    (ii) Scientists hypothesise that there is a positive correlation between temperature
          and the rate of uptake of NAA through the lower leaf surface.
          Use the data in the table to calculate the correlation coefficient, r, to
          test this, where $r = \dfrac{\Sigma(x - \bar{x}) \times (y - \bar{y})}{\sqrt{\Sigma(x - \bar{x})^2 \times \Sigma(y - \bar{y})^2}}$ (4 marks)

    (iii) at 10 degrees of freedom the critical value of r at the
          5% level is 0.576. Comment on the support for the
          scientists' hypothesis.   (2 marks)
    (iv) NAA is a similar substance to IAA. It is used to control the
          growth of cultivated plants. Plant physiologists investigated
          the effect of temperature on the uptake of NAA by leaves.
          They sprayed a solution containing NAA on the upper and
          lower surfaces of a leaf. The graph shows their results.

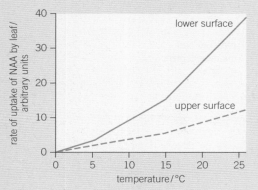

    (c) Explain the effect of temperature on the rate at which
         NAA is taken up by the lower surface of the leaf.   (2 marks)
    (d) There are differences in the properties of the cuticle
         on the upper and lower surfaces of leaves.
         (i)   Suggest how these differences in the cuticle might
               explain the differences in rates of uptake of NAA
               by the two surfaces.   (2 marks)
         (ii) In this investigation, the physiologists investigated
               the leaves of pear trees. Explain why the results
               might be different for other species.   (1 mark)
                                    AQA June 2011 (apart from 1 (b) (ii) and (iii))

2   Scientists investigated the response of lateral roots to gravity. Lateral roots grow from the
    side of main roots.
    The diagrams show four stages, **A** to **D**, in the growth of a lateral root and typical cells
    from the tip of the lateral root in each stage. All of the cells are drawn with the bottom
    of the cell.
    (a) Describe **three** changes in the root tip cells between stages **A** and **D**.   (3 marks)
    (b) The scientists' hypothesis was that there was a relationship between the starch
         grains in the root tip cells and the bending and direction of growth of lateral roots.
         Does the information in the diagram support this hypothesis? Give reasons
         for your answer.   (3 marks)

(c) The diagram shows the distribution of indoleacetic acid (IAA) in the lateral root at Stage **B**.

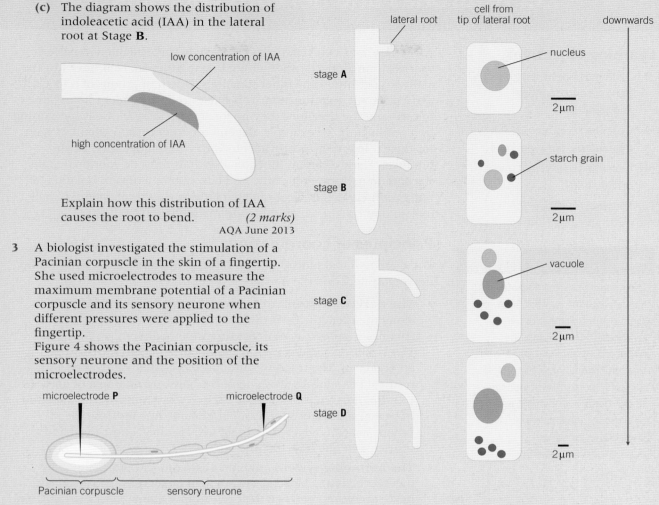

Explain how this distribution of IAA causes the root to bend. *(2 marks)*
AQA June 2013

3 A biologist investigated the stimulation of a Pacinian corpuscle in the skin of a fingertip. She used microelectrodes to measure the maximum membrane potential of a Pacinian corpuscle and its sensory neurone when different pressures were applied to the fingertip.
Figure 4 shows the Pacinian corpuscle, its sensory neurone and the position of the microelectrodes.

▲ **Figure 4**

▼ **Table 2** *shows some of the biologist's results.*

Pressure applied to the fingertip	Membrane potential at P / millivolts	Membrane potential at Q / millivolts
None	−70	−70
Light	−50	−70
Medium	+30	+40
Heavy	+40	+40

(a) Explain how the resting potential of −70 mV is maintained in the sensory neurone when no pressure is applied. *(2 marks)*

(b) Explain how applying pressure to the Pacinian corpuscle produces the changes in membrane potential recorded by microelectrode P. *(3 marks)*

(c) The membrane potential at Q was the same whether medium or heavy pressure was applied to the finger tip. Explain why. *(2 marks)*

(d) Multiple sclerosis is a disease in which parts of the myelin sheaths surrounding neurones are destroyed. Explain how this results in slower responses to stimuli.
(2 marks)
AQA Specimen 2014

Learning objectives

→ Distinguish between how nervous and hormonal coordination.

→ Describe the structure of a myelinated motor neurone.

→ Describe the different types of neurone.

Specification reference: 3.6.2.1

Hint

The nervous system operates like a telephone system, allowing rapid communication between two specific individuals. The hormonal system can be likened to a nationwide mail shot, sending a slower, more general message to everyone, everywhere, but only those individuals who are sensitive to it respond.

As species have evolved, their cells have become adapted to perform specialist functions. By specialising in one function, cells have lost the ability to perform some other functions. Different groups of cells each carry out their own function. This makes cells dependent upon others to carry out the functions they no longer specialise in. Cells specialising in reproduction, for example, depend on other cells to obtain oxygen for their respiration, to provide glucose or to remove their waste products. These different functional systems must be coordinated if they are to perform efficiently. No body system works in isolation, all must be integrated in a coordinated fashion. In this chapter we shall look at one way in which this coordination is achieved.

Principles of coordination

There are two main forms of coordination in animals as a whole – the nervous system and the hormonal system:

- **The nervous system** uses nerve cells to pass electrical impulses along their length. They stimulate their target cells by secreting chemicals, known as **neurotransmitters**, directly on to them. This results in rapid communication between specific parts of an organism. The responses produced are often short-lived and restricted to a localised region of the body. An example of nervous coordination is a reflex action, such as the withdrawal of the hand from an unpleasant stimulus. For obvious reasons this type of action, which is covered in Topic 14.3, is rapid, short-lived and restricted to one region of the body.

- **The hormonal system** produces chemicals (hormones) that are transported in the blood plasma to their target cells. The target cells have specific receptors on their cell-surface membranes and the change in the concentration of hormones stimulates them. This results in a slower, less specific form of communication between parts of an organism. The responses are often long-lasting and widespread. An example of hormonal coordination is the control of blood glucose concentration, which produces a slower response but has a more long term and more widespread effect.

Although different, both systems work together and interact with one another. A comparison of the nervous and hormonal systems is given in Table 1.

▼ **Table 1** *Comparison of hormonal and nervous systems*

Hormonal system	Nervous system
Communication is by chemicals called hormones	Communication is by nerve impulses
Transmission is by the blood system	Transmission is by neurones
Transmission is usually relatively slow	Transmission is very rapid
Hormones travel to all parts of the body, but only target cells respond	Nerve impulses travel to specific parts of the body
Response is widespread	Response is localised
Response is slow	Response is rapid
Response is often long-lasting	Response is short-lived
Effect may be permanent and irreversible	Effect is usually temporary and reversible

Neurones

Neurones (nerve cells) are specialised cells adapted to rapidly carrying electrochemical changes called **nerve impulses** from one part of the body to another.

A mammalian motor neurone is made up of:

- a **cell body**, which contains all the usual cell organelles, including a nucleus and large amounts of rough endoplasmic reticulum. This is associated with the production of proteins and neurotransmitters

- **dendrons**, extensions of the cell body which subdivide into smaller branched fibres, called **dendrites**, that carry nerve impulses towards the cell body

- an **axon**, a single long fibre that carries nerve impulses away from the cell body

- **Schwann cells**, which surround the axon, protecting it and providing electrical insulation. They also carry out phagocytosis (the removal of cell debris) and play a part in nerve regeneration. Schwann cells wrap themselves around the axon many times, so that layers of their membranes build up around it

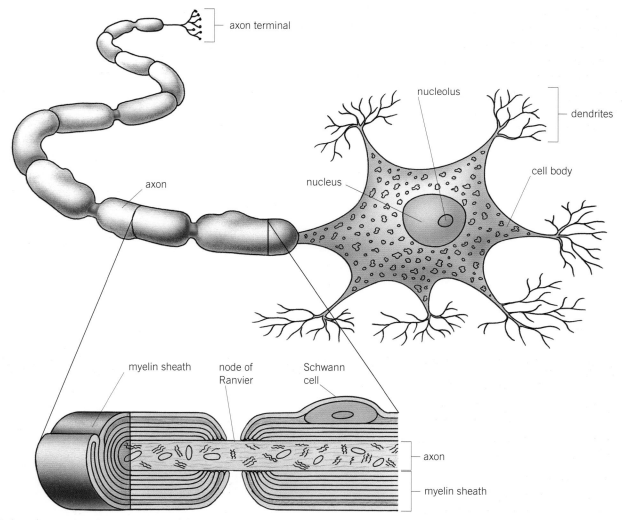

▲ **Figure 1** *Myelinated motor neurone*

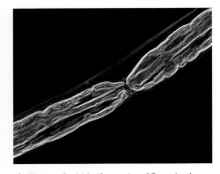

▲ **Figure 2** *LM of a node of Ranvier in a neurone. The node is the constriction in the centre. The constriction is a small area without myelin in an otherwise myelinated nerve fibre*

- a **myelin sheath**, which forms a covering to the axon and is made up of the membranes of the Schwann cells. These membranes are rich in a lipid known as **myelin**. Neurones with a myelin sheath are called myelinated neurones.

- **nodes of Ranvier**, constrictions between adjacent Schwann cells where there is no myelin sheath. The constrictions are 2–3 µm long and occur every 1–3 mm in humans (Figure 2).

The structure of a myelinated motor neurone is illustrated in Figure 1.

Neurones can be classified according to their function:

- **Sensory neurones** transmit nerve impulses from a receptor to an intermediate or motor neurone. They have one dendron that is often very long. It carries the impulse towards the cell body and one axon that carries it away from the cell body.

- **Motor neurones** transmit nerve impulses from an intermediate or relay neurone to an effector, such as a gland or a muscle. Motor neurones have a long axon and many short dendrites.

- **Intermediate or relay neurones** transmit impulses between neurones, for example, from sensory to motor neurones. They have numerous short processes.

Figure 3, shows the structure of all three types of neurone.

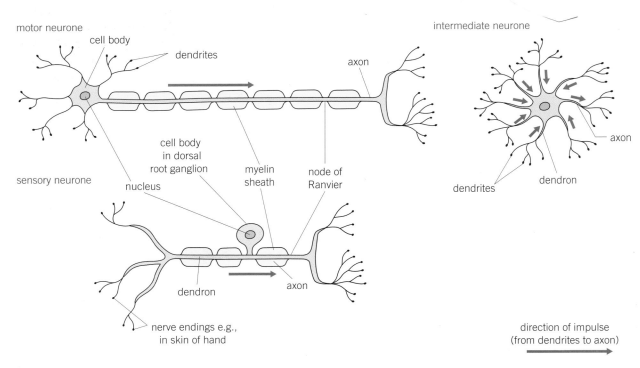

▲ **Figure 3** *Types of neurone*

Summary questions

In the following passage give the word that best replaces the numbers in brackets.

Neurones are adapted to carry electrochemical charges called (1). Each neurone comprises a cell body that contains a (2) and large amounts of (3), which is used in the production of proteins and neurotransmitters. Extending from the cell body is a single long fibre called an axon and smaller branched fibres called (4). Axons are surrounded by (5) cells, which protect and provide (6) because their membranes are rich in a lipid known as (7). There are three main types of neurone. Those that carry nerve impulses to an effector are called (8) neurones. Those that carry impulses from a receptor are called (9) neurones and those that link the other two types are called (10) neurones.

(11) List **three** ways in which a response to a hormone differs from a response to a nerve impulse.

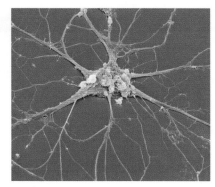

▲ **Figure 4** *SEM of a neurone with the cell body at its centre and dendrites radiating from it*

Maths link $\sqrt{x}$

MS 0.3, see Chapter 22.

Ageing neurones $\sqrt{x}$

Neurones are found throughout the brain. In Figure 5 we see drawings of some neurones from a region of the brain of healthy humans of different ages. The last diagram in the series shows neurones from a 70-year-old with Alzheimer's disease, a condition that causes dementia.

1 Using Figure 5 describe the changes in the neurones that take place when healthy humans age.
2 Comment on the appearance of the neurone from the 70 year old who has Alzheimer's disease.
3 $\sqrt{x}$ After the age of 50 years, humans lose 5% of the neurones in this region of the brain every 10 years. Calculate how many neurones will be left at the age of 70 years from each 2000 neurones present at the age of 50 years.

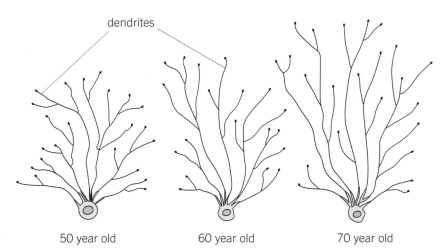

dendrites

| 50 year old | 60 year old | 70 year old | 70 year old with Alzheimer's disease |

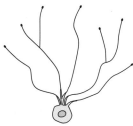

▲ **Figure 5**

15.2 The nerve impulse

Learning objectives

→ Describe the nature of the resting potential.

→ Explain how a resting potential is established in a neurone.

→ Explain what an action potential is.

Specification reference: 3.6.2.1

Synoptic link

To understand the nerve impulse requires a thorough knowledge and understanding of plasma membranes, particularly the structure of plasma membranes and the role of their carrier proteins in the sodium–potassium pump. It would be useful to revise Topics 4.1, 4.2, and 4.4 as a starting point for this section.

Study tip

Where sodium and potassium ions are actively transported through carrier proteins, it is known as the sodium-potassium pump.

Hint

As the phospholipid bilayer prevents diffusion of sodium and potassium ions, they move back across the bilayer by facilitated diffusion through channel proteins that are permanently open. These channel proteins are known as sodium or potassium 'gates', depending on which ion they transport.

A nerve impulse may be defined as a self-propagating wave of electrical activity that travels along the axon membrane. It is a temporary reversal of the electrical potential difference across the axon membrane. This reversal is between two states, called the **resting potential** and the **action potential**.

Resting potential

The movement of ions, such as sodium ions (Na$^+$) and potassium ions (K$^+$), across the axon membrane is controlled in a number of ways:

- The phospholipid bilayer of the axon plasma membrane prevents sodium and potassium ions diffusing across it.
- Proteins, known as **channel proteins**, span this phospholipid bilayer. These proteins have channels, called ion channels, which pass through them. Some of these channels have 'gates', which can be opened or closed so that sodium or potassium ions can move through them by facilitated diffusion at any one time, but not on other occasions. There are different gated channels for sodium and potassium ions. Some channels, however, remain open all the time, so the sodium and potassium ions move unhindered through them by facilitated diffusion.
- Some carrier proteins actively transport potassium ions into the axon and sodium ions out of the axon. This mechanism can be called a **sodium-potassium pump**.

As a result of these various controls, the inside of an axon is negatively charged relative to the outside. This is known as the **resting potential** and ranges from 50 to 90 millivolts (mV), but is usually 65 mV in humans. In this condition the axon is said to be **polarised**. The establishment of this potential difference (the difference in charge between the inside and outside of the axon) is due to the following events:

- Sodium ions are actively transported **out** of the axon by the sodium–potassium pumps.
- Potassium ions are actively transported **into** the axon by the sodium–potassium pumps.
- The active transport of sodium ions is greater than that of potassium ions, so three sodium ions move out for every two potassium ions that move in.
- Although both sodium and potassium ions are positive, the outward movement of sodium ions is greater than the inward movement of potassium ions. As a result, there are more sodium ions in the tissue fluid surrounding the axon than in the cytoplasm, and more potassium ions in the cytoplasm than in the tissue fluid, thus creating an electrochemical gradient.
- The sodium ions begin to diffuse back naturally into the axon while the potassium ions begin to diffuse back out of the axon.

- However, most of the gates in the channels that allow the potassium ions to move through are open, while most of the gates in the channels that allow the sodium ions to move through are closed.

These events are summarised in Figure 1.

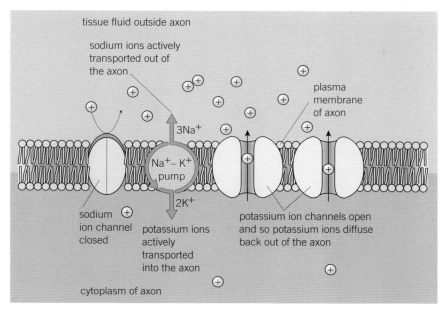

▲ **Figure 1** *Distribution of ions at resting potential*

The action potential

When a stimulus of sufficient size is detected by a receptor in the nervous system, its energy causes a temporary reversal of the charges either side of this part of the axon membrane. If the stimulus is great enough, the negative charge of −65 mV inside the membrane becomes a positive charge of around +40 mV. This is known as the **action potential**, and in this condition this part of the axon membrane is said to be **depolarised**. This depolarisation occurs because the channels in the axon membrane change shape, and hence open or close, depending on the voltage across the membrane. They are therefore called voltage-gated channels. The sequence of events is described on the next page (the numbers relate to the stages illustrated in Figure 2). It is important to stress that the events described relate to a particular point on the axon membrane and not the whole of the membrane.

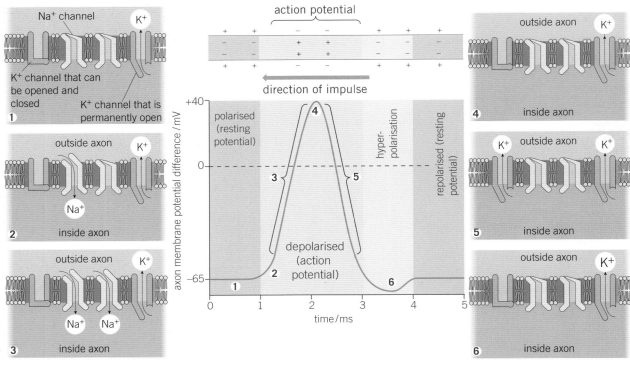

▲ **Figure 2** *The action potential*

Hint

The unit of time given on the y-axis of the graph (see Figure 2) is the millisecond (ms). A millisecond is 0.001 of a second. There are therefore 1000 milliseconds in a second. At 2 ms each, action potentials are very short-lived!

1   At resting potential some potassium voltage-gated channels are open (namely those that are permanently open) but the sodium voltage-gated channels are closed.

2   The energy of the stimulus causes some sodium voltage-gated channels in the axon membrane to open and therefore sodium ions diffuse into the axon through these channels along their electrochemical gradient. Being positively charged, they trigger a reversal in the potential difference across the membrane.

3   As the sodium ions diffuse into the axon, so more sodium channels open, causing an even greater influx of sodium ions by diffusion.

4   Once the action potential of around +40 mV has been established, the voltage gates on the sodium ion channels close (thus preventing further influx of sodium ions) and the voltage gates on the potassium ion channels begin to open.

5   With some potassium voltage-gated channels now open, the electrical gradient that was preventing further outward movement of potassium ions is now reversed, causing more potassium ion channels to open. This means that yet more potassium ions diffuse out, starting repolarisation of the axon.

6   The outward diffusion of these potassium ions causes a temporary overshoot of the electrical gradient, with the inside of the axon being more negative (relative to the outside) than usual (= hyperpolarisation) The closable gates on the potassium ion channels now close and the activities of the sodium–potassium pumps once again cause sodium ions to be pumped out and potassium ions in. The resting potential of −65 mV is re-established and the axon is said to be **repolarised**.

The terms action potential and resting potential can be misleading because the movement of sodium ions inwards during the action potential is purely due to diffusion – which is a passive process – while the resting potential is maintained by active transport – which is an active process. The term action potential simply means that the axon membrane is transmitting a nerve impulse, whereas resting potential means that it is not.

Summary questions

1 Describe how the movement of ions establishes the resting potential in an axon.

2 Table 1 shows the membrane potential of an axon at different stages of an action potential. The table refers to those channels that can be open and closed, not those that remain permanently open. For each of the letters A–F, indicate the state of the relevant channels, that is, open or closed.

▼ Table 1

	Resting	Beginning to depolarise	Repolarising
Membrane potential / mV	−70	−50	−20
Na⁺ channels in axon membrane	A	B	C
K⁺ channels in axon membrane	D	E	F

 ### Measuring action potentials

The plasma membrane of an axon will transmit an action potential when stimulated to do so. The action potential involves changes in the electrical potential across the membrane due to the movement of positive ions.

1 State which **two** positive ions are responsible for this change in electrical potential.

Figure 3 shows two action potentials that were recorded using electrodes and displayed on an instrument called an oscilloscope.

2 Between 0.5 and 2.0 ms there is a considerable change in membrane potential. Explain how this change is brought about.

3 Calculate how many action potentials will occur in 1 second if the frequency shown on the graph is maintained for this period. Show your working.

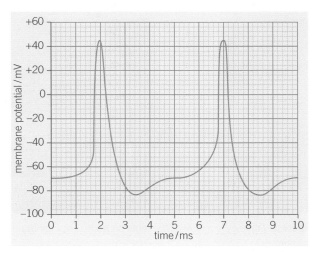

▲ Figure 3

Maths link √x̄

MS 3.1, see Chapter 22.

Learning objectives

→ Explain how an action potential passes along an unmyelinated axon.

→ Explain how an action potential passes along a myelinated axon.

Specification reference: 3.6.2.1

Once it has been created, an action potential moves rapidly along an axon. The size of the action potential remains the same from one end of the axon to the other. Strictly speaking, nothing physically moves from place to place along the axon of the neurone. As one region of the axon produces an action potential and becomes depolarised, it acts as a stimulus for the depolarisation of the next region of the axon. In this manner, action potentials are generated along each small region of the axon membrane. The action potential is therefore a travelling wave of depolarisation. In the meantime, the previous region of the membrane returns to its resting potential, that is, it undergoes repolarisation.

The process can be likened to the Mexican wave that often takes place in a crowded stadium during a sporting event. Although the wave of people standing up and raising their hands (the action potential) moves around the stadium, the people themselves do not move from seat to seat with the wave (i.e., they do not physically pass around the stadium until they reach their original seat again). Instead, their individual actions of standing and raising their hands are stimulated by the action of the person on one side of them and are reproduced by the person on the other side.

Passage of an action potential along an unmyelinated axon

It is easier to understand how a nerve impulse is propagated in a myelinated axon if we first look at how it is propagated in an unmyelinated one. This process is described and illustrated in Figure 2.

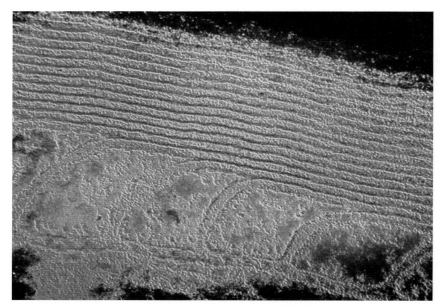

▲ **Figure 1** *False-colour TEM of the myelin sheath (orange bands at top) around the axon (bottom)*

polarised

depolarised

repolarised

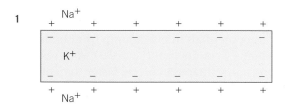

1 At resting potential the concentration of sodium ions outside the axon membrane is high relative to the inside, whereas that of the potassium ions is high inside the membrane relative to the outside. The overall concentration of positive ions is, however, greater on the outside, making this positive compared with the inside. The axon membrane is polarised. In our Mexican wave analogy, this is equivalent to the whole stadium being seated, that is, at rest.

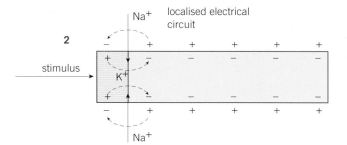

2 A stimulus causes a sudden influx of sodium ions and hence a reversal of charge on the axon membrane. This is the action potential and the membrane is depolarised. In our analogy, a prompt leads a vertical line of people to stand and wave their arms, that is, they are stimulated into action.

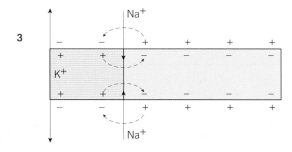

3 The localised electrical currents established by the influx of sodium ions cause the opening of sodium voltage-gated channels a little further along the axon. The resulting influx of sodium ions in this region causes depolarisation. Behind this new region of depolarisation, the sodium voltage-gated channels close and the potassium ones open. Potassium ions begin to leave the axon along their electrochemical gradient. So, once initiated, the depolarisation moves along the membrane. The sight of the person next to them standing and waving prompts the person in the adjacent seat to stand and wave. A new vertical line of people stands and waves, while the original line of

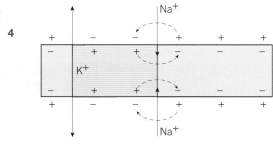

4 The action potential (depolarisation) is propagated in the same way further along the axon. The outward movement of the potassium ions has continued to the extent that the axon membrane behind the action potential has returned to its original charged state (positive outside, negative inside), that is, it has been repolarised. The second line of people standing and waving prompts the third line of people to do the same. Meanwhile, the first line have now resumed their original positions, that is, they are re-seated.

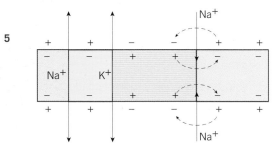

5 Repolarisation of the axon allows sodium ions to be actively transported out, once again returning the axon to its resting potential in readiness for a new stimulus if it comes. The people who have just sat down settle back in their seats and readjust themselves in readiness to repeat the process should they be prompted to do so again.

▲ **Figure 2** *Passage of an impulse along the axon of an unmyelinated neurone*

Passage of an action potential along a myelinated axon

In myelinated axons, the fatty sheath of myelin around the axon acts as an electrical insulator, preventing action potentials from forming. At intervals of 1–3 mm there are breaks in this myelin insulation, called nodes of Ranvier (see Topic 15.1). Action potentials can occur at these points. The localised circuits therefore arise between adjacent nodes of Ranvier and the action potentials in effect jump from node to node in a process known as saltatory conduction (Figure 3). As a result, an action potential passes along a myelinated neurone faster than along the axon of an unmyelinated one of the same diameter. This is because in an unmyelinated neurone, the events of depolarisation have to take place all the way along an axon and this takes more time. In our Mexican wave analogy, this is equivalent to a whole block of spectators leaping up simultaneously, followed by the next block and so on. Instead of the wave passing around the stadium in hundreds of small stages, it passes around in 20 or so large ones and is consequently more rapid.

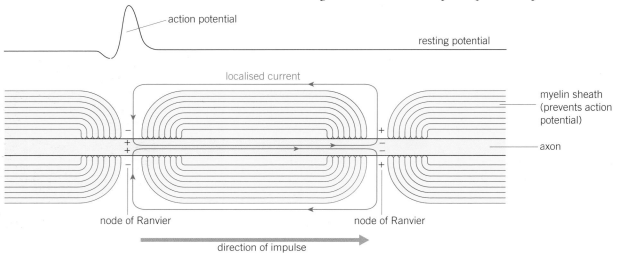

▲ **Figure 3** *Passage of an action potential along a myelinated axon. Action potentials are produced only at nodes of Ranvier. Depolarisation therefore skips from node to node (= saltatory conduction)*

Hint

The term saltatory, in saltatory conduction, comes from the Latin word saltare, meaning to jump.

Summary questions

1   In a myelinated axon, sodium and potassium ions can only be exchanged at certain points along it.

   a   State the name given to these points.

   b   Explain why ions can only be exchanged at these points.

   c   Describe the effect this has on the way an action potential is conducted along the axon.

   d   State the name that is given to this type of conduction.

   e   Describe how it affects the speed with which the action potential is transmitted compared to an unmyelinated axon.

2   Describe what happens to the size of an action potential as it moves along an axon.

Once an action potential has been set up, it moves rapidly from one end of the axon to the other without any decrease in size. In other words, the action potential at the end of the axon is the same size as when it starts. This transmission of an action potential along the axon of a neurone is the **nerve impulse**.

Factors affecting the speed at which an action potential travels

A number of factors affect the speed at which action potentials pass along an axon. Depending upon these factors, action potentials may travel at a speed of as little as $0.5\,\mathrm{m\,s^{-1}}$ or as much as $120\,\mathrm{m\,s^{-1}}$. These factors include:

- **The myelin sheath**. We saw in Topic 15.3, Passage of an action potential, that the myelin sheath acts as an electrical insulator, preventing an action potential forming in the part of the axon covered in myelin. It does, however, jump from one node of Ranvier to another (saltatory conduction). This increases the speed of conductance from $30\,\mathrm{m\,s^{-1}}$ in an unmyelinated neurone to $90\,\mathrm{m\,s^{-1}}$ in a similar myelinated one.

- **The diameter of the axon**. The greater the diameter of an axon, the faster the speed of conductance. This is due to less leakage of ions from a large axon (leakage makes membrane potentials harder to maintain).

- **Temperature**. This affects the rate of diffusion of ions and therefore the higher the temperature the faster the nerve impulse. The energy for active transport comes from respiration. Respiration, like the sodium–potassium pump, is controlled by enzymes. Enzymes function more rapidly at higher temperatures up to a point. Above a certain temperature, enzymes and the plasma membrane proteins are denatured and impulses fail to be conducted at all. Temperature is clearly an important factor in response times in cold-blooded (ectothermic) animals, whose body temperature varies in accordance with the environment. Temperature also affects the speed and strength of muscle contractions.

All-or-nothing principle

Nerve impulses are described as **all-or-nothing** responses. There is a certain level of stimulus, called the **threshold value**, which triggers an action potential. Below the threshold value, no action potential, and therefore no impulse, is generated. Any stimulus, of whatever strength, that is below the threshold value will fail to generate an action potential – this is the nothing part. Any stimulus above the threshold value will succeed in generating an action potential and so a nerve impulse will travel. All action potentials are more or less the same size, and so the strength of a stimulus cannot be detected by the size of the action potentials. How then can an organism perceive the size of a stimulus? This is achieved in two ways:

Learning objectives

→ Describe the factors that affect the speed of conductance of an action potential.

→ Explain what is meant by the refractory period.

→ Explain the role of the refractory period in separating one impulse from the next.

→ Explain the meaning of the all-or-nothing principle.

Specification reference: 3.6.2.1

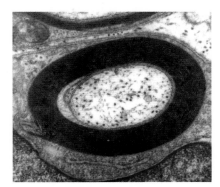

▲ **Figure 1** *False-coloured TEM of a section through a myelinated neurone and Schwann cell. Myelin (black) surrounds the axon (purple), increasing the speed at which nerve impulses travel. It is formed when Schwann cells (green) wrap around the axon, depositing layers of myelin between each coil*

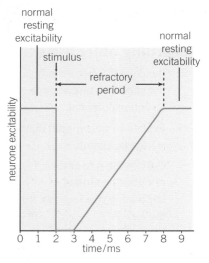

▲ **Figure 2** *Graph illustrating neurone excitability before and after a nerve impulse*

Hint

The brain would be overloaded with information if it became aware of every little stimulus. The all-or-nothing nature of the action potential acts as a filter, preventing minor stimuli from setting up nerve impulses and thus preventing the brain becoming overloaded.

Study tip

The refractory period limits the strength of stimulus that can be detected.

- by the number of impulses passing in a given time. The larger the stimulus, the more impulses that are generated in a given time (Figure 3)
- by having different neurones with different threshold values. The brain interprets the number and type of neurones that pass impulses as a result of a given stimulus and thereby determines its size.

The refractory period

Once an action potential has been created in any region of an axon, there is a period afterwards when inward movement of sodium ions is prevented because the sodium **voltage-gated channels** are closed. During this time it is impossible for a further action potential to be generated. This is known as the **refractory period** (Figure 2).

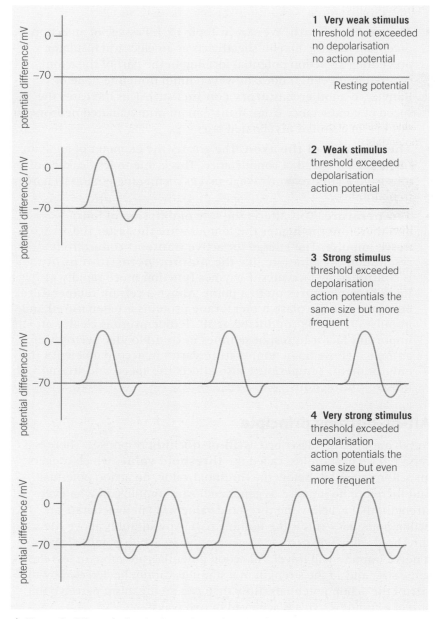

▲ **Figure 3** *Effect of stimulus intensity on impulse frequency*

The refractory period serves three purposes:

- **It ensures that action potentials are propagated in one direction only**. Action potentials can only pass from an active region to a resting region. This is because action potentials cannot be propagated in a region that is refractory, which means that they can only move in a forward direction. This prevents action potentials from spreading out in both directions, which they would otherwise do.

- **It produces discrete impulses**. Due to the refractory period, a new action potential cannot be formed immediately behind the first one. This ensures that action potentials are separated from one another.

- **It limits the number of action potentials**. As action potentials are separated from one another this limits the number of action potentials that can pass along an axon in a given time, and thus limits the strength of stimulus that can be detected.

Summary questions

1 Explain how the refractory period ensures that nerve impulses are kept separate from one another.

2 State the all-or-nothing principle.

3 Earthworms have unmyelinated axons and so to increase the speed of conduction of action potentials these are relatively large in diameter. Suggest **two** reasons why mammals do not require large diameter axons to achieve rapid transmission of action potentials.

 Different axons, different speeds

Table 1 below shows the speeds at which different axons conduct action potentials.

▼ **Table 1**

Axon	Myelin	Axon diameter / μm	Transmission speed / m s^{-1}
Human motor axon to leg muscle	Yes	20	120
Human sensory axon from skin pressure receptor	Yes	10	50
Squid giant axon	No	500	25
Human motor axon to internal organ	No	1	2

1 Using data from the table, describe the effect of axon diameter on the speed of conductance of an action potential.

2 The data show that a myelinated axon conducts an action potential faster than an unmyelinated axon. Explain why this is so.

3 Name the cells whose membranes make up the myelin sheath around some types of axon.

4 State which has the greater effect on the speed of conductance of an action potential: the presence of myelin or the diameter of the axon. Use information from Table 1 to explain your answer.

5 The squid is an ectothermic animal. This means that its body temperature fluctuates with the temperature of the water in which it lives. Suggest how this might affect the speed at which action potentials are conducted along a squid axon.

6 Assuming it is circular in cross section, calculate the surface area of a squid giant axon. Give your answer in mm^2 to five significant figures.

Learning objectives

→ Describe the structure of a synapse.

→ Describe the functions that synapses perform.

Specification reference: 3.6.2.2

Synoptic link

It would help your understanding of synapses to firstly revise protein channels and carrier proteins (Topic 4.1) and facilitated diffusion (Topic 4.2).

A synapse is the point where one **neurone** communicates with another or with an **effector**. They are important in linking different neurones together and therefore coordinating activities.

Structure of a synapse

Synapses transmit information, but not impulses, from one neurone to another by means of chemicals known as **neurotransmitters**. Neurones are separated by a small gap, called the **synaptic cleft**, which is 20–30 nm wide. The neurone that releases the neurotransmitter is called the **presynaptic neurone**. The axon of this neurone ends in a swollen portion known as the **synaptic knob**. This possesses many mitochondria and large amounts of endoplasmic reticulum. These are required in the manufacture of the neurotransmitter which takes place in the axon. The neurotransmitter is stored in the **synaptic vesicles**. Once the neurotransmitter is released from the vesicles it diffuses across to the postsynaptic neurone, which possesses specific receptor proteins on its membrane to receive it.

The structure of a chemical synapse is illustrated in Figure 1.

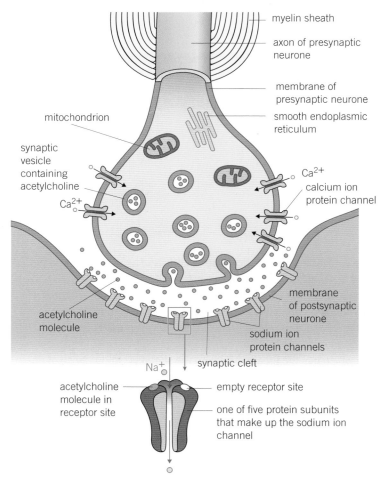

▲ **Figure 1** *Structure of a synapse*

Features of synapses

The basic way in which synapses function means they have a number of different features.

Unidirectionality

Synapses can only pass information in one direction – from the presynaptic neurone to the postsynaptic neurone. In this way, synapses act like valves.

Summation

Low-frequency action potentials often lead to the release of insufficient concentrations of neurotransmitter to trigger a new action potential in the postsynaptic neurone. They can, however, do so in a process called summation. This entails a rapid build-up of neurotransmitter in the synapse by one of two methods:

*   **spatial summation**, in which a number of different presynaptic neurones together release enough neurotransmitter to exceed the threshold value of the postsynaptic neurone. Together they therefore trigger a new action potential.

*   **temporal summation**, in which a single presynaptic neurone releases neurotransmitter many times over a very short period. If the concentration of neurotransmitter exceeds the threshold value of the postsynaptic neurone, then a new action potential is triggered.

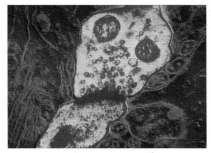

▲ **Figure 2** *TEM of synapse. The synaptic cleft between the two neurones (centre) appears deep red. The cell above the cleft has many small vesicles (red–yellow spheres) containing neurotransmitter, whereas the two larger spheres above the vesicles are mitochondria*

spatial summation

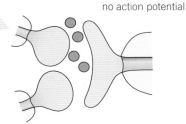

no action potential

Neurone A releases neurotransmitter but concentration is below threshold to trigger action potential in postsynaptic neurone.

presynaptic neurone A

presynaptic neurone B

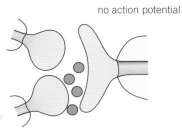

no action potential

Neurone B releases neurotransmitter but concentration is below threshold to trigger action potential in postsynaptic neurone.

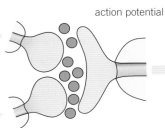

action potential

Neurone A and B release neurotransmitter. Concentration is above threshold and so an action potential is triggered in the postsynaptic neurone.

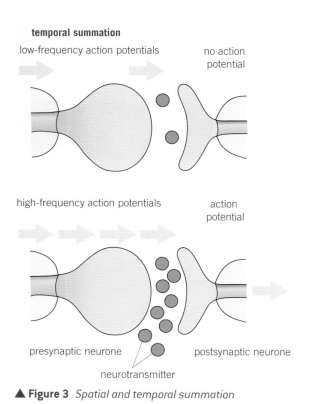

temporal summation

low-frequency action potentials

no action potential

Low-frequency action potentials lead to neurotransmitter being broken down rapidly. Concentration is below the threshold to trigger an action potential in the postsynaptic neurone.

high-frequency action potentials

action potential

High-frequency action potentials lead to release of of neurotransmitter in a short time. Concentration is above the threshold to trigger an action.potential in the postsynaptic neurone.

presynaptic neurone

postsynaptic neurone

neurotransmitter

▲ **Figure 3** *Spatial and temporal summation*

Inhibition

Some synapses make it less likely that a new action potential will be created on the postsynaptic neurone. These are known as **inhibitory synapses**. They operate as follows:

- The presynaptic neurone releases a type of neurotransmitter that binds to chloride ion protein channels on the postsynaptic neurone.
- The neurotransmitter causes the chloride ion protein channels to open.
- Chloride ions (Cl^-) move into the postsynaptic neurone by facilitated diffusion.
- The binding of the neurotransmitter causes the opening of nearby potassium (K^+) protein channels.
- Potassium ions move out of the postsynaptic neurone into the synapse.
- The combined effect of negatively charged chloride ions moving in and positively charged potassium ions moving out is to make the inside of the postsynaptic membrane more negative and the outside more positive.
- The membrane potential increases to as much as $-80\,mV$ compared with the usual $-65\,mV$ at resting potential.
- This is called hyperpolarisation and makes it less likely that a new action potential will be created because a larger influx of sodium ions is needed to produce one.

Maths link $\sqrt{x}$

MS 0.3, see Chapter 22.

Functions of synapses

Synapses transmit information from one neurone to another. In so doing, they act as junctions, allowing:

- a single impulse along one neurone to initiate new impulses in a number of different neurones at a synapse. This allows a single stimulus to create a number of simultaneous responses
- a number of impulses to be combined at a synapse. This allows nerve impulses from receptors reacting to different stimuli to contribute to a single response.

We shall look in more detail at how synapses transmit information in Topic 15.6, Transmission across a synapse. However, to understand the basic functioning of synapses as described here, it is sufficient to appreciate the following:

- A chemical (the neurotransmitter) is made **only** in the presynaptic neurone and not in the postsynaptic neurone.
- The neurotransmitter is stored in synaptic vesicles. When an **action potential** reaches the synaptic knob the membranes of these vesicles fuse with the pre-synaptic membrane to release the neurotransmitter.
- When released, the neurotransmitter diffuses across the synaptic cleft to bind to specific receptor proteins which are found **only** on the postsynaptic neurone.
- The neurotransmitter binds with the receptor proteins and this leads to a new action potential in the postsynaptic neurone. Synapses that produce new action potentials in this way are called **excitatory synapses**.

Summary questions $\sqrt{x}$

1 Explain how a presynaptic neurone is adapted for the manufacture of neurotransmitter.

2 Explain how the postsynaptic neurone is adapted to receive the neurotransmitter.

3 Outline the events in the transmission of information from one neurone to another across a synapse.

4 If a neurone is stimulated in the middle of its axon, an action potential will pass both ways along it to the synapses at each end of the neurone. However, the action potential will only pass across the synapse at one end. Explain why.

5 When walking along a street we barely notice the background noise of traffic. However, we often respond to louder traffic noises, such as the sound of a horn.

   a From your knowledge of summation, explain this difference.

   b Suggest an advantage in responding to high-level stimuli but not to low-level ones.

6 Explain why hyperpolarisation reduces the likelihood of a new action potential being created.

7 $\sqrt{x}$ Table 1 compares the number of synapses with the speed of transmission in three neural pathways: A, B and C.

   a Calculate the percentage increase in transmission speed when the number of synapses is reduced from 13 to 9.

   b Explain why the neural pathways of reflex arcs have very few synapses.

▼ Table 1

Neural pathway	Number of synapses	Speed of transmission /m s^{-1}
A	13	40
B	9	64
C	5	93

15.6 Transmission across a synapse

Learning objectives

→ Explain how information is transmitted across a synapse.

Specification reference: 3.6.2.2

In Topic 15.5 we outlined how neurotransmitters transmit information from one neurone to another. Let us now consider this in more detail by looking at a cholinergic synapse.

A **cholinergic** synapse is one in which the neurotransmitter is a chemical called **acetylcholine**. Acetylcholine is made up of two parts: acetyl (more precisely ethanoic acid) and choline. Cholinergic synapses are common in vertebrates, where they occur in the central nervous system and at neuromuscular junctions (junctions between neurones and muscles). Details of the neuromuscular junction are given in Topic 15.7.

The process of transmission across a cholinergic synapse is described in the series of diagrams in Figure 2 later in this topic. To simplify matters, only the relevant structures are shown on each diagram. Each receptor is a protein that binds specifically to a neurotransmitter because they have complementary shapes.

Hint

You need to think in terms of separate bursts of neurotransmitter release from the presynaptic knob. Each one relates to the arrival of an action potential along the neurone.

Synoptic link

In this topic, there are many possibilities for synoptic questions that bring together other topics. These include membrane structure, enzyme action, mitochondria and ATP production, diffusion (across the synaptic cleft and down channels in membrane proteins) and how molecular shapes fit one another (e.g., a neurotransmitter fitting into receptors on the postsynaptic neurone).

Effects of drugs on synapses

There are many different neurotransmitters responsible for the exchange of information across a synapse. There are also many different types of receptor on the postsynaptic neurone. Each receptor is a protein that binds specifically to a neurotransmitter because they have complementary shapes. Some of these neurotransmitters and receptors are excitatory, that is, they lead to a new action potential in the postsynaptic neurone. Others are inhibitory, that is, they make it less likely that a new action potential will be created in the postsynaptic neurone. Overall, the action of a specific neurotransmitter depends on the specific receptor to which it binds.

Given that our perception of the world is through stimuli detected by receptors and information transferred to the brain as nerve impulses by neurones that connect via synapses, it is not surprising that the effects of many medicinal and recreational drugs are due to their actions on synapses. Drugs act on synapses in two main ways:

- **They stimulate the nervous system by creating more action potentials in postsynaptic neurones**. A drug may do this by mimicking a neurotransmitter, stimulating the release of more neurotransmitter, or inhibiting the enzyme that breaks down the neurotransmitter. The outcome is to enhance the body's responses to impulses passed along the postsynaptic neurone. For example, if the neurone transmits impulses from sound receptors, a person will perceive the sound as being louder.

- **They inhibit the nervous system by creating fewer action potentials in postsynaptic neurones**. A drug may do this by inhibiting the release of neurotransmitter or blocking receptors on sodium/potassium ion channels on the postsynaptic neurone. The outcome is to reduce the impulses passed along the postsynaptic neurone. In this case, if the neurone transmits impulses from sound receptors, a person will perceive the sound as being quieter.

The effects of a drug on the synapse depend on the type of transmitter. For example, a drug that inhibits the action of an excitatory neurotransmitter will reduce a particular effect, but a drug that inhibits an inhibitory neurotransmitter will enhance a particular effect. Let us look at some examples of the effects of drugs on synapses.

Endorphins are neurotransmitters used by certain sensory nerve pathways, especially pain pathways. Endorphins block the sensation of pain. Drugs such as morphine and codeine bind to specific receptors in the brain used by endorphins and so mimic the effects of endorphins.

1   Suggest the likely effect of drugs like morphine and codeine on the body.
2   Explain how the effect you suggest might be brought about.

Serotonin is a neurotransmitter involved in the regulation of sleep and certain emotional states. Reduced activity of the neurones that release serotonins is thought to be one cause of clinical depression. **Prozac** is an antidepressant drug that affects serotonin within synaptic clefts.

3   Suggest a way that the drug Prozac might affect serotonin within synaptic clefts.
4   Explain how the effect you suggest makes Prozac an effective antidepressant.

GABA is a neurotransmitter that inhibits the formation of action potentials when it binds to postsynaptic neurones. **Valium** is a drug that enhances the binding of GABA to its receptors.

5   Suggest the likely effect of Valium on the nerve pathways that cause muscle contractions.
6   Explain the reasoning for your answer.
7   Epilepsy can be the result of an increase in the activity of neurones in the brain due to insufficient GABA. An enzyme breaks down GABA on the postsynaptic membrane. A drug called Vigabatrin has a molecular structure similar to GABA and is used to treat epilepsy. Suggest a way in which Vigabatrin might be effective in treating epilepsy.

Summary questions

1   For each of the following, state as accurately as possible the name of the substance described.

a   They diffuse into the postsynaptic neurone where they generate an action potential.

b   A neurotransmitter found in a cholinergic synapse.

c   It is released by mitochondria to enable the neurotransmitter to be reformed.

d   Their influx into the presynaptic neurone causes synaptic vesicles to release their neurotransmitter.

2   State why it is necessary for acetylcholine to be hydrolysed by acetylcholinesterase.

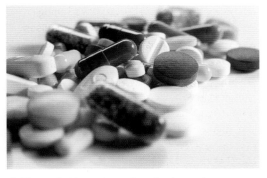

▲ **Figure 1** *Many drugs function by acting on synapses*

1 The arrival of an action potential at the end of the presynaptic neurone causes calcium ion protein channels to open and calcium ions (Ca²⁺) enter the synaptic knob by facilitated diffusion.

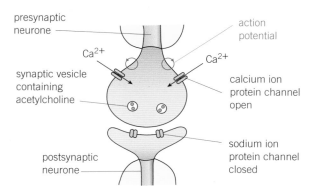

4 The influx of sodium ions generates a new action potential in the postsynaptic neurone.

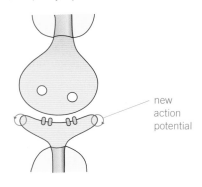

2 The influx of calcium ions into the presynaptic neurone causes synaptic vesicles to fuse with the presynaptic membrane, releasing acetylcholine into the synaptic cleft.

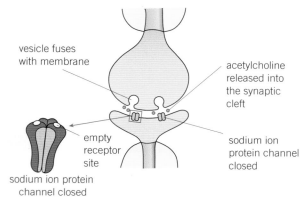

5 Acetylcholinesterase hydrolyses acetylcholine into choline and ethanoic acid (acetyl), which diffuse back across the synaptic cleft into the presynaptic neurone (= recycling). In addition to recycling the choline and ethanoic acid, the rapid breakdown of acetylcholine also prevents it from continuously generating a new action potential in the postsynaptic neurone, and so leads to discrete transfer of information across synapses

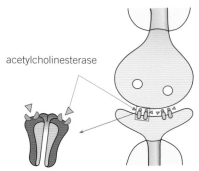

3 Acetylcholine molecules diffuse across the narrow synaptic cleft very quickly because the diffusion pathway is short. Acetylcholine then binds to receptor sites on sodium ion protein channels in the membrane of the postsynaptic neurone.
This causes the sodium ion protein channels to open, allowing sodium ions (Na⁺) to diffuse in rapidly along a concentration gradient.

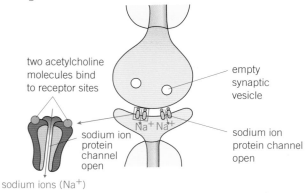

6 ATP released by mitochondria is used to recombine choline and ethanoic acid into acetycholine. This is stored in synaptic vesicles for future use. Sodium ion protein channels close in the absence of acetylcholine in the receptor sites.

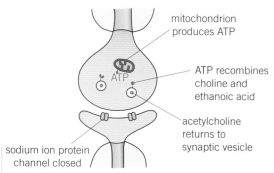

▲ **Figure 2** *Mechanism of transmission across a cholinergic synapse*

Muscles are effector organs that respond to nervous stimulation by contracting and so bring about movement. There are three types of muscle in the body. **Cardiac muscle** is found exclusively in the heart while **smooth muscle** is found in the walls of blood vessels and the gut. Neither of these types of muscle is under conscious control and we remain largely unaware of their contractions. The third type, **skeletal muscle**, makes up the bulk of body muscle in vertebrates. It is attached to bone and acts under voluntary, conscious control.

A rope is made up of millions of separate threads. Each thread has very little individual strength and can easily be snapped. Yet grouped together in a rope, these threads can support a mass running into hundreds of tonnes. In the same way, individual muscles are made up of millions of tiny muscle fibres called **myofibrils**. In themselves, they produce almost no force while collectively they can be extremely powerful. Just as the threads in a rope are lined up parallel to each other in order to maximise its strength, so the myofibrils are arranged in order to give maximum force. And just as the threads of a rope are grouped into strings, the strings are grouped into small ropes and small ropes are grouped into bigger ropes, so muscle is composed of smaller units bundled into progressively larger ones (Figure 1).

If muscle was made up of individual cells joined end to end it would not be able to perform the function of contraction very efficiently. This is partly because the junction between adjacent cells would be a point of weakness that would reduce the overall strength of the muscle. To overcome this, muscles have a different structure. The separate cells have become fused together into muscle fibres. These muscle fibres share nuclei and also cytoplasm, called **sarcoplasm**, which is mostly found around the circumference of the fibre (Figure 1). Within the sarcoplasm is a large concentration of mitochondria and endoplasmic reticulum.

Learning objectives

→ Describe the gross and microscopic structure of a skeletal muscle.

→ Describe the ultrastructure of a myofibril.

→ Explain how actin and myosin are arranged within a myofibril.

Specification reference: 3.6.3

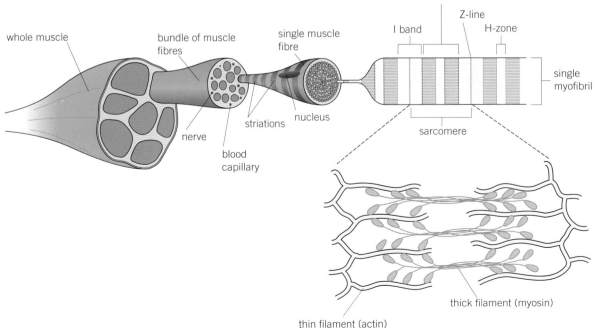

▲ **Figure 1** *The gross and microscopic structure of skeletal muscle*

Microscopic structure of skeletal muscle

We can see from Figure 1 that each muscle fibre is made up of myofibrils. Myofibrils are made up mainly of two types of protein filament:

- **actin**, which is thinner and consists of two strands twisted around one another
- **myosin**, which is thicker and consists of long rod-shaped tails with bulbous heads that project to the side.

The structure of these filaments and their constituent molecules shown in Figure 2.

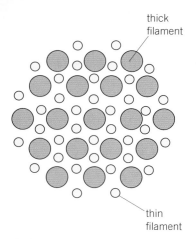

▲ **Figure 3** *Transverse section through part of a myofibril showing the arrangement of thick and thin filaments*

▶ **Figure 2** *Structure of actin and myosin molecules and their arrangement into thick and thin filaments*

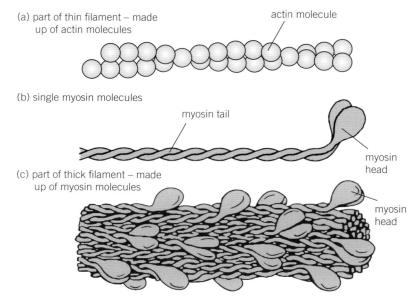

(a) part of thin filament – made up of actin molecules

actin molecule

(b) single myosin molecules

myosin tail

myosin head

myosin head

(c) part of thick filament – made up of myosin molecules

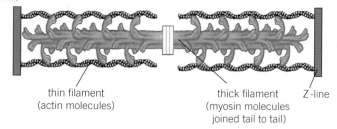

(d) arrangement of thick and thin filaments in muscle. Note that the myosin molecules are joined tail to tail in the thick filament

Z-line

thin filament (actin molecules)

thick filament (myosin molecules joined tail to tail)

Z-line

Hint

To help you remember which band is the dark band and which is the light band, look at the vowels in the words light and dark. This vowel is the first letter of the relevant band. Therefore the d**a**rk band is the **A**-band and the l**i**ght band is the **I**-band.

Hint

The arrangement of sarcomeres into a long line means that, when one sarcomere contracts a little, the line as a whole contracts a lot! In addition, having the lines of sarcomeres running parallel to each other means that all the force is generated in one direction.

Myofibrils appear striped due to their alternating light-coloured and dark-coloured bands. The light bands are called **I bands** (isotropic bands). They appear lighter because the thick and thin filaments do not overlap in this region. The dark bands are called **A bands** (anisotropic bands). They appear darker because the thick and thin filaments overlap in this region.

At the centre of each A band is a lighter-coloured region called the **H-zone**. At the centre of each I band is a line called the **Z-line**. The distance between adjacent Z-lines is called a **sarcomere** (Figure 1). When a muscle contracts, these sarcomeres shorten and the pattern of light and dark bands changes (Topic 15.8).

Another important protein found in muscle is **tropomyosin**, which forms a fibrous strand around the actin filament.

Types of muscle fibre

There are two types of muscle fibre, the proportions of which vary from muscle to muscle and person to person. The two types are:

- **slow-twitch fibres**. These contract more slowly than fast-twitch fibres and provide less powerful contractions but over a longer period. They are therefore adapted to endurance work, such as running a marathon. In humans they are more common in muscles like the calf muscle, which must contract constantly to maintain the body in an upright position. They are suited to this role by being adapted for aerobic respiration in order to avoid a build-up of lactic acid, which would cause them to function less effectively and prevent long-duration contraction. These adaptations include having:
  - a large store of myoglobin (a bright red molecule that stores oxygen, which accounts for the red colour of slow-twitch fibres)
  - a rich supply of blood vessels to deliver oxygen and glucose for aerobic respiration
  - numerous mitochondria to produce ATP.

- **fast-twitch fibres**. These contract more rapidly and produce powerful contractions but only for a short period. They are therefore adapted to intense exercise, such as weight-lifting. As a result they are more common in muscles which need to do short bursts of intense activity, like the biceps muscle of the upper arm. Fast-twitch fibres are adapted to their role by having:
  - thicker and more numerous myosin filaments
  - a high concentration of glycogen
  - a high concentration of enzymes involved in anaerobic respiration which provides ATP rapidly
  - a store of phosphocreatine, a molecule that can rapidly generate ATP from ADP in anaerobic conditions and so provide energy for muscle contraction.

Before we look at how muscle contracts in the next topic, let us explore how the muscle is stimulated. To do so, we must first look at where neurones meet muscle – the neuromuscular junction.

Neuromuscular junctions

A neuromuscular junction is the point where a motor neurone meets a skeletal muscle fibre. There are many such junctions along the muscle. If there were only one junction of this type it would take time for a wave of contraction to travel across the muscle, in which case not all the fibres would contract simultaneously and the movement would be slow. As rapid and coordinated muscle contraction is frequently essential for survival there are many neuromuscular junctions spread throughout the muscle. This ensures that contraction of a muscle is rapid and powerful when it is simultaneously stimulated by action potentials. All muscle fibres supplied by a single motor neurone act together as a single functional unit and are known as a motor unit. This arrangement gives control over the force that the muscle exerts. If only slight force is needed, only a few units are stimulated. If a greater force is required, a larger number of units are stimulated.

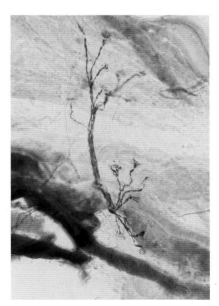

▲ **Figure 4** *Light micrograph of a neuromuscular junction*

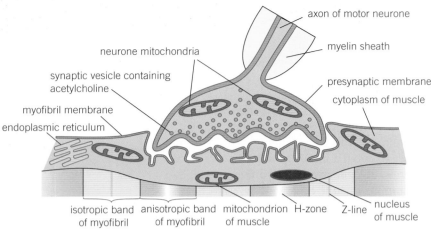

▲ **Figure 5** *The neuromuscular junction*

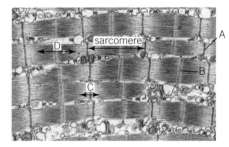

▲ **Figure 6** *TEM of skeletal muscle*

When a nerve impulse is received at the neuromuscular junction, the synaptic vesicles fuse with the presynaptic membrane and release their acetylcholine. The acetylcholine diffuses to the postsynaptic membrane (which is the membrane of the muscle fibre), altering its permeability to sodium ions (Na^+), which enter rapidly, depolarising the membrane. A description of how this leads to the contraction of the muscle is given in Topic 15.8.

The acetylcholine is broken down by acetylcholinesterase to ensure that the muscle is not over-stimulated. The resulting choline and ethanoic acid (acetyl) diffuse back into the neurone, where they are recombined to form acetylcholine using energy provided by the mitochondria found there.

The structure of a neuromuscular junction is shown in Figure 5 and an account of how it functions is provided in Topic 15.8.

Comparison of the neuromuscular junction and a synapse

The neuromuscular junction has some similarities with a cholinergic synapse but also differs in some respects. Their similarities include that both:

- have neurotransmitters that are transported by diffusion
- have receptors, that on binding with the neurotransmitter, cause an influx of sodium ions
- use a sodium–potassium pump to repolarise the axon
- use enzymes to breakdown the neurotransmitter.

Some of the differences are shown in Table 1.

▼ **Table 1** *Differences between a neuromuscular junction and cholinergic synapse*

Neuromuscular junction	Cholinergic synapse
Only excitatory (Topic 15.5)	May be excitatory or inhibitory (Topic 15.5)
Only links neurones to muscles	Links neurones to neurones, or neurones to other effector organs
Only motor neurones are involved	Motor, sensory and intermediate neurones may be involved
The action potential ends here (it is the end of a neural pathway)	A new action potential may be produced along another neurone (the postsynaptic neurone)
Acetylcholine binds to receptors on membrane of muscle fibre	Acetylcholine binds to receptors on membrane of post-synaptic neurone

To understand how skeletal muscles bring about movement, it is important to appreciate that these muscles are attached to the skeleton. In humans, this skeleton is made up of bone which is incompressible. Therefore, if muscle exerts a force, via tendons the bone moves rather than the muscle changing shape. The different parts of the skeleton can be moved relative to one another around a series of points called joints.

The contraction of a skeletal muscle will move a part of the skeleton, for example, a limb, in one direction but the same muscle cannot move it in the opposite direction. Muscles cannot push they can only pull. To move the limb in the opposite direct requires a second muscle that works antagonistically to the first one, i.e. in the opposite direction. In doing so it stretches its partner muscle (which has relaxed) returning it to its original state ready to contract again.

Skeletal muscles therefore occur, and act, in antagonistic pairs. These pairs pull in opposite directions and when one is contracted the other is relaxed.

We have looked at the structure of skeletal muscle in Topic 15.7, now let us turn our attention to how exactly the arrangement of the various proteins brings about contraction of the muscle fibre. The process involves the actin and myosin filaments sliding past one another and is therefore called the **sliding filament mechanism**.

Learning objectives

→ Explain what is meant by antagonistic muscles and how they operate.

→ Summarise the evidence that supports the sliding filament mechanism of muscle contraction.

→ Explain how the sliding filament mechanism causes a muscle to contract and relax.

→ State where the energy for muscle contraction comes from.

Specification reference: 3.6.3

Synoptic link

The functioning of muscle depends on the molecular shapes of the four main proteins involved. The importance of shape on the functioning of proteins is covered in Topic 1.6.

This Figure is simplified - see Figure 2 in Topic 15.7 for the detailed arrangement of myosin filaments

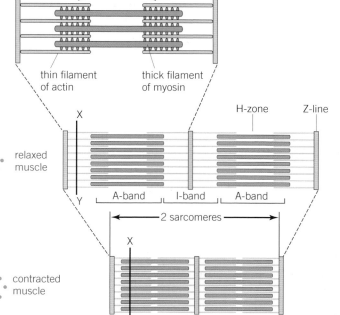

▲ **Figure 1** *Comparison of two sarcomeres in a relaxed and a contracted muscle*

Evidence for the sliding filament mechanism

In Topic 15.7, we saw that myofibrils appear darker in colour where the actin and myosin filaments overlap and lighter where they do not. If the sliding filament mechanism is correct, then there will be more overlap of actin and myosin in a contracted muscle than in a relaxed one. If you look at Figure 1, you will see that, when a muscle contracts, the following changes occur to a **sarcomere**:

*   The I-band becomes narrower.
*   The Z-lines move closer together or, in other words, the sarcomere shortens.
*   The H-zone becomes narrower.

The A-band remains the same width. As the width of this band is determined by the length of the myosin filaments, it follows that the myosin filaments have not become shorter. This discounts the theory that muscle contraction is due to the filaments themselves shortening.

Before we look at how the sliding filament mechanism works, let us take a closer look at the three main proteins involved in the process:

*   Myosin is made up of two types of protein:
    *   a fibrous protein arranged into a filament made up of several hundred molecules (the tail)
    *   a globular protein formed into two bulbous structures at one end (the head).
*   Actin is a globular protein whose molecules are arranged into long chains that are twisted around one another to form a helical strand.
*   Tropomyosin forms long thin threads that are wound around actin filaments.

The arrangement of the molecules of actin and tropomyosin are shown in Figure 2.

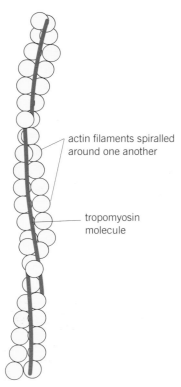

actin filaments spiralled around one another

tropomyosin molecule

▲ **Figure 2** *The relationship of tropomyosin to an actin filament*

The sliding filament mechanism of muscle contraction

The theory that actin and myosin filaments slide past one another during muscle contraction is supported by the changes seen in the band pattern on myofibrils. The next question for the scientists was, by what mechanism do the filaments slide past one another? Clues to the answer lie in the shape of the various proteins involved.

The bulbous heads of the myosin filaments form cross-bridges with the actin filaments. They do this by attaching themselves to binding sites on the actin filaments, and then flexing in unison, pulling the actin filaments along the myosin filaments. They then become detached and, using ATP as a source of energy, return to their original angle and re-attach themselves further along the actin filaments. This process is repeated up to 100 times per second. The action is similar to the way a ratchet operates. This process is illustrated in Figure 4.

Hint

The action of the myosin heads is similar to the rowing action of oarsmen in a boat. The oars (myosin heads) are dipped into the water, flexed as the oarsmen pull on them, removed from the water and then dipped back into the water further along. The oarsmen work in unison and the boat and water move relative to one another.

The following account describes the sliding filament mechanism of muscle contraction in detail. The process is continuous but, for ease of understanding, has been divided into stimulation, contraction, and relaxation.

Muscle stimulation

- An **action potential** reaches many **neuromuscular junctions** simultaneously, causing calcium ion protein channels to open and calcium ions to diffuse into the synaptic knob.

- The calcium ions cause the synaptic vesicles to fuse with the presynaptic membrane and release their **acetylcholine** into the synaptic cleft.

- Acetylcholine diffuses across the synaptic cleft and binds with receptors on the muscle cell-surface membrane, causing it to depolarise.

Muscle contraction

- The action potential travels deep into the fibre through a system of tubules (T-tubules) that are extensions of the cell-surface membrane and branch throughout the cytoplasm of the muscle (sarcoplasm).

- The tubules are in contact with the endoplasmic reticulum of the muscle (sarcoplasmic reticulum) which has actively transported calcium ions from the cytoplasm of the muscle leading to very low Ca^{2+} concentration in cytoplasm.

- The action potential opens the calcium ion protein channels on the endoplasmic reticulum and calcium ions diffuse into the muscle cytoplasm down a concentration gradient.

- The calcium ions cause the tropomyosin molecules that were blocking the binding sites on the actin filament to pull away (Figure 4, stages 1 and 2).

- ADP molecules attached to the myosin heads mean they are in a state to bind to the actin filament and form a cross-bridge (Figure 4, stage 3).

- Once attached to the actin filament, the myosin heads change their angle, pulling the actin filament along as they do so and releasing a molecule of ADP (Figure 4, stage 4).

- An ATP molecule attaches to each myosin head, causing it to become detached from the actin filament (Figure 4, stage 5).

- The calcium ions then activate the enzyme ATPase, which hydrolyses the ATP to ADP. The hydrolysis of ATP to ADP provides the energy for the myosin head to return to its original position (Figure 4, stage 6).

- The myosin head, once more with an attached ADP molecule, then reattaches itself further along the actin filament and the cycle is repeated as long as the concentration of calcium ions in the myofibril remains high (Figure 4, stage 7).

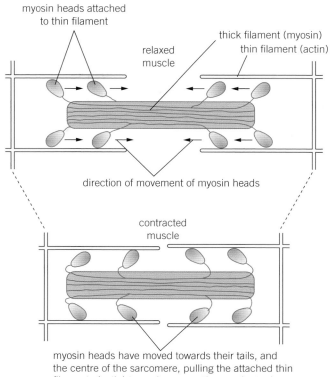

▲ Figure 3 *The action of myosin in shortening a sarcomere and causing muscle contraction*

> **Hint**
> The contraction of muscle is the result of a wave of excitation that spreads across it.

> **Hint**
> The hydrolysis of ATP releases energy and produces ADP and inorganic phosphate (P_i). For simplicity, this account just refers to ADP as the product.

1 Tropomyosin molecule prevents myosin head from attaching to the binding site on the actin molecule.

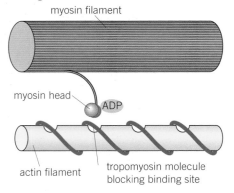

2 Calcium ions released from the endoplasmic reticulum cause the tropomyosin molecule to change shape and so pull away from the binding sites on the actin molecule.

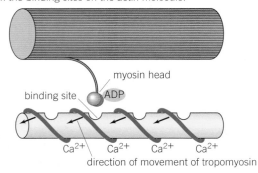

3 Myosin head now attaches to the binding site on the actin filament.

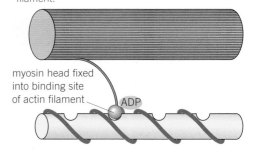

4 Head of myosin changes angle, moving the actin filament along as it does so. The ADP molecule is released.

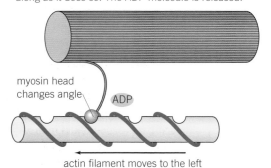

5 ATP molecule fixes to myosin head, causing it to detach from the actin filament.

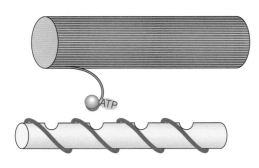

6 Hydrolysis of ATP to ADP by ATPase provides the energy for the myosin head to resume its normal position.

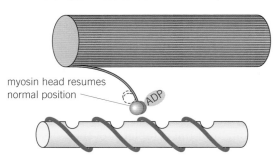

7 Head of myosin reattaches to a binding site further along the actin filament and the cycle is repeated.

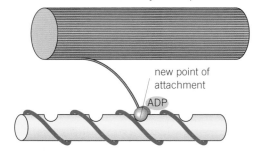

Synoptic link

The presence of calcium ions changes the environment of the protein tropomyosin leading to a change in its tertiary structure. We first met this idea in Topic 1.8.

▲ **Figure 4** *Sliding filament mechanism of muscle contraction (showing only one myosin head throughout)*

- As the myosin molecules are joined tail to tail in two oppositely facing sets, the movement of one set of myosin heads is in the opposite direction to the other set. This means that actin filaments to which they are attached also move in opposite directions.

- The movement of actin filaments in opposite directions pulls them towards each other, shortening the distance between the two adjacent Z-lines. The process is illustrated in Figure 4. The overall effect of this process taking place repeatedly and simultaneously throughout a muscle is to shorten it and so bring about movement of a part of the body.

Muscle relaxation

- When nervous stimulation ceases, calcium ions are actively transported back into the endoplasmic reticulum using energy from the hydrolysis of ATP.

- This reabsorption of the calcium ions allows tropomyosin to block the actin filament again.

- Myosin heads are now unable to bind to actin filaments and contraction ceases, that is, the muscle relaxes.

- In this state force from antagonistic muscles can pull actin filaments out from between myosin (to a point).

Energy supply during muscle contraction

Muscle contraction requires considerable energy. This is supplied by the hydrolysis of ATP to ADP and inorganic phosphate (P_i). The energy released is needed for:

- the movement of the myosin heads
- the reabsorption of calcium ions into the endoplasmic reticulum by active transport.

In an active muscle, there is clearly a great demand for ATP. In some circumstances, for example, escaping from danger, the ability of muscles to work intensely can be life-saving. Most ATP is regenerated from ADP during the respiration of pyruvate in the mitochondria, which are particularly plentiful in the muscle. However, this process requires oxygen. In a very active muscle the demand for ATP, and therefore oxygen, is greater than the rate at which the blood can supply oxygen. Therefore a means of rapidly generating ATP anaerobically is also required. This is partly achieved using a chemical called **phosphocreatine** and partly by more glycolysis.

Phosphocreatine cannot supply energy directly to the muscle, so instead it regenerates ATP, which can. Phosphocreatine is stored in muscle and acts as a reserve supply of phosphate, which is available immediately to combine with ADP and so re-form ATP. The phosphocreatine store is replenished using phosphate from ATP when the muscle is relaxed.

▲ **Figure 5** *Marathon runners undergoing strenuous exercise*

Maths link √x̄

MS 0.1 and 2.4, see Chapter 22.

Summary questions √x̄

1 Explain how the shape of the myosin molecule is adapted to its role in muscle contraction.

2 Trained sprinters have high levels of phosphocreatine in the muscles. Explain the advantage of this.

3 √x̄ During the contraction of a muscle sarcomere, a single actin filament moves 0.8 μm. If the hydrolysis of a single ATP molecule provides enough energy to move an actin filament 40 nm, calculate how many ATP molecules are needed to move the actin filament 0.8 μm. Show your working.

4 Dead cells can no longer produce ATP. Soon after death, muscles contract, making the body stiff – a state known as rigor mortis. From your knowledge of muscle contraction, explain the reasons why rigor mortis occurs after death.

1 During an action potential, the permeability of the cell-surface membrane of an axon changes. The graph shows changes in permeability of the membrane to sodium ions (Na⁺) and to potassium ions (K⁺) during a single action potential.

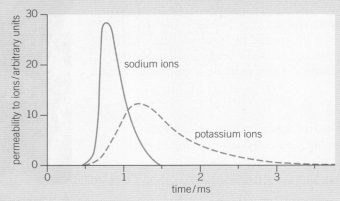

(a) Explain the shape of the curve for sodium ions between 0.5 ms and 0.7 ms. (*3 marks*)

(b) (i) During an action potential, the membrane potential rises to +40 mV and then falls. Use information from the graph to explain the fall in membrane potential. (*3 marks*)

(ii) The refractory period of a neurone has two components, absolute and relative. Calculate the maximum number of impulses that can be generated in a neurone when the total refractory period is 5 ms. (*2 marks*)

(iii) Calculate the percentage increase in nerve impulses if the stimulus intensity is raised high enough to overcome the relative refractory period of 4 ms. (*3 marks*)

(c) After exercise, some ATP is used to re-establish the resting potential in axons. Explain how the resting potential is re-established. (*2 marks*)

AQA June 2010 (apart from 1 (b) (ii) and (iii))

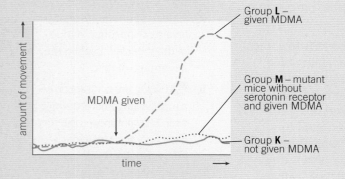

2 Serotonin is a neurotransmitter released in some synapses in the brain. It is transported back out of the synaptic gap by a transport protein in the pre-synaptic membrane.

(a) Serotonin diffuses across the synaptic gap and binds to a receptor on the post-synaptic membrane. Describe how this causes depolarisation of the post-synaptic membrane. (*2 marks*)

(b) It is important that a neurotransmitter such as serotonin is transported back out of synapses. Explain why. (*2 marks*)

(c) Scientists investigated the effect of a drug called MDMA on movement of mice. They measured the amount of movement of three groups of mice, **K**, **L** and **M**.

• Group **K**, mice not given MDMA.
• Group **L**, mice given MDMA.
• Group **M**, mutant mice that did not produce a serotonin receptor on their post-synaptic membranes and were given MDMA.

The graph shows their results.

The scientists concluded that MDMA affects movement by binding to serotonin receptors.

How do these results support this conclusion? (*3 marks*)

AQA June 2012

Feature	Fast muscle fibre	Slow muscle fibre
Type of respiration	Mainly anaerobic	Mainly aerobic
Glycogen	High concentration	Low concentration
Capillaries	Few	Many

3 **(a)** Describe the part played by each of the following in myofibril contraction.
   (i) Tropomyosin *(2 marks)*
   (ii) Myosin *(2 marks)*
(b) The table shows features of fast and slow muscle fibres.
Use information from the table to suggest and explain **one** advantage of:
   (i) the high glycogen content of fast muscle fibres *(2 marks)*
   (ii) the number of capillaries supplying slow muscle fibres. *(2 marks)*

AQA June 2013

4 The diagram shows two relaxed sarcomeres from skeletal muscle.

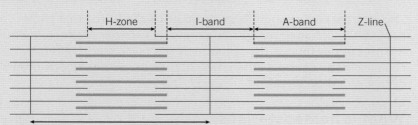

length of sarcomere = 2.2 μm

(a) When the sarcomeres contract, what happens to the length of
   (i) the I-band *(1 mark)*
   (ii) the A-band? *(1 mark)*
(b) The length of each sarcomere in the diagram is 2.2 μm. Use this information to calculate the magnification of the diagram. Show your working. *(2 marks)*
(c) People who have McArdle's disease produce less ATP than healthy people. As a result, they are not able to maintain strong muscle contraction during exercise. Use your knowledge of the sliding filament theory to suggest why. *(3 marks)*

AQA June 2012

5 The drawing is a tracing of a cross-section through skeletal muscle tissue. This muscle contains fast muscle fibres and slow muscle fibres. The section has been stained to show the distribution of the enzyme succinate dehydrogenase. This enzyme is found in mitochondria.
(a) (i) Succinate dehydrogenase catalyses one of the reactions in the Krebs cycle. What is the evidence from the drawing that muscle fibre S is a slow muscle fibre? Explain your answer. *(2 marks)*
   (ii) Use evidence from the diagram to describe the distribution of mitochondria inside the slow muscle fibres. Explain the importance of this distribution. *(3 marks)*
(b) (i) You could use an optical microscope and a slide of stained muscle tissue to find the diameter of one of the muscle fibres. Explain how. *(2 marks)*
   (ii) A student found the mean diameter for the slow muscle fibres in a section. Give two precautions that she should have taken when sampling the fibres. Give a reason for each precaution. *(2 marks)*

muscle fibre **S**

AQA June 2010

Learning objectives

→ Describe the nature of homeostasis.

→ Explain the importance of homeostasis.

→ Explain how control mechanisms work.

→ Explain how control mechanisms are coordinated.

Specification reference: 3.6.4.1

In the previous chapter we looked at how complex organisms control and coordinate their activities. In particular we considered the way in which such organisms respond rapidly to environmental changes using their nervous system. A feature of an increase in complexity is the ability of organisms to control their internal environment. By maintaining a relatively constant internal environment for their cells, organisms can limit the external changes these cells experience. This maintenance of a constant internal environment is called **homeostasis**. In this chapter we shall learn about homeostasis and the role of the other coordination system, hormonal coordination, in an organism's physiological control.

The internal environment is made up of **tissue fluids** that bathe each cell, supplying nutrients and removing wastes. Maintaining the features of this fluid at the optimum levels protects the cells from changes in the external environment, thereby giving the organism a degree of independence.

What is homeostasis?

Homeostasis is the maintenance of an internal environment within restricted limits in organisms. It involves trying to maintain the chemical make-up, volume and other features of blood and tissue fluid within restricted limits. Homeostasis ensures that the cells of the body are in an environment that meets their requirements and allows them to function normally despite external changes. This does not mean that there are no changes. On the contrary, there are continuous fluctuations brought about by variations in internal and external conditions, such as changes in temperature, pH and water potential. These changes, however, occur around an optimum point. Homeostasis is the ability to return to that optimum point and so maintain organisms in a balanced equilibrium.

The importance of homeostasis

Homeostasis is essential for the proper functioning of organisms for the following reasons amongst others:

- The enzymes that control the biochemical reactions within cells, and other proteins, such as channel proteins, are sensitive to changes in pH and temperature. Any change to these factors reduces the rate of reaction of enzymes or may even prevent them working altogether, for example, by denaturing them. Even small fluctuations in temperature or pH can impair the ability of enzymes to carry out their roles effectively. Maintaining a fairly constant internal environment means that reactions take place at a suitable rate.

- Changes to the **water potential** of the blood and tissue fluids may cause cells to shrink and expand (even to bursting point) as a result of water leaving or entering by osmosis. In both instances the cells cannot operate normally. The maintenance of a constant blood glucose concentration is essential in ensuring a constant water potential. A constant blood glucose concentration also ensures a reliable source of glucose for respiration by cells.

Synoptic link

The importance of temperature and pH in relation to enzyme activity (Topic 1.8) and water potential in relation to cells (Topic 3.7) make useful background reading for homeostasis.

Hint

A change in water potential may affect the concentration of substrates and enzymes and therefore the rate of reactions. See Topic 1.8.

- Organisms with the ability to maintain a constant internal environment are more independent of changes in the external environment. They may have a wider geographical range and therefore have a greater chance of finding food, shelter, etc. Mammals, for example, with their ability to maintain a constant temperature, are found in most habitats, ranging from hot arid deserts to cold, frozen polar regions.

Control mechanisms

The control of any self-regulating system involves a series of stages that feature:

- the **optimum point**, the point at which the system operates best. This is monitored by a …
- **receptor**, which detects any deviation from the optimum point (ie., a stimulus) and informs the …
- **coordinator**, which coordinates information from receptors and sends instructions to an appropriate …
- **effector**, often a muscle or gland, which brings about the changes needed to return the system to the optimum point. This return to normality creates a …
- **feedback mechanism**, by which a receptor responds to a stimulus created by the change to the system brought about by the effector.

Figure 2 illustrates the relationship between these stages using the everyday example of controlling a central heating system.

▲ **Figure 1** *Homeostasis allows animals such as these camels in the desert (top) and these penguins in the Antarctic (bottom) to survive in extreme environments*

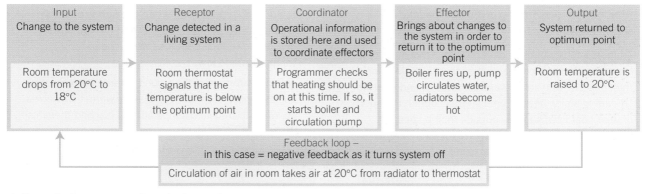

Input Change to the system	Receptor Change detected in a living system	Coordinator Operational information is stored here and used to coordinate effectors	Effector Brings about changes to the system in order to return it to the optimum point	Output System returned to optimum point
Room temperature drops from 20°C to 18°C	Room thermostat signals that the temperature is below the optimum point	Programmer checks that heating should be on at this time. If so, it starts boiler and circulation pump	Boiler fires up, pump circulates water, radiators become hot	Room temperature is raised to 20°C

Feedback loop –
in this case = negative feedback as it turns system off
Circulation of air in room takes air at 20°C from radiator to thermostat

▲ **Figure 2** *Components of a typical control system*

Coordination of control mechanisms

Most systems, including biological ones, use **negative feedback** Negative feedback is when the change produced by the control system leads to a change in the stimulus detected by the receptor and turns the system off. We shall meet an example of negative feedback when we look at the regulation of blood glucose in Topic 16.3.

Positive feedback occurs when a deviation from an optimum causes changes that result in an even greater deviation from the normal. One example occurs in neurones where a stimulus leads to a small influx of sodium ions. This influx increases the permeability of the neurone membrane to sodium ions, more ions enter, causing a further increase

Summary questions

1 Describe homeostasis.

2 Explain why maintaining a constant temperature is important in mammals.

3 Suggest why maintaining a constant blood glucose concentration might be important in mammals.

Maths link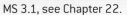

MS 3.1, see Chapter 22.

in permeability and even more rapid entry of ions. In this way, a small stimulus can bring about a large and rapid response.

Control systems normally have many receptors and effectors. This allows them to have separate mechanisms taht each produce a positive movement towards an optimum. This allows a greater degree of control of the particular factor being regulated. Having separate mechanisms that controls departures in different directions from the original state is a general feature of homeostasis. It is important to ensure that the information provided by receptors is analysed by the coordinator before action is taken. For example, temperature receptors in the skin may signal that the skin itself is cold and that the body temperature should be raised. However, information from regions in the hypothalamus in the brain may indicate that blood temperature is already above normal. This situation might arise during strenuous exercise when blood temperature rises but sweating cools the skin. By analysing the information from all detectors, the brain can decide the best course of action – in this case not to raise the body temperature further. In the same way, the control centre must coordinate the action of the effectors so that they operate harmoniously. For example, sweating would be less effective in cooling the body if it were not accompanied by **vasodilation**.

Comparing thermoregulation in ectotherms and endotherms ✓x̄

As with all extension boxes, the material here is to broaden understanding of material beyond the specification.

Animals such as birds and mammals derive most of their heat from the metabolic activities that take place inside their bodies. They are therefore known as **endotherms** (meaning inside heat). Some animals obtain a proportion of their heat from sources outside their bodies, namely the environment. They are therefore known as **ectotherms** (meaning outside heat).

Regulation of body temperature in ectotherms

Many ectotherms gain heat from the environment, so their body temperature fluctuates with that of the environment. They therefore control their body temperature by adapting their behaviour to changes in the external temperature. Reptiles, such as lizards, are ectotherms. They control their body temperature by:

- **exposing themselves to the Sun**. In order to gain heat lizards orientate themselves so that the maximum surface area of their body is exposed to the warming rays of the Sun.

- **taking shelter**. Lizards will shelter in the shade to prevent over-heating when the Sun's radiation is at its peak. At night they retreat into burrows in order to reduce heat loss when the external temperature is low.

- **gaining warmth from the ground**. Lizards will press their bodies against areas of hot ground to warm themselves up. When the required temperature is reached, they raise themselves off the ground on their legs.

▲ **Figure 3** *A lizard showing thermoregulatory behaviour by gaining heat both from the sun and the warm rock*

Regulation of body temperature in endotherms

Endotherms gain most of their heat from internal metabolic activities. Their body temperature remains relatively constant despite fluctuations in the external temperature. Like ectotherms, endothermic animals use behaviour to maintain a constant body temperature. Unlike ectotherms, however, they also use a wide range of physiological mechanisms to regulate their temperature.

Conserving and gaining heat in response to a cold environment

Mammals and birds that live in cold climates have evolved a number of adaptations in order to survive in these environments. One of the most important is having a body with a small surface area to volume ratio. It is from within the volume that heat is produced and from the surface area that heat is lost. Mammals and birds in cold climates therefore tend to be relatively large, for example, the polar bear and penguin. Compared with animals in warmer climates they also have smaller extremities, such as ears, and thick fur, feathers, or fat layers to insulate the body.

To make more rapid body temperature changes, mammals use one or more of the following mechanisms:

- **vasoconstriction**. The diameter of the arterioles near the surface of the skin is made smaller. This reduces the volume of blood reaching the skin surface through the capillaries. Most of the blood entering the skin passes beneath the insulating layer of fat and so loses little heat to the environment (Figure 5).

▲ **Figure 4** *The penguin and the polar bear both have large compact bodies with a small surface area to volume ratio. This helps them to conserve heat in the cold environments of the South and North Poles where they live*

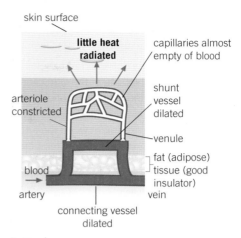

▲ **Figure 5** *Vasoconstriction*

- **shivering**. The muscles of the body undergo involuntary rhythmic contractions that produce metabolic heat.

- **raising of hair**. The hair erector muscles in the skin contract, raising the hairs on the body. This enables a thicker layer of still air, which is a good insulator, to be trapped next to the skin, insulation and conserving heat in mammals with thick fur.

- **increased metabolic rate**. In cold conditions more of the hormones that increase metabolic rate are produced. As a result metabolic activity, including respiration, is increased and so more heat is generated.

- **decrease in sweating**. Sweating is reduced, or ceases altogether, in cold conditions.

- **behavioural mechanisms**. Sheltering from the wind, basking in the sun and huddling together all help animals to maintain their core body temperature.

Losing heat in response to a warm environment

Long-term adaptations to life in a warm climate include having a large surface area to volume ratio and lighter coloured fur to reflect heat. Rapid responses that enable heat to be lost when the environmental temperature is high include:

- **vasodilation**. The diameter of the arterioles near the surface of the skin becomes larger. This allows warm blood to pass close to the skin surface through the capillaries. The heat from this blood is then radiated away from the body (Figure 6).

- **increased sweating**. To evaporate water from the skin surface requires energy in the form of heat. In relatively hairless mammals, such as humans, sweating is a highly effective means of losing heat. In mammals with fur, cooling is achieved by the evaporation of water from the mouth and tongue, during panting. The high latent heat of vaporisation of water makes sweating an efficient way of losing heat.

* **lowering of body hair**. The hair erector muscles in the skin relax and the elasticity of the skin causes them to flatten against the body. This reduces the thickness of the insulating layer and allows more heat to be lost to the environment when the internal temperature is higher than the external temperature.

* **behavioural mechanisms**. Avoiding the heat of the day by sheltering in burrows and seeking out shade help to prevent the body temperature from rising.

The graphs shown in Figure 7 compare the rates of metabolic heat generation and evaporative heat loss in a mammal and a reptile as the environmental temperature changes.

1 Give a reason why the values for heat generation and heat loss are measured per gram of body mass.

2 a Describe the relationship between metabolic heat generation and evaporative heat loss in a reptile.

   b Explain how this relationship differs in a mammal.

3 Reptiles frequently seek shade when the environmental temperature rises above 25°C. Use the graphs to explain this type of behaviour.

4 Suggest a reason for the change in the evaporative heat loss in the mammal at point A on the graph.

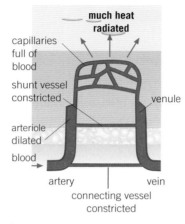

▲ **Figure 6** *Vasodilation*

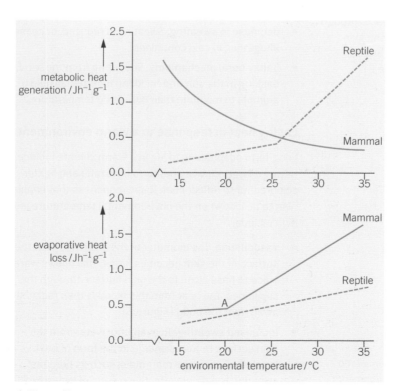

▲ **Figure 7**

16.2 Feedback mechanisms

We saw in Topic 16.1, that the homeostatic control of any system involves a series of stages featuring:

- **the optimum point**, or desired level (norm), at which the system operates
- **a receptor**, which detects the stimulus of any deviation from the set point (norm)
- **a coordinator**, which coordinates information from various sources
- **an effector**, which brings about the corrective measures needed to return the system to the optimum point (norm)
- **a feedback mechanism**, by which a receptor detects a stimulus created by the change to the system and the effector brings about the appropriate response.

Let us now look in more detail at the last stage in the list – the feedback mechanism. When an effector has corrected any deviation and returned the system to the optimum point, it is important that this information is fed back to the receptor. If the information is not fed back, the receptor will continue to stimulate the effector, leading to an over-correction and causing a deviation in the opposite direction. There are two types of feedback – negative feedback and positive feedback.

Negative feedback

Negative feedback occurs when the stimulus causes the corrective measures to be turned off. In doing so this tends to return the system to its original (optimum) level (and prevents any overshoot).

There are separate negative feedback mechanisms to regulate departures from the norm in each direction.

An example is in the control of blood glucose that is covered in more detail in Topic 16.3. If there is a fall in the concentration of glucose in the blood this stimulus is detected by receptors on the cell-surface membrane of the α cells (coordinator) in the pancreas. These α cells secrete the hormone glucagon. Glucagon causes liver cells (effectors) to convert glycogen to glucose which is released into the blood raising the blood glucose concentration. As this blood with a raised glucose concentration circulates back to the pancreas there is reduced stimulation of α cells which therefore secrete less glucagon. So the secretion of glucagon leads to a reduction in its own secretion (=negative feedback). These events are illustrated in Figure 1.

In the same way, if the blood glucose concentration rises, rather than falls, insulin will be produced from the β cells in the pancreas. Insulin increases the uptake of glucose by cells and its conversion to glycogen and fat. The fall in blood glucose concentration that results reduces insulin production once blood glucose concentrations return to their optimum (= negative feedback).

Having separate negative feedback mechanisms that control departures from the norm in either direction gives a greater degree of homeostatic control. This is because there are positive actions in both directions.

Learning objectives

→ Explain what negative feedback is.

→ Explain how negative feedback helps to control homeostatic processes.

→ Distinguish between negative feedback and positive feedback.

Specification reference: 3.6.4.1

Synoptic link

The concept of stimulus → receptor → coordinator → effector → response is a recurring theme in biology. For example, we met it throughout Chapter 14.

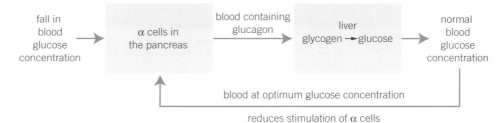

▲ **Figure 1** *Negative feedback in the control of blood glucose levels*

For example, if glucagon raised the blood sugar concentration above the optimum, it would take some time for it to fall again if the only way of lowering it was through metabolic activity. However, by having a second hormone, insulin, that lowers blood sugar concentration, its secretion brings about a return to optimum blood sugar concentration far more rapidly.

Summary questions

1 Explain why negative feedback is important in maintaining a system at a set point.

2 Explain the advantage of having separate negative feedback mechanisms to control deviations away from normal.

Positive feedback

Positive feedback occurs when the feedback causes the corrective measures to remain turned on. In doing so it causes the system to deviate even more from the original (normal) level. Examples are less common, but one occurs in neurones when a stimulus causes a small influx of sodium ions. This influx increases the permeability of the neurone to sodium ions so more ions enter, causing a further increase in permeability and even more rapid entry of ions. This results in a very rapid build-up of an action potential that allows an equally rapid response to a stimulus.

Positive feedback occurs more often when there is a breakdown of control systems. In certain diseases, for example typhoid fever, there is a breakdown of temperature regulation resulting in a rise in body temperature leading to hyperthermia. In the same way, when the body gets too cold (hypothermia) the temperature control system tends to break down, leading to positive feedback resulting in the body temperature dropping even lower.

1 Oxytocin is a hormone that causes contractions of the uterus at childbirth. The contractions produce a positive feedback loop that results in the release of more oxytocin. Explain the advantage of positive feedback rather than negative feedback in this situation.

Negative feedback in temperature control

If the temperature of the blood increases, thermoreceptors in a region of the brain called the hypothalamus send more nerve impulses to the heat loss centre, which is also in the hypothalamus. This in turn sends impulses to the skin (effector organ). Vasodilation, sweating and lowering of body hairs all lead to a reduction in blood temperature. If the fact that blood temperature has returned to normal is not fed back to the hypothalamus, it will continue to stimulate the skin to lose body heat. Blood temperature will then fall below normal and may continue to do so causing hypothermia and the death of the organism.

What happens in practice is that the cooler blood returning from the skin passes through the hypothalamus. As a result thermoreceptors send fewer impulses to the heat loss centre. This in turn stops sending impulses to the skin and so vasodilation, sweating, etc. cease, and blood temperature remains at its normal level rather than continuing to fall. The blood, having been cooled to its normal temperature, has resulted in turning *off* the effector (the skin) that was correcting the rise in temperature. This is therefore negative feedback and is illustrated in Figure 2.

1   State what would happen to the temperature of the blood if the feedback was positive rather than negative.

2   Cutting the nerves connecting the thermoreceptors to the heat loss centre in the hypothalamus might cause the death of the individual. In terms of the information in Figure 2, explain precisely why this action might cause death.

3   Cutting the nerves connecting the heat loss centre to the skin would be less likely to cause death than if those between the thermoreceptors and the heat loss centre were cut. Suggest why.

4   Negative feedback in temperature regulation occurs as a result of blood passing from the skin to the brain. In doing so, this blood passes through the heart. List, in sequence, the major vessels joined to the heart that this blood would pass through on its journey.

▲ **Figure 3** *Control of body temperature involves negative feedback mechanisms*

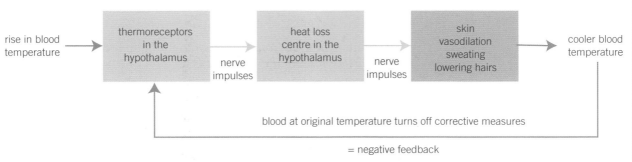

▲ **Figure 2** *Negative feedback in the control of body temperature*

Hormones and the regulation of blood glucose concentration

We saw in Topic 15.1 that animals possess two principal coordinating systems: the nervous system, which communicates rapidly, and the hormonal system, which usually communicates more slowly. Both systems interact in order to maintain the constancy of the internal environment while also being responsive to changes in the external environment. Both systems also use chemical messengers – the hormonal system exclusively so, and the nervous system through the use of neurotransmitters in chemical synapses.

The regulation of blood glucose is an example of how different hormones interact in achieving homeostasis. However, let us first look at what hormones are and how they work.

Hormones and their mode of action

Hormones differ from one another chemically but they all have certain characteristics in common. Hormones are:

- produced in glands, which secrete the hormone directly into the blood (endocrine glands)
- carried in the blood plasma to the cells on which they act – known as **target cells** – which have specific receptors on their cell-surface membranes that are complementary to a specific hormone
- are effective in very low concentrations, but often have widespread and long-lasting effects.

One mechanism of hormone action is known as the **second messenger model**. This mechanism is used by two hormones involved in the regulation of blood glucose concentration, namely adrenaline and glucagon.

The mechanism involving adrenaline is detailed below and illustrated in Figure 1.

- Adrenaline binds to a transmembrane protein receptor within the cell-surface membrane of a liver cell.
- The binding of adrenaline causes the protein to change shape on the inside of the membrane.
- This change of protein shape leads to the activation of an enzyme called adenyl cyclase. The activated adenyl cyclase converts ATP to cyclic AMP (cAMP).
- The cAMP acts as a second messenger that binds to protein kinase enzyme, changing its shape and therefore activating it.
- The active protein kinase enzyme catalyses the conversion of glycogen to glucose which moves out of the liver cell by facilitated diffusion and into the blood, through channel proteins.

The role of the pancreas in regulating blood glucose

The pancreas is a large, pale-coloured gland that is situated in the upper abdomen, behind the stomach. It produces enzymes (protease, amylase and lipase) for digestion and hormones (insulin and glucagon) for regulating blood glucose concentration.

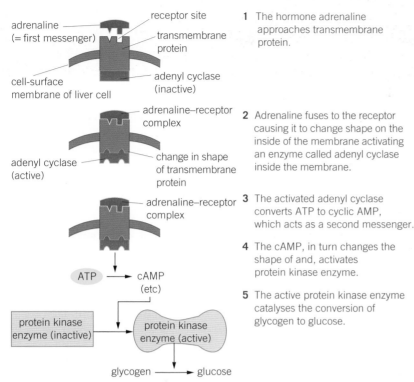

1. The hormone adrenaline approaches transmembrane protein.

2. Adrenaline fuses to the receptor causing it to change shape on the inside of the membrane activating an enzyme called adenyl cyclase inside the membrane.

3. The activated adenyl cyclase converts ATP to cyclic AMP, which acts as a second messenger.

4. The cAMP, in turn changes the shape of and, activates protein kinase enzyme.

5. The active protein kinase enzyme catalyses the conversion of glycogen to glucose.

◀ **Figure 1** *Second messenger model of hormone action as illustrated by the action of adrenaline in regulating blood sugar*

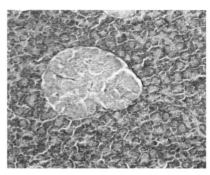

▲ **Figure 2** *LM of the pancreas showing an islet of Langerhans (centre) containing α cells and β cells. The meshworks of blue and white in the islet are blood capillaries. Around the islet are the enzyme-producing pancreatic cells*

When examined microscopically, the pancreas is made up largely of the cells that produce its digestive enzymes. Scattered throughout these cells are groups of hormone-producing cells known as **islets of Langerhans**. The cells of the islets of Langerhans include:

* **α cells**, which are the larger and produce the hormone **glucagon**
* **β cells**, which are smaller and produce the hormone **insulin**.

The role of the liver in regulating blood sugar

The liver is located immediately below the diaphragm, has a mass of up to 1.5 kg and is made up of cells called hepatocytes. It serves a large variety of roles including regulating blood glucose concentration. While the pancreas produces the hormones insulin and glucagon, it is in the liver where they have their effects. There are three important processes associated with regulating blood sugar which take place in the liver.

* **Glycogenesis** is the conversion of glucose into glycogen. When blood glucose concentration is higher than normal the liver removes glucose from the blood and converts it to glycogen. It can store 75–100 g of glycogen, which is sufficient to maintain a human's blood glucose concentration for about 12 hours when at rest, in the absence of other sources.
* **Glycogenolysis** is the breakdown of glycogen to glucose. When blood glucose concentration is lower than normal, the liver can convert stored glycogen back into glucose which diffuses into the blood to restore the normal blood glucose concentration.
* **Gluconeogenesis** is the production of glucose from sources other than carbohydrate. When its supply of glycogen is exhausted, the liver can produce glucose from non-carbohydrate sources such as glycerol and amino acids.

Practical link

Required practical 11. Production of a dilution series of a glucose solution and use of colorimetric technique to produce a calibration curve with which to identify the concentration of glucose in an unknown 'urine' sample.

Study tip

Hormones only affect their target cells and not other cells because only target cells have the specific protein receptors that are complementary to the shape of that specific hormone.

Synoptic link

It would be useful at this stage to recall information on respiration by reviewing Topic 12.1, Topic 12.2 and Topic 2.3.

Regulation of blood glucose concentration

Glucose is a substrate for respiration, providing the source of energy for almost all organisms. It is therefore essential that the blood of mammals contains a relatively constant concentration of glucose for respiration. If the concentration falls too low, cells will be deprived of energy and die – brain cells are especially sensitive in this respect because they can only respire glucose. If the concentration rises too high, it lowers the water potential of the blood and creates osmotic problems that can cause dehydration and be equally dangerous. Homeostatic control (Topic 16.1) of blood glucose is therefore essential.

Factors that influence blood glucose concentration

The normal concentration of blood glucose is $5\,\text{mmol}\,\text{dm}^{-3}$. Blood glucose comes from three sources:

- **directly from the diet** in the form of glucose absorbed following hydrolysis of other carbohydrates such as starch, maltose, lactose, and sucrose
- **from the hydrolysis in the small intestine of glycogen = glycogenolysis** stored in the liver and muscle cells
- **from gluconeogenesis**, which is the production of glucose from sources other than carbohydrate.

As animals do not eat continuously and their diet varies, their intake of glucose fluctuates. Likewise, glucose is used during respiration at different rates depending on the level of mental and physical activity. It is against these changes in supply and demand that the three main hormones, **insulin**, **glucagon** and **adrenaline**, operate to maintain a constant blood glucose concentration.

Insulin and the β cells of the pancreas

The β cells of the islets of Langerhans in the pancreas have receptors that detect the stimulus of a rise in blood glucose concentration and respond by secreting the hormone insulin directly into the blood plasma. Insulin is a globular protein made up of 51 amino acids.

Almost all body cells (red blood cells being a notable exception) have glycoprotein receptors on their cell-surface membranes that bind specifically with insulin molecules. When it combines with the receptors, insulin brings about:

- a change in the tertiary structure of the glucose transport carrier proteins, causing them to change shape and open, allowing more glucose into the cells by facilitated diffusion
- an increase in the number of the carrier proteins responsible for glucose transport in the cell-surface membrane. At low insulin concentrations, the protein from which these channels are made is part of the membrane of vesicles. A rise in insulin concentration results in these vesicles fusing with the cell-surface membrane so increasing the number of glucose transport channels
- activation of the enzymes that convert glucose to glycogen and fat.

Hint

You will find it easier to understand the terms used in this topic if you remember the following:

gluco / glyco = glucose

glycogen = glycogen

neo = new

lysis = splitting

genesis = birth / origin

Therefore:

glycogen – o – lysis = splitting of glycogen

gluco – neo – genesis = formation of new glucose

As a result, the blood glucose concentration is lowered in one or more of the following ways:

- by increasing the rate of absorption of glucose into the cells, especially in muscle cells
- by increasing the respiratory rate of the cells, which therefore use up more glucose, thus increasing their uptake of glucose from the blood
- by increasing the rate of conversion of glucose into glycogen (glycogenesis) in the cells of the liver and muscles
- by increasing the rate of conversion of glucose to fat.

The effect of these processes is to remove glucose from the blood and so return its concentration to the optimum. This lowering of the blood glucose concentration causes the β cells to reduce their secretion of insulin (= negative feedback).

Glucagon and the α cells of the pancreas

The α cells of the islets of Langerhans detect a fall in blood glucose concentration and respond by secreting the hormone glucagon directly into the blood plasma. Glucagon's actions include:

- attaching to specific protein receptors on the cell-surface membrane of liver cells
- activating enzymes that convert glycogen to glucose
- activating enzymes involved in the conversion of amino acids and glycerol into glucose (= gluconeogenesis).

The overall effect is therefore to increase the concentration of glucose in the blood and return it to its optimum concentration. This raising of the blood glucose concentration causes the α cells to reduce the secretion of glucagon (= negative feedback).

Role of adrenaline in regulating the blood glucose level

There are at least four other hormones apart from glucagon that can increase blood glucose concentration. The best known of these is adrenaline. At times of excitement or stress, adrenaline is produced by the adrenal glands that lie above the kidneys. Adrenaline raises the blood glucose concentration by:

- attaching to protein receptors on the cell-surface membrane of target cells
- activating enzymes that causes the breakdown of glycogen to glucose in the liver.

Hormone interaction in regulating blood glucose

The two hormones insulin and glucagon act in opposite directions. Insulin lowers the blood glucose concentration, whereas glucagon increases it. The two hormones are said to act antagonistically. The system is self-regulating through negative feedback in that it is the concentration of glucose in the blood that determines the quantity of insulin and glucagon produced. In this way the interaction of

> **Study tip**
>
> When writing about negative feedback it is important to mention that the secretion of a hormone such as insulin results in a reduction of its own secretion.

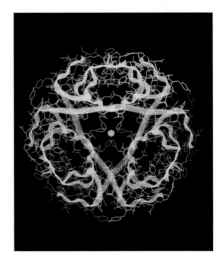

▲ **Figure 3** *Molecular graphic of an insulin molecule. Insulin is made up of 51 amino acids arranged in two chains (shown here as yellow and green ribbons)*

Hint

There is almost always a time lag between a hormone being produced and the response to it. This is because it takes time to produce it, transport it in the blood and for it to affect the enzyme or transport protein of the target cell.

these two hormones allows highly sensitive control of the blood glucose concentration. The concentration of glucose is not, however, constant, but fluctuates around an optimum point. This is because of the way negative feedback mechanisms work. Only when the blood glucose concentration falls below the set point is insulin secretion reduced (negative feedback), leading to a rise in blood glucose concentration. In the same way, only when the concentration exceeds the set point is glucagon secretion reduced (negative feedback), causing a fall in the blood glucose concentration.

The control of blood glucose concentration is summarised in Figure 4.

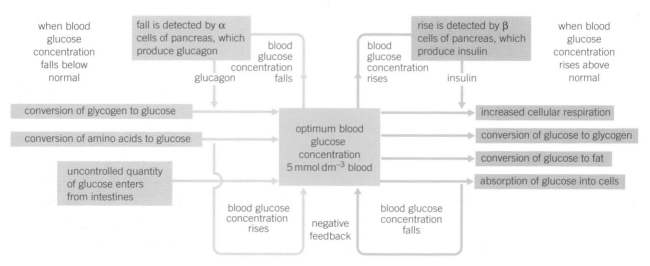

▲ **Figure 4** *Summary of regulation of blood glucose concentration*

Summary questions

In the following passage, state the most suitable word to replace the numbers in brackets.

The chemical energy in glucose is released by cells during the process known as (1). It is therefore important that the blood glucose concentration is maintained at a constant level because if it falls too low cells are deprived of energy, and (2) cells are especially sensitive in this respect. If it gets too high (3) problems occur that may cause dehydration. Blood glucose is formed directly from (4) in the diet or from the breakdown of (5), which is stored in the cells of the liver and (6). The liver can also increase the blood glucose concentration by making glucose from other sources, such as glycerol and (7), in a process known as (8). Blood glucose is used up when it is absorbed into cells, converted into fat or (9) for storage, or used up during (10) by cells. In order to maintain a constant concentration of blood glucose the pancreas produces two hormones from clusters of cells within it called (11). The β cells produce the hormone (12), which causes the blood glucose concentration to fall. The α cells produce the hormone (13), which has the opposite effect. Another hormone, called (14), can also raise blood glucose concentration.

16.4 Diabetes and its control

Diabetes is a disease in which a person is unable to metabolise carbohydrate, especially glucose, properly. There are around 350 million people worldwide with diabetes, 3.2 million of whom are in the UK. In addition, a further 1 million people in the UK are thought to have the disease but are currently unaware of it. One form of diabetes is diabetes mellitus, or 'sugar diabetes'.

Types of sugar diabetes

Diabetes is a metabolic disorder caused by an inability to control blood glucose concentration due to a lack of the hormone insulin or a loss of responsiveness to insulin.

There are two forms of diabetes:

- **Type I (insulin dependent)** is due to the body being unable to produce insulin. It normally begins in childhood. It may be the result of an autoimmune response whereby the body's immune system attacks its own cells, in this case the β cells of the islets of Langerhans. Type I diabetes develops quickly, usually over a few weeks, and the signs and symptoms (see Hint) are normally obvious.

- **Type II (insulin independent)** is normally due to glycoprotein receptors on body cells being lost or losing their reponsiveness to insulin. However, it may also be due to an inadequate supply of insulin from the pancreas. Type II diabetes usually develops in people over the age of 40 years. There is, however, an increasing number of cases of obesity and poor diet leading to type II diabetes in adolescents. It develops slowly, and the symptoms are normally less severe and may go unnoticed. People who are overweight are particularly likely to develop type II diabetes. About 90% of people with diabetes have type II.

Figure 2 illustrates the differences in blood glucose concentration between people with and without diabetes who have swallowed a glucose solution.

Control of diabetes

Although diabetes cannot be cured, recent trials in transplanting insulin-producing cells have shown promise. Diabetes can also be successfully treated. Treatment varies depending on the type of diabetes.

- **Type I diabetes** is controlled by injections of insulin. This cannot be taken by mouth because, being a protein, it would be digested in the alimentary canal. It is therefore injected, typically either two or four times a day. The dose of insulin must be matched exactly to the glucose intake. If a person with diabetes takes too much insulin, he or she will experience a low blood glucose concentration that can result in unconsciousness. To ensure the correct dose, blood glucose concentration is monitored using biosensors. By injecting insulin and managing their carbohydrate intake and exercise carefully, people with diabetes can lead normal lives.

- **Type II diabetes** is usually controlled by regulating the intake of carbohydrate in the diet and matching this to the amount

Learning objectives

→ Describe the two main types of diabetes and how they differ.

→ Explain how each type of diabetes can be controlled.

Specification reference: 3.6.4.2

Hint

Signs of diabetes:

- high blood glucose concentration
- presence of glucose in urine
- need to urinate excessively
- genital itching or regular episodes of thrush
- weight loss
- blurred vision

Symptoms of diabetes:

- tiredness
- increased thirst and hunger

▲ **Figure 1** *A person with diabetes injecting insulin*

127

Hint

Blood glucose concentration can be controlled by changing the uptake of glucose from the gut (diet) and by changing the rate at which glucose is removed from the blood (exercise and insulin).

of exercise taken. In some cases, this may be supplemented by injections of insulin or by the use of drugs that stimulate insulin production. Other drugs can slow down the rate at which the body absorbs glucose from the intestine.

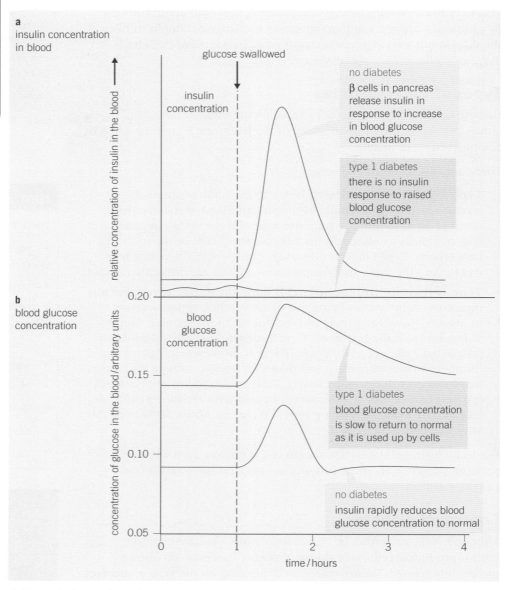

▲ **Figure 2** *Comparison of blood glucose and insulin concentrations in a person with type I diabetes and a person without diabetes after each has swallowed a glucose solution*

Summary questions

1 State **one** difference between the causes of type I and type II diabetes.

2 State **one** difference between the main ways of controlling type I and type II diabetes.

3 Suggest an explanation for why tiredness is a symptom of diabetes.

4 Suggest what lifestyle advice you might give someone in order to help them avoid developing type II diabetes.

Effects of diabetes on substance concentrations in the blood

An experiment was carried out with two groups of people. Group X had type 1 diabetes while group Y did not (control group). Every 15 minutes blood samples were taken from all members of both groups and the mean concentrations of insulin, glucagon and glucose were determined. After an hour, each person was given a glucose drink. The results are shown in the graphs below.

1 Name a hormone other than insulin and glucagon that is involved in regulating blood glucose concentration.
2 State *two* differences between groups X and Y in the way insulin secretion responds to the drinking of glucose.

3 Suggest a reason why the glucose concentration falls in both groups during the first hour.
4 Using information from the graphs, explain the changes in the blood glucose concentration in group Y after drinking the glucose.
5 Explain the difference in blood glucose concentration of group X compared with group Y.
6 Suggest what might happen to the blood glucose concentration of group X if they have no food over the next 24 hours.

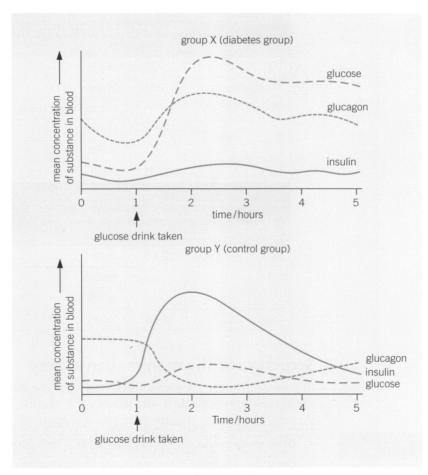

▲ Figure 3

Learning objectives

→ Describe the structure of the mammalian kidney.

→ Describe the structure of the nephron.

Specification reference: 3.6.4.3

The amount of water and mineral ions we take in varies from day to day, as does the quantity we lose. Table 1 shows the daily balance between loss and gain of salts and water for a typical human. In the blood, however, an optimum concentration of water and salts is maintained to ensure a fairly constant water potential of blood plasma and tissue fluid. The homeostatic control of the water potential of the blood is called **osmoregulation**. To understand osmoregulation, we must first understand the structure of the organ that carries it out, the kidney, and in particular its functional unit – the **nephron**.

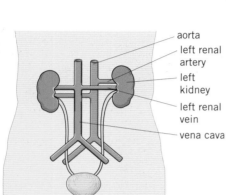

▲ **Figure 1** *Position of the kidneys in humans*

aorta
left renal artery
left kidney
left renal vein
vena cava

▼ **Table 1** *Daily water and sodium chloride balance in a typical human*

Water			
Volume of water / $cm^3\ day^{-1}$			
Water gain		Water loss	
Diet	2300	Urine	1500
Metabolism (e.g., respiration)	200	Expired air	400
		Evaporation from skin	350
		Faeces	150
		Sweat	100
Total	2500	Total	2500
Sodium chloride			
Mass of sodium chloride / $g\ day^{-1}$			
Salt gain		Salt loss	
Diet	10.50	Urine	10.00
		Faeces	0.25
		Sweat	0.25
Total	10.50	Total	10.50

Structure of the mammalian kidney

In mammals there are two kidneys found at the back of the abdominal cavity, one on each side of the spinal cord (Figure 1). A section through the kidney (Figure 2) shows it is made up of the:

- **fibrous capsule** – an outer membrane that protects the kidney
- **cortex** – a lighter coloured outer region made up of renal (Bowman's) capsules, convoluted tubules and blood vessels
- **medulla** – a darker coloured inner region made up of loops of Henle, collecting ducts and blood vessels

- **renal pelvis** – a funnel-shaped cavity that collects urine into the ureter
- **ureter** – a tube that carries urine to the bladder
- **renal artery** – supplies the kidney with blood from the heart via the aorta
- **renal vein** – returns blood to the heart via the vena cava.

A microscopic examination of the cortex and medulla reveals around one million tiny tubular structures in each kidney. These are the basic structural and functional units of the kidney – the nephrons.

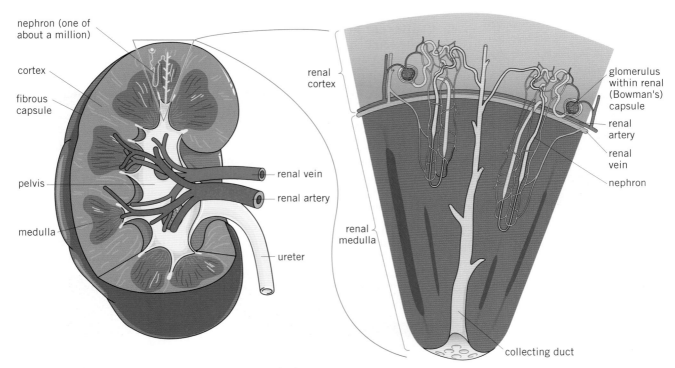

▲ **Figure 2** *Detailed structure of mammalian kidney (LS) showing the position of two of the million or more nephrons in each kidney*

The structure of the nephron

The nephron is the functional unit of the kidney. It is a narrow tube up to 14 mm long, closed at one end, with two twisted regions separated by a long hairpin loop. Each nephron is made up of a:

- **Renal (Bowman's) capsule** – the closed end at the start of the nephron. It is cup-shaped and surrounds a mass of blood capillaries known as the glomerulus. The inner layer of the renal capsule is made up of specialised cells called **podocytes**.
- **Proximal convoluted tubule** – a series of loops surrounded by blood capillaries. Its walls are made of epithelial cells which have microvilli.

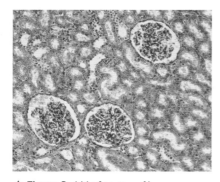

▲ **Figure 3** *LM of cortex of human kidney (TS). Three glomeruli are seen as regions of small cells surrounded by a clear space – the lumen of the renal capsule. The background shows convoluted tubules*

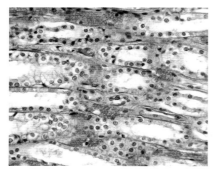

▲ **Figure 4** *LM of medulla of human kidney showing loops of Henle (white tubes). Around them are blood capillaries containing red blood cells (red)*

- **Loop of Henle** – a long, hairpin loop that extends from the cortex into the medulla of the kidney and back again. It is surrounded by blood capillaries.
- **Distal convoluted tubule** – a series of loops surrounded by blood capillaries. Its walls are made of epithelial cells, but it is surrounded by fewer capillaries than the proximal tubule.
- **Collecting duct** – a tube into which a number of distal convoluted tubules from a number of nephrons empty. It is lined by epithelial cells and becomes increasingly wide as it empties into the pelvis of the kidney.

Associated with each nephron are a number of blood vessels (Figure 6), namely:

- **afferent arteriole** – a tiny vessel that ultimately arises from the renal artery and supplies the nephron with blood. The afferent arteriole enters the renal capsule of the nephron where it forms the
- **glomerulus** – a many-branched knot of capillaries from which fluid is forced out of the blood. The glomerular capillaries recombine to form the
- **efferent arteriole** – a tiny vessel that leaves the renal capsule. It has a smaller diameter than the afferent arteriole and so causes an increase in blood pressure within the glomerulus. The efferent arteriole carries blood away from the renal capsule and later branches to form the
- **blood capillaries** – a concentrated network of capillaries that surrounds the proximal convoluted tubule, the loop of Henle and the distal convoluted tubule and from where they reabsorb mineral salts, glucose and water. These capillaries merge together into venules (tiny veins) that in turn merge together to form the renal vein.

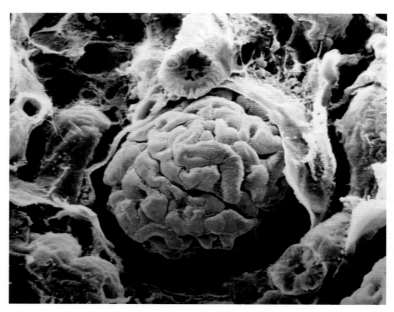

▲ **Figure 5** *Colourised SEM of a glomerulus (centre) surrounded by the renal capsule, seen as a white-brown membrane at centre right. Part of the proximal convoluted tubule is seen, coloured blue*

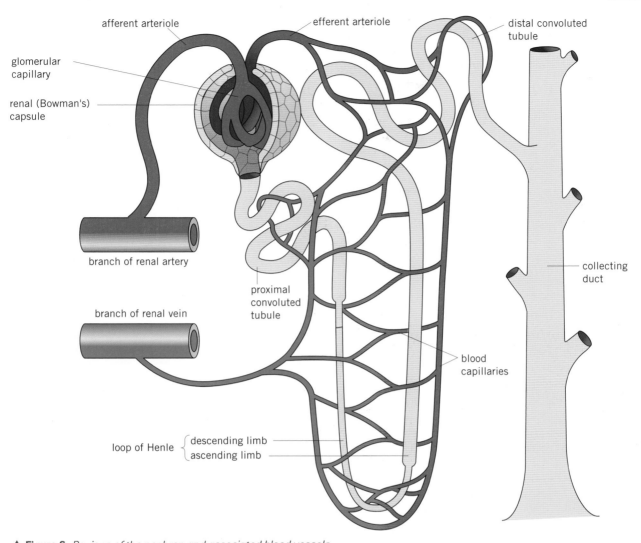

▲ **Figure 6** *Regions of the nephron and associated blood vessels*

Summary questions

Complete the passage below by stating the word or words that best replace the numbers in brackets.

The nephron is the structural unit of the kidney. It comprises a cup-shaped structure called the (**1**) that contains a knot of blood vessels called the (**2**) which receives its blood from a vessel called the (**3**) arteriole. The inner wall of this cup-shaped structure is lined with specialised cells called (**4**) and from it extends the first, or (**5**), convoluted tubule whose walls are lined with (**6**) that have (**7**) to increase their surface area. The next region of the nephron is a hairpin loop called the (**8**) which then leads onto the second, or (**9**), convoluted tubule. This in turn leads onto the (**10**) which empties into the renal pelvis. Around much of the nephron is a dense network of blood vessels called the (**11**) capillaries.

Control of blood water potential

Figure 7 shows some of the changes that occur as a result of water being lost from the blood due to sweating.

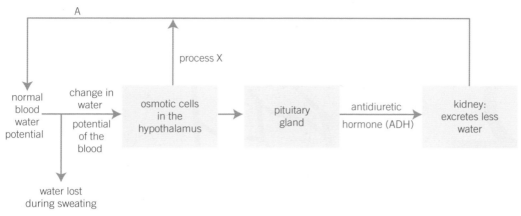

▲ Figure 7

1 Describe the change in water potential that occurs in the blood as a result of sweating.
2 Which of the structures shown in Figure 7 acts as:
  a a receptor
  b an effector?
3 Describe how ADH gets from the pituitary gland to the kidney.
4 The kidney conserves the water that is already in the blood. Given that the water potential of the blood returns to its normal level prior to sweating, suggest what is happening in process X.
5 State as concisely as possible what mechanism is shown by the line labelled A.

 ## The glomerulus – a unique capillary network

In mammals, the glomerulus is the only capillary bed in which an arteriole (the afferent arteriole) supplies it with blood and an arteriole (the efferent arteriole) also drains blood away. In all other mammalian capillary beds it is a venule that drains away the blood.

1 By reference to Figure 6, suggest a reason why the efferent arteriole is not called a venule.
2 Suggest another way in which you could show that the afferent arteriole was not a venule.

One important function of the kidney is to maintain the water potential of plasma and hence tissue fluid (osmoregulation).

The nephron carries out its role of osmoregulation in a series of stages. These are:

- the formation of glomerular filtrate by ultrafiltration
- reabsorption of glucose and water by the proximal convoluted tubule
- maintenance of a gradient of sodium ions in the medulla by the loop of Henle
- reabsorption of water by the distal convoluted tubule and collecting ducts.

Let us now look at each stage in detail.

Formation of glomerular filtrate by ultrafiltration

Blood enters the kidney through the renal artery, which branches frequently to give around one million tiny arterioles, each of which enters a **renal (Bowman's) capsule** of a nephron. This arteriole is called the **afferent arteriole** and it divides to give a complex of capillaries known as the glomerulus. The **glomerular** capillaries later merge to form the **efferent arteriole**, which then sub-divides again into capillaries, which wind their way around the various tubules of the nephron before combining to form the renal vein. The walls of the glomerular capillaries are made up of epithelial cells with pores between them. As the diameter of the afferent arteriole is greater than that of the efferent arteriole, there is a build up of hydrostatic pressure within the glomerulus. As a result, water, glucose and mineral ions are squeezed out of the capillary to form the **glomerular filtrate**. Blood cells and proteins cannot pass across into the renal capsule as they are too large. The movement of this filtrate out of the glomerulus is resisted by the:

- capillary epithelial cells
- connective tissue and epithelial cells of the blood capillary
- epithelial cells of the renal capsule
- the hydrostatic pressure of the fluid in the renal capsule space
- the low water potential of the blood in the glomerulus.

This total resistance would be sufficient to prevent filtrate leaving the glomerular capillaries, but for some modifications to reduce this barrier to the flow of filtrate:

- The inner layer of the renal capsule is made up of highly specialised cells called **podocytes**. These cells, which are illustrated in Figure 1, have spaces between them. This allows filtrate to pass beneath them and through gaps between their branches. Filtrate passes between these cells rather than through them.
- The endothelium of the glomerular capillaries has spaces up to 100 nm wide between its cells (Figure 1). Again, fluid can therefore pass between, rather than through, these cells.

Learning objectives

→ Describe ultrafiltration and the production of glomerular filtrate.

→ Explain reabsorption of water by the proximal convoluted tubule.

→ Explain how a gradient of sodium ions in the medulla of the loop of Henle is maintained.

→ Explain the role of the distal convoluted tubule and collecting duct in the reabsorption of water.

Specification reference: 3.6.4.3

As a result, the hydrostatic pressure of the blood in the glomerulus is sufficient to overcome the resistance and so filtrate passes from the blood into the renal capsule. The filtrate, which contains urea, does not contain cells or plasma proteins which are too large to pass across the connective tissue. Many of the substances in the 125 cm^3 of filtrate passing out of blood each minute are extremely useful to the body and are reabsorbed.

Reabsorption of glucose and water by the proximal convoluted tubule

In the proximal convoluted tubule nearly 85% of the filtrate is reabsorbed back into the blood. Ultrafiltration operates on the basis of size of molecule – small ones are removed. Some, such as urea, are wastes, but most are useful and so are reabsorbed.

The proximal convoluted tubules are adapted to reabsorb substances into the blood by having epithelial cells that have:

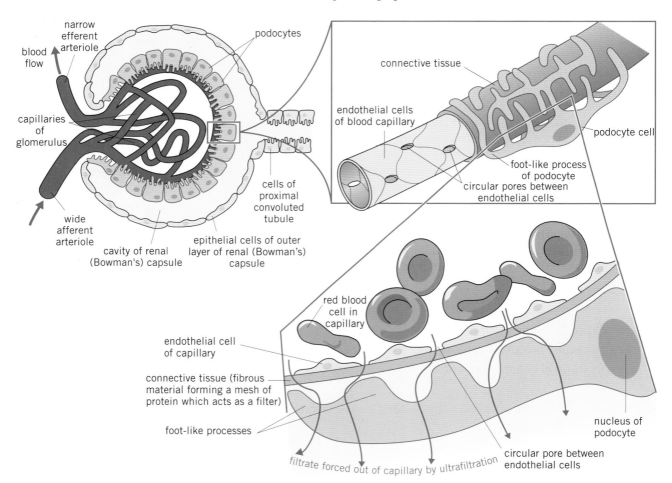

▲ **Figure 1** *Podocyte and ultrafiltration*

- microvilli to provide a large surface area to reabsorb substances from the filtrate
- infoldings at their bases to give a large surface area to transfer reabsorbed substances into blood capillaries
- a high density of mitochondria to provide ATP for active transport.

The process is as follows:

- Sodium ions are actively transported out of the cells lining the proximal convoluted tubule into blood capillaries which carry them away. The sodium ion concentration of these cells is therefore lowered.
- Sodium ions now diffuse down a concentration gradient from the lumen of the proximal convoluted tubule into the epithelial lining cells but only through special carrier proteins by facilitated diffusion.
- These carrier proteins are of specific types, each of which carries another molecule (glucose or amino acids or chloride ions, etc.) along with the sodium ions. This is known as co-transport.
- The molecules which have been co-transported into the cells of the proximal convoluted tubule then diffuse into the blood. As a result, all the glucose and most other valuable molecules are reabsorbed as well as water.

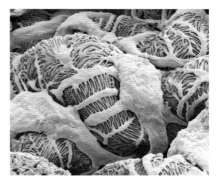

▲ **Figure 2** *Colourised SEM of podocyte cells around a glomerulus in a human kidney*

Synoptic link

Your understanding of reabsorption will be improved if you first revise Topics 4.1, 4.2, 4.4, and 4.5.

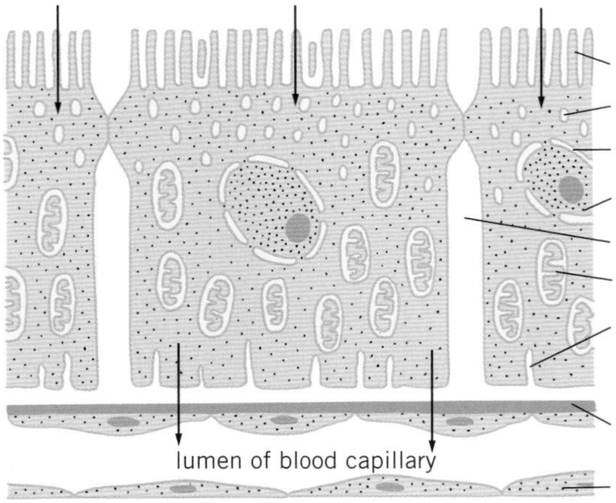

lumen of proximal convoluted tubule
reabsorption of material by facilitated diffusion, active transport and pinocytosis

- microvillus
- pinocytic vesicle
- nuclear envelope
- nucleoplasm
- intercellular space
- mitochondrion
- infolding of basal membrane
- basement membrane
- endothelial cell

lumen of blood capillary

▲ **Figure 3** *Details of cells from the wall of the proximal convoluted tubule*

About $180\,dm^3$ of water enters the nephrons each day. Of this volume, only about $1\,dm^3$ leaves the body as urine. 85% of the reabsorption of water occurs in the proximal convoluted tubule. The remainder is reabsorbed from the collecting duct as a result of the functioning of the loop of Henle.

Maintenance of a gradient of sodium ions by the loop of Henle

The loop of Henle is a hairpin-shaped tubule that extends into the medulla of the kidney. It is responsible for water being reabsorbed from the collecting duct, thereby concentrating the urine so that it has a lower **water potential** than the blood. The concentration of the urine produced is directly related to the length of the loop of Henle.

The loop of Henle has two regions:

- The descending limb, which is narrow, with thin walls that are highly permeable to water.
- The ascending limb, which is wider, with thick walls that are impermeable to water.

The loop of Henle acts as a counter-current multiplier. To understand how this works it is necessary to consider the following sequence of events in conjunction with Figure 4, to which the numbers refer.

1 Sodium ions are actively transported out of the ascending limb of the loop of Henle using ATP provided by the many mitochondria in the cells of its wall.

2 This creates a low water potential (high ion concentration) in the region of the medulla between the two limbs (called the interstitial region). In normal circumstances water would pass out of the ascending limb by osmosis. However, the thick walls are almost impermeable to water and so very little, if any, escapes.

3 The walls of the descending limb are, however, very permeable to water and so it passes out of the filtrate, by osmosis, into the

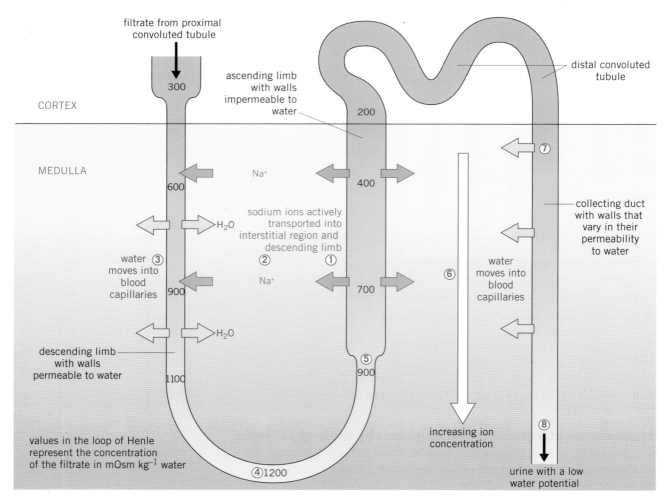

▲ **Figure 4** *Counter-current multiplier of the loop of Henle*

interstitial space. This water enters the blood capillaries in this region by osmosis and is carried away.

4 The filtrate progressively loses water in this way as it moves down the descending limb lowering its water potential. It reaches its lowest water potential at the tip of the hairpin.

5 At the base of the ascending limb, sodium ions diffuse out of the filtrate and as it moves up the ascending limb these ions are also actively pumped out (see point 1) and therefore the filtrate develops a progressively higher water potential.

6 In the interstitial space between the ascending limb and the collecting duct there is a gradient of water potential with the highest water potential (lowest concentration of ions) in the cortex and an increasingly lower water potential (higher concentration of ions) the further into the medulla one goes.

7 The collecting duct is permeable to water and so, as the filtrate moves down it, water passes out of it by osmosis. This water passes by osmosis into the blood vessels that occupy this space, and is carried away.

8 As water passes out of the filtrate its water potential is lowered. However, the water potential is also lowered in the interstitial space and so water continues to move out by osmosis down the whole length of the collecting duct. The counter-current multiplier ensures that there is always a water potential gradient drawing water out of the tubule.

The water that passes out of the collecting duct by osmosis does so through channel proteins that are specific to water (aquaporins). Antidiuretic hormone (ADH) (Topic 16.7) can alter the number of these channels and so control water loss. By the time the filtrate, now called urine, leaves the collecting duct on its way to the bladder, it has lost most of its water and so it has a lower water potential (is more concentrated) than the blood.

The distal convoluted tubule
The cells that make up the walls of the distal convoluted tubule have microvilli and many mitochondria that allow them to reabsorb material rapidly from the filtrate, by active transport. The main role of the distal tubule is to make final adjustments to the water and salts that are reabsorbed and to control the pH of the blood by selecting which ions to reabsorb. To achieve this, the permeability of its walls becomes altered under the influence of various hormones (Topic 16.7).

Counter-current multiplier
You may remember that when two liquids flow in opposite directions past one another, the exchange of substances (or heat) between them is greater than if they flowed in the same direction next to each other. In the case of the loop of Henle, the counter-current flow means that the filtrate in the collecting duct with a lower water potential meets interstitial fluid that has an even lower water potential. This means that, although the water potential gradient between the collecting duct and interstitial fluid is small, it exists for the whole length of the collecting duct. There is therefore a steady flow of water into the interstitial fluid, so that around 80% of the water enters the interstitial fluid and hence the blood. If the two flows were in the same direction (parallel) less of the water would enter the blood.

Synoptic link
To help you understand how the loop of Henle concentrates urine, you should look back at the counter-current exchange principle in Topic 6.3 Gas exchange in fish.

Summary questions
1 Name the structure in the nephron where the majority of water is reabsorbed.

2 The following is a list of the various parts of a nephron: distal convoluted tubule, glomerulus, loop of Henle, collecting duct, distal convoluted tubule; renal capsule.

List the sequence of structures that a molecule of water which is excreted from the body passes through on its journey to the bladder.

3 Describe how the proximal convoluted tubule is adapted to its function.

4 The length of the loop of Henle in animals living in dry environments is different from the length in those living in environments where water is abundant. Suggest if the length is longer or shorter and explain how it helps animals in dry areas to survive.

16.7 The role of hormones in osmoregulation

16.7

Learning objectives

→ Explain how the water potential of the blood is regulated.

→ Describe the roles of the hypothalamus, posterior pituitary and antidiuretic hormone (ADH) in osmoregulation.

Specification reference: 3.6.4.3

Hint

The name antidiuretic hormone (ADH) may, at first, seem unusual. However, it describes its function precisely. **Diuresis** is the production of large volumes of dilute urine. As the effect of ADH is to increase the permeability of collecting ducts so that more water is reabsorbed into the blood, it causes the production of small volumes of concentrated urine.

This is the opposite of diuresis – hence the name **anti**diuretic hormone.

Hint

The pituitary gland has two parts – the anterior and the posterior part.

Synoptic link

The regulation of water potential of the blood is another example of the stimulus → receptor → coordinator → effector → response pathway that we have seen many times before.

The **homeostatic** control of osmoregulation in the blood is achieved by a hormone that acts on the distal convoluted tubule and the collecting duct.

Regulation of the water potential of the blood

The **water potential** of the blood depends on the concentration of solutes like glucose, proteins, sodium chloride, and other mineral ions as well as the volume of water in the body. A rise in solute concentration lowers its water potential. This may be caused by:

- too little water being consumed
- much sweating occurring
- large amounts of ions, for example, sodium chloride, being taken in.

The body responds to this fall in water potential as follows:

- Cells called **osmoreceptors** in the **hypothalamus** of the brain detect the fall in water potential.
- It is thought that, when the water potential of the blood is low, water is lost from these osmoreceptor cells by osmosis.
- Due to this water loss the osmoreceptor cells shrink, a change that causes the hypothalamus to produce a hormone called **antidiuretic hormone (ADH)**.
- ADH passes to the posterior **pituitary gland**, from where it is secreted into the capillaries.
- ADH passes in the blood to the kidney, where it increases the permeability to water of the cell-surface membrane of the cells that make up the walls of the distal convoluted tubule and the collecting duct.
- Specific protein receptors on the cell-surface membrane of these cells bind to ADH molecules, leading to activation of an enzyme called phosphorylase within the cell.
- The activation of phosphorylase causes vesicles within the cell to move to, and fuse with, its cell-surface membrane.
- These vesicles contain pieces of plasma membrane that have numerous water channel proteins (aquaporins) and so when they fuse with the membrane the number of water channels is considerably increased, making the cell-surface membrane much more permeable to water.
- ADH increases the permeability of the collecting duct to urea, which therefore passes out, further lowering the water potential of the fluid around the duct.
- The combined effect is that more water leaves the collecting duct by osmosis, down a water potential gradient, and re-enters the blood.
- As the reabsorbed water came from the blood in the first place, this will not, in itself, increase the water potential of the blood, but merely prevent it getting lower. The osmoreceptors also send nerve impulses to the thirst centre of the brain, to encourage the individual to seek out and drink more water.

- The osmoreceptors in the hypothalamus detect the rise in water potential and send fewer impulses to the pituitary gland.
- The pituitary gland reduces the release of ADH and the permeability of the collecting ducts to water and urea reverts to its former state. This is an example of homeostasis and the principle of negative feedback (Topic 16.1).

A fall in the solute concentration of the blood raises its water potential. This may be caused by:

- large volumes of water being consumed
- salts used in metabolism or excreted not being replaced in the diet.

The body responds to this rise in water potential as follows:

- The osmoreceptors in the hypothalamus detect the rise in water potential and increase the frequency of nerve impulses to the pituitary gland to reduce its release of ADH.
- Less ADH, via the blood, leads to a decrease in the permeability of the collecting ducts to water and urea.
- Less water is reabsorbed into the blood from the collecting duct.
- More dilute urine is produced and the water potential of the blood falls.
- When the water potential of the blood has returned to normal, the osmoreceptors in the hypothalamus cause the pituitary to raise its ADH release back to normal levels (= negative feedback).

These events are summarised in Figure 1.

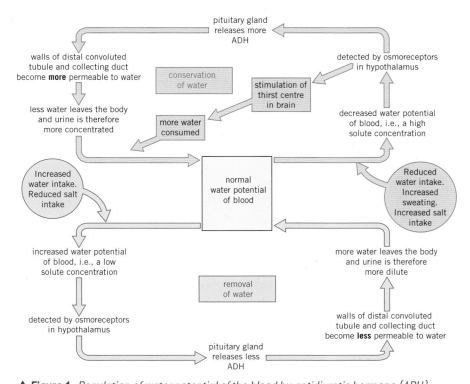

▲ **Figure 1** *Regulation of water potential of the blood by antidiuretic hormone (ADH)*

Summary questions

1 State where the cells which monitor the water potential of the blood are located.

2 In each of the following situations deduce whether more or less ADH would be produced by the body:

   a drinking a large volume of water in a short time.

   b exercising intensely for 30 minutes.

3 Explain how ADH causes the collecting ducts to reabsorb more water.

4 √x̄ The concentration of proteins in a sample of glomerular filtrate is 0.0625% of their concentration in the blood plasma. If the concentration of proteins in the blood plasma is 80 g dm^{-3}, calculate their concentration in the glomerular filtrate.

The significance of glucose in the urine √x̄

The presence of glucose in urine may indicate a clinical disorder. The kidney should reabsorb all glucose from the filtrate leaving the urine free of it. The presence of glucose in urine (glucosuria) suggests:

- There may be so much glucose in the filtrate that the kidney is overwhelmed and cannot reabsorb it all.

This is often an indication of a high blood glucose concentration due to diabetes mellitus.

- More rarely, it may be that the kidney is not functioning properly and not reabsorbing glucose. This could be due to a disease of the kidneys.

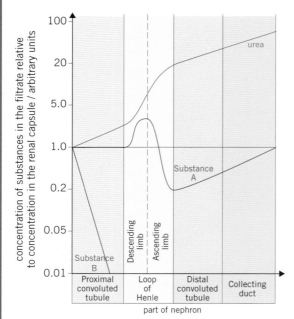

▲ **Figure 2** *Relative concentrations of three substances in the filtrate as it passes along a nephron. NB Scale is not linear*

1 Name three hormones involved in controlling the level of glucose in the blood and state the effect of each on the level of glucose.

   Figure 2 shows the relative concentrations of three substances in the filtrate as it passes along the nephron. (NB the scale is not linear.)

2 Urea is a waste product of the body that is removed by the kidneys. The quantity of urea in the filtrate entering the nephron does not change significantly as the filtrate passes along the nephron. Explain why the level of urea shown in Figure 2 rises considerably as the filtrate passes along the nephron.

3 Suggest the name of substance A and explain the reasons for your answer.

4 Suggest the name of substance B and explain the reasons for your answer.

5 One symptom of diabetes is dehydration. From your knowledge of how water is reabsorbed in the collecting ducts, explain why diabetes might cause dehydration.

1 The release of a substance called dopamine in some areas of the brain increases the desire to eat. Scientists measured increases in the release of dopamine in the brains of rats given different concentrations of sucrose solution to drink.

Sucrose stimulates taste receptors on the tongue.

The graph shows their results. Each point is the result for one rat.

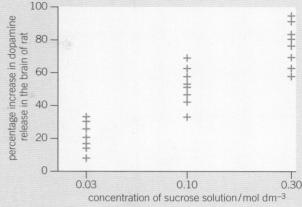

(a) The scientists concluded that drinking a sucrose solution had a positive feedback effect on the rats' desire to eat.

How do these data support this conclusion?

(3 marks)

(b) In this investigation, the higher the concentration of sucrose in a rat's mouth, the higher the frequency of nerve impulses from each taste receptor to the brain. If rats are given very high concentrations of sucrose solution to drink, the refractory period makes it impossible for information about the differences in concentration to reach the brain. Explain why. *(2 marks)*

(c) In humans, when the stomach starts to become full of food, receptors in the wall of the stomach are stimulated. This leads to negative feedback on the desire to eat. Suggest why this negative feedback is important. *(3 marks)*

AQA June 2013

2 Different substances are involved in coordinating responses in animals.

(a) Hormones are different from local chemical mediators such as histamine in the cells they affect.

(i) Describe how hormones are different in the cells they affect. *(1 mark)*

(ii) Describe how hormones and local chemical mediators reach the cells they affect. *(2 marks)*

(b) Synapses are unidirectional. Explain how acetylcholine contributes to a synapse being unidirectional. *(2 marks)*

(c) Cells in the stomach wall release gastric juice after a meal. The graph shows how the volumes of gastric juice produced by nervous stimulation and by hormonal stimulation change after a meal.

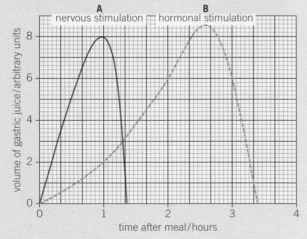

(i) Describe the evidence from the graph that curve **A** represents the volume of gastric juice produced by nervous stimulation. *(2 marks)*

(ii) Copy and complete the table to show the percentage of gastric juice produced by nervous stimulation at the times shown.

	Time after meal / hours		
	1	2	3
Percentage of gastric juice produced by nervous stimulation			

(1 mark)
AQA June 2011

3 **(a)** Adrenaline binds to receptors in the plasma membranes of liver cells. Explain how this causes the blood glucose concentration to increase. *(2 marks)*

(b) Scientists made an artificial gene which codes for insulin. They put the gene into a virus which was then injected into rats with type I diabetes. The virus was harmless to the rats but carried the gene into the cells of the rats. The treated rats produced insulin for up to 8 months and showed no side-effects. The scientists measured the blood glucose concentrations of the rats at regular intervals. While the rats were producing the insulin, their blood glucose concentrations were normal.

(i) The rats were not fed for at least 6 hours before their blood glucose concentration was measured. Explain why. *(1 mark)*

(ii) The rats used in the investigation had type I diabetes. This form of gene therapy may be less effective in treating rats that have type II diabetes. Explain why. *(1 mark)*

(iii) Research workers have suggested that treating diabetes in humans by this method of gene therapy would be better than injecting insulin. Evaluate this suggestion. *(4 marks)*
AQA June 2012

4 **(a)** Technicians in a hospital laboratory tested urine and blood samples from a girl with diabetes at intervals over a one-year period. Each time the technicians tested her urine, they also measured her blood glucose concentration. Their results are shown in the graph.

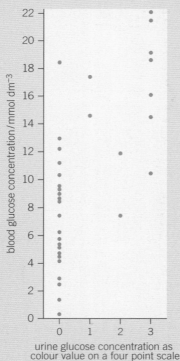

(i) The girl who took part in this investigation was being successfully treated with insulin. The graph shows that on some occasions, the concentration of glucose in her blood was very high. Suggest why. *(2 marks)*

(ii) Use the graph to evaluate the use of the urine test as a measure of blood glucose concentration. *(3 marks)*

(b) Diabetic people who do not control their blood glucose concentration may become unconscious and go into a coma. A doctor may inject a diabetic person who is in a coma with glucagon. Explain how the glucagon would affect the person's blood glucose concentration. *(2 marks)*

AQA June 2010

5 Osmoreceptors are specialised cells that respond to changes in the water potential of the blood.

(a) Give the location of osmoreceptors in the body of a mammal. *(1 mark)*

(b) When a person is dehydrated, the cell volume of an osmoreceptor decreases. Explain why. *(2 marks)*

(c) Stimulation of osmoreceptors can lead to secretion of the hormone ADH. Describe and explain how the secretion of ADH affects urine produced by the kidneys. *(4 marks)*

The efficiency with which the kidneys filter the blood can be measured by the rate at which they remove a substance called creatinine from the blood. The rate at which they filter the blood is called the glomerular filtration rate (GFR).

In 24 hours, a person excreted 1660 mg of creatinine in his urine. The concentration of creatinine in the blood entering his kidneys was constant at 0.01 mg cm^{-3}.

(d) Calculate the GFR in cm^3 minute^{-1} *(1 mark)*

(e) Creatinine is a breakdown product of creatine found in muscle tissues. Apart from age and gender, give **two** factors that could affect the concentration of creatinine in the blood. *(1 mark)*

AQA Specimen 2014

6 In an investigation of blood glucose levels, colorimetry was used to find the absorbance of blue light by plasma samples that had been reacted with Benedict's Reagent. The % absorbance was converted to glucose concentration using a standard calibration curve, drawn using the following data.

Glucose concentration (mMol dm^{-3})	Absorbance of blue light (%)
0.001	10
0.01	22
0.1	40
1	58
10	80

(a) Produce recipes for each of the five glucose concentrations, assuming a supply of 10mMol dm^{-3} glucose and distilled water. *(3 marks)*

(b) Plot a calibration curve of % absorbance against log$_{10}$ glucose concentration. *(3 marks)*

(c) Why is a log scale appropriate in this case? *(1 mark)*

(d) Use your graph to estimate the concentration of glucose that would correspond with a % absorbance of 45%? *(2 marks)*

Section 6 Summary

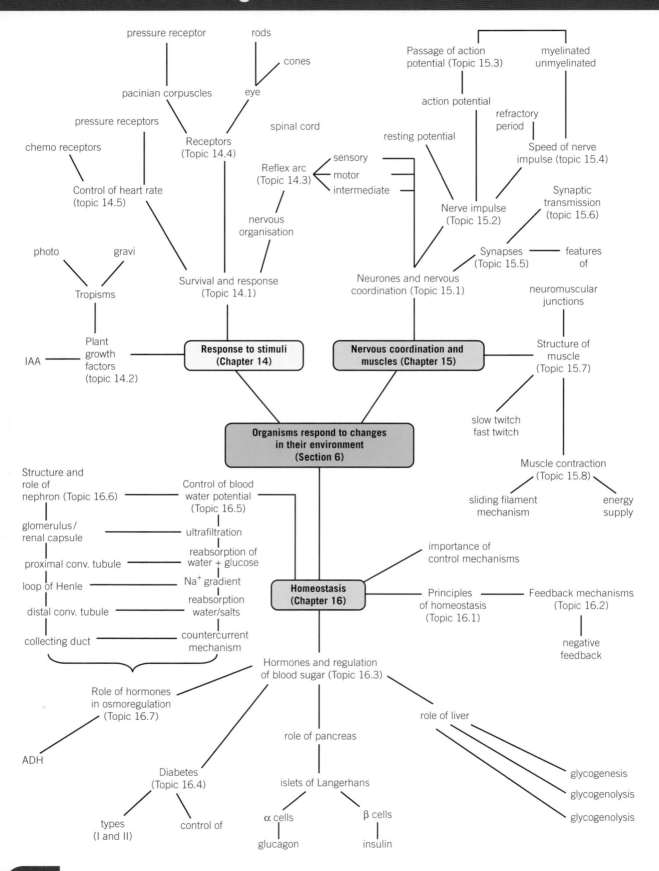

Practical skills

In this section you have met the following practical skills:

- How to carry out experiments to determine how plant growth factors such as auxins like IAA have their effects on cell growth and elongation.

- How to carry out an experiment to investigate the effects on blood sugar levels of consuming a glucose drink by diabetics and non-diabetics.

Maths skills

In this section you have met the following maths skills:

- Calculating percentage change in transmission speeds.

- Solving algebraic equations to determine the number of ATP molecules needed to contract a muscle fibre a specified distance.

- Translating information between graphical and numerical forms in calculating the number of action potentials in a given time.

Extension task

Using only the technique of a person catching a 30 cm ruler between his/her thumb and forefinger when the ruler is dropped, design and carry out a series of experiments to determine the distance travelled by the ruler before it is caught using:

1. the stimulus of sight only,

2. the stimulus of sound only,

3. the stimulus of touch only.

Using textbooks or the internet, research how to convert the distance travelled by the ruler before it is caught to the time taken for it to fall that distance. Calculate the average reaction time for each type of stimulus.

Suggest reasons for any differences you found between the reaction times for the three different stimuli.

You could also devise and carry out an experiment using the same technique to compare the reaction times between a person's dominant and non-dominant hand.

1   Scientists investigated the response of the roots of pea seedlings to gravity.
    They took three samples of seedlings, **A**, **B**, and **C**, and placed them so that their roots
    were growing horizontally. The root tips of each sample had been given different
    treatments. After a set time, the scientists recorded whether the roots of the seedlings had
    grown upwards or downwards and the amount of curvature. Table 1 shows the treatment
    they gave to each sample and their results.

▼ Table 1

Treatment	Results	
	Direction of growth	Mean amount of curvature / degrees
A None	Downwards	60
B Root tip removed	Continues to grow horizontally	0
C Upper half of root tip removed	Downwards	30

(a)   The pea seedlings were kept in the dark after each treatment. Explain why
      this was necessary.   *(1 mark)*
(b)   What conclusion can be made from the results for treatment **B**?   *(1 mark)*
(c)   Suggest how indoleacetic acid (IAA) could have caused the results for
      (i)   treatment **A**   *(2 marks)*
      (ii)   treatment **C**   *(2 marks)*

AQA June 2012

2   Figure 2 shows a diagram of part of a muscle myofibril.

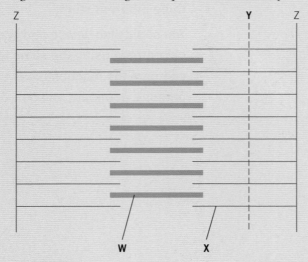

▲ Figure 2

(a) Name the protein present in the filaments labelled **W** and **X**. *(1 mark)*

(b) Figure 3 shows the cut ends of the protein filaments when the myofibril was cut at position Y. Figure 4 shows the protein filaments when the myofibril was cut at the same distance from a Z line at a different stage of contraction.

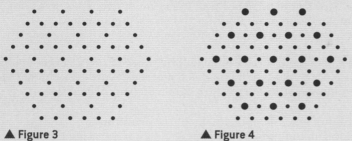

▲ **Figure 3**   ▲ **Figure 4**

Explain why the pattern of protein filaments differs in Figure 3 and 4. *(2 marks)*

(c) Describe the role of calcium ions in the concentration of a sarcomere. *(4 marks)*

AQA Jan 2004

3 (a) Describe how insulin reduces the concentration of glucose in the blood. *(3 marks)*

Some people produce no insulin. As a result they have a condition called diabetes. In an investigation, a man with diabetes drank a glucose solution. The concentration of glucose in his blood was measured at regular intervals. The results are shown in Figure 5.

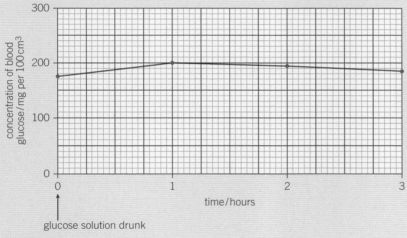

▲ **Figure 5**

(b) Suggest **two** reasons why the concentration of glucose decreased after 1 hour even though this man's blood contained no insulin. *(2 marks)*

(c) The investigation was repeated on a man who did not have diabetes. The concentration of glucose in his blood before drinking the glucose solution was 80 mg per 100 cm³. Copy the graph roughly and sketch a curve on it to show the results you would expect. *(1 mark)*

(d) The diabetic man adopted a daily routine to stabilise his blood glucose concentration within narrow limits. He ate three meals a day: breakfast, a midday meal, and an evening meal. He injected insulin once before breakfast and once before the evening meal.
The injection he used before breakfast was a mixture of two types of insulin.
The mixture contained slow-acting insulin and fast-acting insulin.
(i) Explain the advantage of injecting both types of insulin before breakfast.

(2 marks)

(ii) One day, the man did not eat a midday meal. Suggest one reason why his blood glucose concentration did not fall dangerously low even though he had injected himself with the mixture of insulin before breakfast. *(1 mark)*
AQA Jan 2004

4 **(a)** Explain how a resting potential is maintained in a neurone. *(4 marks)*
(b) In an investigation, an impulse was generated in a neurone using electrodes. During transmission along the neurone, an action potential was recorded at one point on the neurone. When the impulse reached the neuromuscular junction, it stimulated a muscle cell to contract. The force generated by the contraction was measured. The results are shown in the graph.
The distance between the point on the neurone where the action potential was measured and the neuromuscular junction was exactly 18 mm.

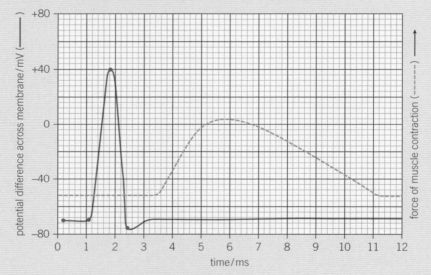

(i) Use the graph to estimate the time between the maximum depolarisation and the start of contraction by the muscle cell. *(1 mark)*
(ii) Use the answer to part (i) to calculate the speed of transmission along this neurone to the muscle cell. Give your answer in mm per second ($mm\,s^{-1}$). Show your working. *(2 marks)*
(iii) Give one reason why the value calculated in part (ii) would be an underestimate of the speed of transmission of an impulse along a neurone. *(1 mark)*
Acetylcholine is the neurotransmitter at neuromuscular junctions.

(c) Describe how the release of acetylcholine into a neuromuscular junction causes the cell membrane of a muscle fibre to depolarise. (*3 marks*)

(d) Use your knowledge of the processes occurring at a neuromuscular junction to explain each of the following.

    (i) The cobra is a very poisonous snake. The molecular structure of the cobra toxin is similar to the molecular structure of acetylcholine. The toxin permanently prevents muscle contraction. (*2 marks*)

    (ii) The insecticide DFP combines with the active site of the enzyme acetylcholinesterase. The muscles stay contracted until the insecticide is lost from the neuromuscular junction. (*2 marks*)

AQA Jan 2004

Section 7
Genetics, populations, evolution, and ecosystems

Introduction

The individuals of a species share the same genes but usually have different combinations of alleles of these genes. An individual inherits alleles from their parent or parents. While this process is universal, the way in which the alleles interact to produce the characteristics of the new individual depends on the type of inheritance involved. Sometimes one allele is dominant to another and so expresses itself, at other times the two alleles are equally dominant and the offspring have intermediate features. A characteristic is sometimes inherited along with the sex of an individual.

Populations of different species live in communities. Competition occurs within and between these populations for the means of survival. Populations within communities are also affected by, and in turn affect, the abiotic factors in an ecosystem. A species exists as one or more populations. The phenotypes of organisms in a population vary due to both genetic and environmental factors. Two forces affect genetic variety within a population – genetic drift and natural selection. Genetic drift can cause changes in allele frequency in small populations. Natural selection occurs when alleles that enhance the survival chances of the individuals that carry them rise in frequency. This change in the allele frequency of a population is known as evolution.

Different populations of the same species can sometimes become isolated from one another. This can be because they are geographically separated and therefore cannot interbreed. When this happens there is no flow of genes between the isolated populations. This may lead to the accumulation of genetic differences between each of these populations. These differences may ultimately lead to organisms in one population becoming unable to breed and produce fertile offspring with organisms from the other populations. This reproductive isolation means that a new species has evolved.

The theory of evolution is fundamental to biology. It states that all new species arise from existing ones by the process of natural selection. This means species, however different, have a common ancestry. This ancestry is represented in the phylogenetic classification of species. Common ancestry explains the similarities between all living organisms. These similarities include common chemistry such as all proteins having the same 20 or so amino acids, the same physiological pathways, e.g. anaerobic respiration, similar cell structure, DNA, and genetic material as well as a 'universal' genetic code.

Working scientifically

In studying this unit there will be opportunities to perform practical exercises and so develop practical skills. A required practical activity is to carry out an investigation into the effect of a named environmental factor on the distribution of a given species. In performing this activity you will have the chance to develop practical skills such as: safely and ethically use organisms in investigations, using microbiological aseptic techniques such as using agar broth, using sampling techniques in fieldwork, using ICT such as computer modelling.

You will be able to develop a range of mathematical skills. In particular the ability to use ratios, fractions, logarithms, and percentages, find arithmetical means, understand simple probability, understand the principles of sampling when applied to scientific data, select and use a statistical test and solve algebraic problems,

What you already know

The material in this unit is intended to be self-explanatory, but there is certain knowledge from GCSE that will be useful to the understanding of this section. This information includes:

- ◯ When a cell divides by meiosis to form gametes copies of the genetic information are made and then the cell divides twice to form four gametes, each with a single set of chromosomes.

- ◯ When gametes join at fertilisation, a single body cell with new pairs of chromosomes is formed.

- ◯ In human body cells, one of the 23 pairs of chromosomes carries the genes that determine sex. In females the sex chromosomes are the same (XX); in males the sex chromosomes are different (XY).

- ◯ Some characteristics are controlled by a single gene. Each gene may have different forms called alleles.

- ◯ An allele that controls the development of characteristics only if the dominant allele is not present, is a recessive allele.

- ◯ A gene is a small section of DNA and each gene codes for a particular combination of amino acids which make a specific protein.

- ◯ How to interpret genetic diagrams, including family trees and how to construct genetic diagrams of monohybrid crosses and predict the outcomes of monohybrid crosses.

- ◯ Understanding the terms homozygous, heterozygous, phenotype, and genotype.

- ◯ Individuals with characteristics most suited to the environment are more likely to survive to breed. The genes that have enabled these individuals to survive are then passed on to the next generation.

- ◯ New species arise as a result of isolation, genetic variation, natural selection, and speciation.

- ◯ Quantitative data on the distribution of organisms can be obtained by random sampling with quadrats and sampling along a transect.

Learning objectives

→ Define the meaning of the terms genotype and phenotype.

→ Define the terms dominant, recessive and codominant alleles.

→ Explain the nature of multiple alleles.

Specification reference: 3.7.1

Synoptic link

An understanding of inheritance depends on an understanding of the way chromosomes behave during meiosis and mutations. It would therefore be beneficial to study Topic 9.1 and Topic 9.2, again before starting this chapter.

Study tip

Not all genes code for a polypeptide, some code for ribosomal RNA or transfer RNAs

Hint

All individuals of the same species have the same genes, but not necessarily the same alleles of these genes.

The fact that children resemble both their parents to a greater or lesser degree and yet are identical to neither has long been recognised. However, it took the re-discovery, at the beginning of the last century, of the work of a scientist and monk, called Gregor Mendel, to establish the basic laws by which characteristics are inherited. In this chapter we shall look at the way in which characteristics are inherited from one generation to the next and how this can produce genetic variety within a population. Let us begin by looking at some of the terms and conventions that are used in studying inheritance.

Genotype and phenotype

Genotype is the genetic constitution (make-up) of an organism. It describes all the alleles that an organism has. The genotype determines the limits within which the characteristics of an individual may vary. It may determine that a human baby could grow to be 1.8 m tall, but the actual height that this individual reaches is affected by other factors, such as diet. A lack of an element like calcium (for the growth of bone) at a particular stage of development could mean that the individual never reaches his/her potential maximum height.

Phenotype is the observable or biochemical characteristics of an organism. It is the result of the interaction between the expression of the genotype and the environment. The environment can alter an organism's phenotype.

Genes and alleles

A **gene** is a length of DNA, that is, a sequence of nucleotide bases, that normally code for a particular polypeptide. A gene does this by coding for a particular polypeptide. This polypeptide may be an enzyme that is needed in the biochemical pathway that leads to the production of the characteristic (for example, a gene could code for a brown pigment in the iris of the eye). Genes exist in two, or more, different forms called alleles. The position of a gene on a particular DNA molecule is known as the **locus**.

An allele is one of the different forms of a gene. In pea plants, for example, there is a gene for the colour of the seed pod. This gene has two different forms, or alleles, an allele for a green pod and another allele for a yellow pod.

Only one allele of a gene can occur at the locus of any one chromosome. However, in diploid organisms the chromosomes occur in pairs called **homologous chromosomes** (Topic 8.2). There are therefore two loci that each carry one allele of a gene. If the allele on each of the chromosomes is the same (for example, both alleles for green pods are present) then the organism is said to be **homozygous** for the character. If the two alleles are different (for example, one chromosome has an allele for green pods and the other chromosome has an allele for yellow pods) then the organism is said to be **heterozygous** for the characteristic.

In most cases where two different alleles are present in the genotype (heterozygous state) only one of them shows itself in the phenotype. For instance, in our example where the alleles for green pods and yellow pods are present in the genotype, the phenotype is always green pods. The allele of the heterozygote that expresses itself in the phenotype is said to be **dominant**, while the one that is not expressed is said to be **recessive**. A homozygous organism with two dominant alleles is called **homozygous dominant**, whereas one with two recessive alleles is called **homozygous recessive**. The effect of a recessive allele is apparent in the phenotype of a *diploid* organism only when it occurs in the presence of another identical allele, that is, when it is in the homozygous state.

These different genetic types are shown in Figure 1.

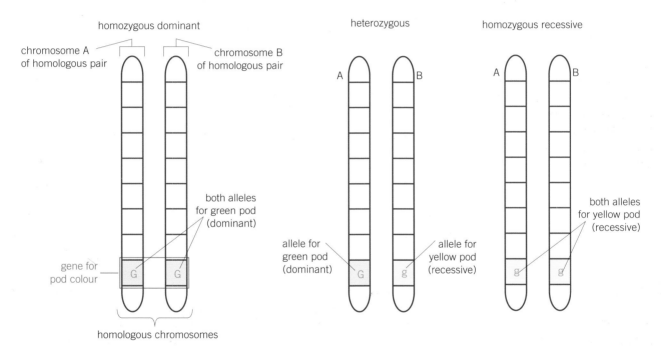

▲ **Figure 1** *Pair of homologous chromosomes showing different possible pairings of dominant and recessive alleles*

In some cases, two alleles both contribute to the phenotype, in which case they are referred to as **codominant**. In this situation when both alleles occur together, the phenotype is either a blend of both features (for example, snapdragons with pink flowers resulting from an allele for red-coloured flowers and an allele for white-coloured flowers) or both features are represented (for example, the presence of both A and B antigens in blood group AB). We will learn more about codominance in Topic 17.5.

Sometimes a gene has more than two allelic forms. In this case, the organism is said to have **multiple alleles** for the character. However, as there are always only two chromosomes in a homologous pair, it follows that only two of the three or more alleles in existence can be present in a single organism. Multiple alleles occur in the human ABO blood grouping system. Again we shall learn more about multiple alleles in Topic 17.5.

Figure 3 summarises the different terms used in genetics.

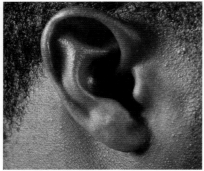

▲ **Figure 2** *Attached earlobe. As with other inherited physical body characteristics, the earlobe is subject to genetic differences. Here, the earlobe is firmly attached to the facial skin rather than hanging freely*

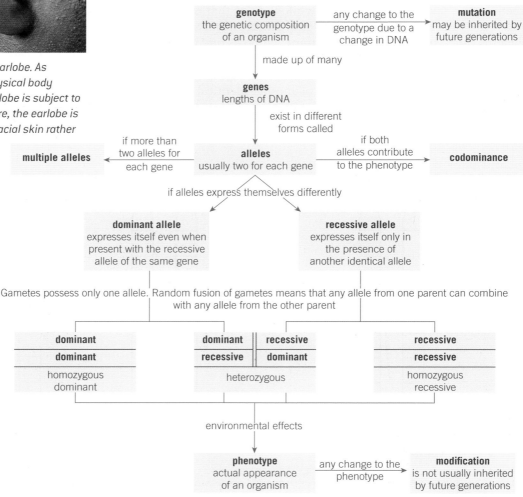

▲ **Figure 3** *Summary of genetic terms*

Summary questions

In the following passage, give the word that best replaces the number in brackets.

The genetic composition of an organism is called the (1) and any change to it is called a (2) and may be inherited by future generations. The actual appearance of an organism is called the (3). A gene is a sequence of (4) along a section of DNA that determines a single characteristic of an organism. It does this by coding for particular (5) that make up the enzymes needed in a biochemical pathway. The position of a gene on the DNA of a chromosome is called the (6). Each gene has two or more different forms called alleles. If the two alleles on a homologous pair of chromosomes are the same they are said to be (7), but if they are different, they are said to be (8). An allele that is not apparent in the phenotype when paired with a dominant allele is said to be (9). Two alleles are called (10) where they contribute equally to the appearance of a characteristic.

Representing genetic crosses

Genetic crosses are usually represented in a standard form of shorthand. This shorthand form is described in Table 1. Although you may occasionally come across variations to this scheme, that outlined in Table 1 is the one normally used. Once you have practised a number of crosses, you may be tempted to miss out stages or explanations. Not only is this likely to lead to errors, it often makes your explanations difficult for others to follow.

Learning objectives

→ Explain how to make labelled genetic diagrams.

→ Explain how a single gene is inherited.

Specification reference: 3.7.1

▼ **Table 1** *Representing genetic crosses*

Instruction	Reason/notes	Example [green pod and yellow pod]
Questions usually give the symbols to be used, in which case always use the ones provided. Choose a single letter to represent each characteristic.	An easy form of shorthand.	—
Choose the first letter of one of the contrasting features.	When more than one character is considered at one time such a logical choice means it is easy to identify which letter refers to which character.	Choose **G** (green) or **Y** (yellow).
If possible, choose the letter in which the higher and lower case forms differ in shape as well as size.	If the higher and lower case forms differ it is almost impossible to confuse them, regardless of their size.	Choose **G** because the higher case form (**G**) differs in shape from the lower case from (**g**) whereas **Y** and **y** are very similar and are likely to be confused.
Let the higher case letter represent the dominant feature and the lower case letter the recessive one. Never use two different letters where one character is dominant.	The dominant and recessive feature can easily be identified. Do not use two different letters as this indicates codominance.	Let **G** = green and **g** = yellow. Do not use **G** for green and **Y** for yellow.
Represent the parents with the appropriate pairs of letters. Label them clearly as 'parents' and state their phenotypes.	This makes it clear to any reader what the symbols refer to.	green pod × yellow pod parents **GG** **gg**
State the gametes produced by each parent. Label them clearly, and encircle them.	Encircling them reinforces the idea that they are separate.	gametes Ⓖ Ⓖ ⓖ ⓖ
Use a type of chequerboard or matrix, called a Punnett square, to show the results of the random crossing of the gametes. Label male and female gametes even though this may not affect the results.	This method is less liable to error than drawing lines between the gametes and the offspring. Labelling the sexes is a good habit to acquire – it has considerable relevance in certain types of crosses, e.g., sex-linked crosses.	♂ Gametes: Ⓖ Ⓖ ; ♀ Gametes: ⓖ → Gg, Gg ; ⓖ → Gg, Gg
State the phenotypes of each different genotype and indicate the numbers of each type. Always put the higher case (dominant) letter first when writing out the genotype.	Always putting the dominant feature first can reduce errors in cases where it is not possible to avoid using symbols with the higher and lower case letters of the same shape.	All offspring are plants producing green pods (**Gg**).

Inheritance of pod colour in peas

Monohybrid inheritance is the inheritance of a single gene. To take a simple example we will look at one of the features Gregor Mendel studied–the colour of the pods of pea plants. Pea pods come in two basic colours–green and yellow.

If pea plants with green pods are bred repeatedly with each other so that they consistently give rise to plants with green pods, they are said to be **pure-breeding** for the character of green pods. Pure-breeding strains can be bred for almost any character. This means that the organisms are homozygous (that is, they have two alleles that are the same) for that particular gene.

If these pure-breeding green-pod plants are then crossed with pure-breeding yellow-pod plants, all the offspring, known as the **first filial**, or F_1, **generation**, produce green pods. This means that the allele for green pods is dominant to the allele for yellow pods, which is therefore recessive. This cross is shown in Figure 1.

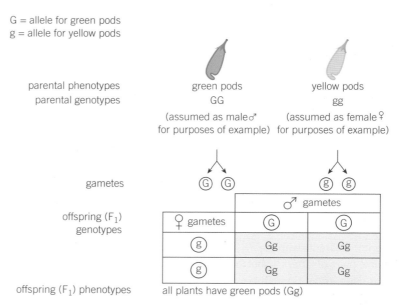

G = allele for green pods
g = allele for yellow pods

▲ **Figure 1** *Cross between a pea plant that is pure breeding for green pods and one that is pure breeding for yellow pods*

When the heterozygous plants (**Gg**) of the F_1 generation are crossed with one another (= F_1 intercross), the offspring (known as the second filial, or F_2, generation) are always in an approximate ratio of three plants with green pods to each one plant with yellow pods. This cross is shown in Figure 2.

These observed facts led to the formation of a basic law of genetics (the law of segregation). This states,

In diploid organisms, characteristics are determined by alleles that occur in pairs. Only one of each pair of alleles can be present in a single gamete.

Study tip

The larger the number of offspring the more likely that the ratio will be 3:1. If the sample is very small it is much less likely that the ratio will be 3:1.

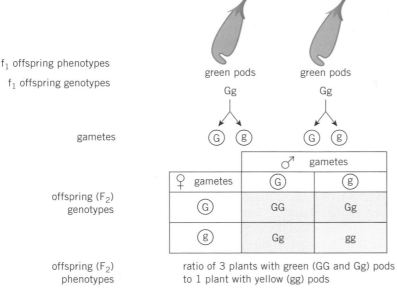

f₁ offspring phenotypes

f₁ offspring genotypes

gametes

offspring (F₂)
genotypes

offspring (F₂)
phenotypes

ratio of 3 plants with green (GG and Gg) pods
to 1 plant with yellow (gg) pods

▲ **Figure 2** *F₁ intercross between pea plants that are heterozygous for green pods*

Summary questions

1   In humans, Huntington's disease is caused by a dominant, mutant allele of a gene. Draw a genetic diagram to show the possible genotypes and phenotypes of the offspring produced by a man with one allele for the disease and a woman who does not suffer from the disease.

2   In cocker spaniels, black coat colour is the result of a dominant allele and red coat colour is the result of a corresponding recessive allele.

   a   Draw a genetic diagram to show a cross between a pure-breeding bitch with a black coat and a pure-breeding dog with a red coat.

   b   If the offspring of this first cross are interbred, calculate the probability that any one of the offspring will have a red coat. Use a genetic diagram to show your working.

Learning objectives

→ Explain why results of genetic crosses often differ from predicted results.

Specification reference: 3.7.1

In Topic 17.2 we looked at monohybrid crosses and we saw that the result of any such cross theoretically produced offspring in the ratio of three offspring with one or more dominant allele to one offspring with only recessive alleles. In practice this ratio is rarely achieved exactly. Before we look at why, let us first be clear what is meant by a ratio.

Ratios

A ratio is a measure of the relative size of two classes (groups) that is expressed as a proportion. For example, any group of humans can be divided into two classes, male and female. If in a group of 60 humans there are 40 males and 20 females, then the ratio of males to females is 40 to 20. This is usually expressed as a ratio which is simplified to 2 to 1 and is written 2:1. If our group of 60 humans comprised 35 males and 25 females the ratio of males to females would be 7:5. For easy comparisons, ratios are often obtained by dividing the value of the smallest group into the value of each larger group. In which case, all ratios have their smallest value as one. For example, our ratio of 7:5 would be 7 ÷ 5 = 1.4 which is written as 1.4:1.

Why actual results of genetic crosses are rarely the same as the predicted results

If you look at Table 1, you will see the results that Gregor Mendel actually obtained in his experiments. Our knowledge of genetics tells us that for each cross we would expect that, in the F_2 generation, there would be three offspring showing the dominant feature to every one showing the recessive feature. However, in no case did Mendel obtain an exact 3:1 ratio. The same is true of almost any genetic cross. These discrepancies are due to statistical error.

Imagine tossing a coin 20 times. In theory you would expect it to come down heads on 10 occasions and tails on 10 occasions. In practice it rarely does – try it. This is because each toss of the coin is an independent event that is not affected by what went before. If the coin has come down heads nine times out of 19 tosses, there is still a 50% chance it will come down tails, rather than the head needed to complete the 1:1 ratio. The coin does not know it is expected to come down heads.

The same is true of gametes. It is chance that determines which ones fuse with which. In our cross between the heterozygote (**Gg**) and the homozygous recessive (**gg**), all the gametes of the homozygous parent are recessive (**g**), whereas the heterozygote parent produces gametes of which half are dominant (**G**) and half are recessive (**g**). If it is the dominant gamete that combines with the recessive one, plants with green pods are produced (**Gg**). If it is the recessive gamete, the plants have yellow pods. The larger the sample, the more likely the actual results are to come near to matching the theoretical ones. It is therefore important to use large numbers of organisms in genetic crosses if representative results are to be obtained. It is no coincidence that the two ratios nearest to the theoretical value of 3:1 in Mendel's

Hint

Take care that you express a ratio the correct way round.

In the example opposite the ratio of males to females is 7:5. However, if you are asked for the ratio of females to males you should give the answer as 5:7.

experiments were those with the largest sample size, whereas the ratio furthest from the theoretical value had the smallest sample size (Table 1).

▼ **Table 1** *Actual results of Mendel's crosses in pea plants*

Character	F₂ results		Ratio
Cotyledon colour	6020 yellow	2001 green	3.01:1
Seed type	5474 smooth	1850 wrinkled	2.96:1
Pod type	882 inflated	299 constricted	2.95:1
Flower position	651 axial	207 terminal	3.14:1
Petal colour	705 purple	224 white	3.15:1
Stem height	787 long	277 short	2.84:1
Pod colour	428 green	152 yellow	2.82:1

Maths link $\sqrt{x}$

MS 0.3, see Chapter 22.

Summary questions $\sqrt{x}$

1 A cross was carried out between a pea plant producing green pods and one producing yellow pods. The seeds from this cross were germinated and, of the 63 plants grown, all produced green pods.

  a State the probable genotype of the parent plant with green pods.

  b Explain why we cannot be absolutely certain of the parent plant's genotype.

2 In a cross between a different pea plant with green pods and a pea plant with yellow pods, 96 plants were produced. 89 of these had green pods and 7 had yellow pods.

  a State the probable genotype of the parent plant with green pods.

  b Evaluate how certain we can be of the genotype of the parent plant with green pods.

  c In the cross described, state the chance of any one of the offspring plants having yellow pods.

  d $\sqrt{x}$ Calculate, to three significant figures, the percentage of offspring plants with yellow pods that were actually produced in the cross described.

In Topic 17.2 we saw how a single character is passed on from one generation to the next (monohybrid inheritance). In practice, many thousands of characters are inherited together. In this topic we shall look at how two characters, determined by two different **genes** located on different chromosomes, are inherited. This is referred to as **dihybrid inheritance**.

An example of dihybrid inheritance

In one of his experiments, Gregor Mendel investigated the inheritance of two characters of a pea plant at the same time. These were:

- **seed shape** – where round shape is dominant to wrinkled shape
- **seed colour** – where yellow-coloured seeds are dominant to green-coloured ones.

He carried out a cross between the following two pure breeding types of plants:

- one always producing round-shaped, yellow-coloured seeds (both dominant features)
- one always producing wrinkled-shaped, green-coloured seeds (both recessive features).

In the F$_1$ generation he obtained plants all of which produced round-shaped, yellow-coloured seeds, that is, dominant features.

He then raised the plants from these seeds and crossed them with one another to obtain the results shown in Table 1.

The explanation for these results is given in Figures 1 and 2.

▼ **Table 1** *Results obtained by Gregor Mendel when he crossed F$_1$ generation plants with round shaped, yellow coloured seeds*

Appearance of seeds	Condition	Number produced
Round Yellow	Dominant Dominant	315
Round Green	Dominant Recessive	108
Wrinkled Yellow	Recessive Dominant	101
Wrinkled Green	Recessive Recessive	32

R = allele for round shaped seeds   r = allele for wrinkled shaped seeds

G = allele for yellow coloured seeds   g = allele for green coloured seeds

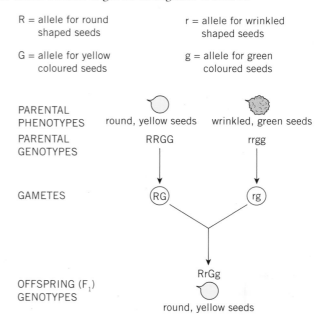

▲ **Figure 1** *Genetic explanation of Mendel's cross between a pure breeding plant for round, yellow seeds and a pure breeding one for wrinkled, green seeds*

From Figure 2 it can be seen that the plants of the F_1 generation produce four types of gamete (**RG, Rg, rG, rg**). This is because the gene for seed colour and the gene for seed shape are on separate chromosomes. As the chromosomes arrange themselves at random on the equator during meiosis, any one of the two **alleles** of the gene for seed colour (**G** and **g**) can combine with any one of the alleles for seed shape (**R** and **r**). Fertilisation is also random, so that any of the four types of gamete (with respect to seed colour and seed shape) of one plant can combine with any of the four types from the other plant.

Synoptic link

To understand much of genetics you need to appreciate what exactly is happening to alleles during a genetic cross. These alleles are attached to chromosomes and so you need to be clear about the behaviour of these chromosomes during meiosis. You can do this by recapping Topic 9.2, Meiosis and genetic variation.

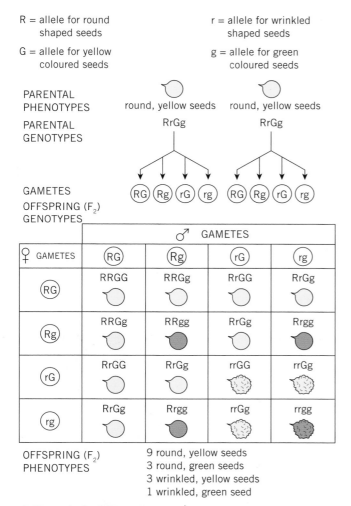

R = allele for round shaped seeds

r = allele for wrinkled shaped seeds

G = allele for yellow coloured seeds

g = allele for green coloured seeds

PARENTAL PHENOTYPES — round, yellow seeds | round, yellow seeds

PARENTAL GENOTYPES — RrGg | RrGg

GAMETES — RG Rg rG rg | RG Rg rG rg

OFFSPRING (F_2) GENOTYPES

	♂ GAMETES			
♀ GAMETES	RG	Rg	rG	rg
RG	RRGG	RRGg	RrGG	RrGg
Rg	RRGg	RRgg	RrGg	Rrgg
rG	RrGG	RrGg	rrGG	rrGg
rg	RrGg	Rrgg	rrGg	rrgg

OFFSPRING (F_2) PHENOTYPES

9 round, yellow seeds
3 round, green seeds
3 wrinkled, yellow seeds
1 wrinkled, green seed

▲ **Figure 2** *Genetic explanation of Mendel's intercross between plants of the F_1 generation*

The theoretical ratio produced of 9:3:3:1 is close enough, allowing for statistical error (Topic 17.3), to Mendel's observed results of 315:108:101:32. Mendel's observations led him to formulate his **law of independent assortment** which, written in today's biological language states–**Each member of a pair of alleles may combine randomly with either of another pair**.

Summary questions

In fruit flies, a pure breeding variety with red eyes and vestigial (tiny) wings was crossed with a pure breeding variety with pink eyes and normal wings. All the F_1 flies had red eyes and normal wings. When these F_1 flies were bred with one another, the F_2 generation produced the following types and numbers:

red eyes with vestigial wings	125
red eyes with normal wings	376
pink eyes with vestigial wings	41
pink eyes with normal wings	117

1 Which characteristics are dominant and which are recessive? Explain your answer.

2 Suggest suitable symbols to represent the alleles of the genes involved.

3 Draw two suitable genetic diagrams to explain the results of this experiment.

+ Better late than never

Gregor Mendel was a monk with some scientific training. This training proved invaluable and led to a scientific approach to his genetic experiments. In particular he:

- chose pea plants to experiment on because they were easy to grow and possessed many contrasting features that could easily be observed

- carefully controlled pollination and hence fertilisation, by accurately transferring pollen from one plant to another with a paint brush

- ensured the plants he used were pure breeding for each feature by self-pollinating them for many generations

- produced quantitative and not just qualitative results

- counted many offspring to ensure his results were reliable and relatively free from statistical error.

In just a nine-year period, Mendel planned a well-organised programme of research, performed it accurately, painstakingly recorded masses of data and analysed his results with precision and insight. He had worked out the basic principles of inheritance by 1865, when the nature of genetics was unknown. He effectively predicted the existence of genes and meiosis long before they were discovered.

Although he circulated his work to libraries and scholars of his day, his theories were not accepted – indeed they were ignored. Even Darwin, whose theory of evolution is based on genetic variation, failed to understand the significance of Mendel's research. It took 35 years before other geneticists rediscovered his work and the significance of his remarkable experiments was at last appreciated.

Some genetic crosses may, at first glance, appear not to follow Mendel's laws because they provide unfamiliar ratios. Consider the following example of a plant that has two different varieties. The crosses made and the results they produced are shown here:

- **Cross 1** – each variety is self-fertilised and in both cases the offspring occur in the ratio 3 green-leaved plants to 1 white-leaved plant.

- **Cross 2** – the two varieties are cross-fertilised and the offspring are all green-leaved.

- **Cross 3** – some plants of the F_1 offspring of the cross between the two varieties are self-fertilised. The F_2 generation produced is in the ratio 9 green-leaved to 7 white-leaved.

Green colour is due to the presence of chlorophyll. In its absence, the plant is white.

1 The production of chlorophyll is controlled in a normal Mendelian way. Justify this statement from the information provided.

If you are having problems answering the rest of this question – try reading topic 17.8 before having another attempt.

The formation of chlorophyll in this plant is controlled by two separate genes A and B. The dominant allele of both genes is required for chlorophyll synthesis.

2 Using this information and the results of cross 1, deduce the genotypes of each variety. Show your reasoning using genetic diagrams.

3 Draw a genetic diagram to explain cross 2.

4 Draw a genetic diagram to explain cross 3.

5 Explain how the presence of both allele **A** and allele **B** might be required for the biochemical synthesis of chlorophyll.

In Topic 17.2 we dealt with straightforward situations in which there were two possible alleles at each locus on a chromosome, one of which was dominant and the other recessive. We shall now look at two different situations.

- **codominance**, in which both alleles are expressed in the phenotype
- **multiple alleles**, where there are more than two alleles, of which only two may be present at the loci of an individual's homologous chromosomes.

Codominance

Codominance occurs where instead of one allele being dominant and the other recessive, both alleles are equally dominant. This means that both alleles of a gene are expressed in the phenotype.

Learning objectives
→ Explain how codominance affects the inheritance of characteristics.
→ Explain how multiple alleles affect inheritance.
→ Explain how blood groups in humans are inherited.

Specification reference: 3.7.1

▲ **Figure 1** *Snapdragons*

One example occurs in the snapdragon plant, in which one allele codes for an enzyme that catalyses the formation of a red pigment in flowers. The other allele codes for an altered enzyme that lacks this catalytic activity and so does not produce the pigment. If these alleles showed the usual pattern of one dominant and one recessive, the flowers would have just two colours–red and white. As they are codominant, however, three colours of flower are found:

- In plants that are **homozygous** for the first allele, both alleles code for the enzyme, and hence pigment, production. These plants have red flowers.
- In plants that are **homozygous** for the other allele, no enzyme and hence no pigment is produced. These plants have white flowers.
- **Heterozygous** plants, with their single allele for the functional enzyme, produce just sufficient red pigment to produce pink flowers.

Hint

Remember to use different letters, such as R and W, when writing about codominance and to put these as superscripts attached to the letter of the gene (C), for example, C^R and C^W.

If a snapdragon with red flowers is crossed with one with white flowers, the resulting seeds give rise to plants with pink flowers. Note that we cannot use upper and lower case letters for the alleles, as this would imply that one (the upper case) was dominant to the other (the lower case). We therefore use different letters – in this case **R** for red and **W** for white – and put them as superscripts on a letter that represents the gene, in this case **C** for colour. Hence the allele for red pigment is written as C^R and the allele for no pigment as C^W. Figure 2 shows a cross between a red and a white snapdragon while Figure 3 shows a cross between the resultant pink-flowered plants.

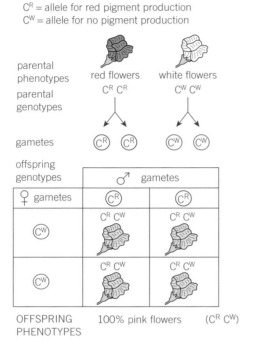

▲ **Figure 2** *Cross between a snapdragon with red flowers and one with white flowers*

▲ **Figure 3** *Cross between two snapdragons with pink flowers*

Synoptic link

We studied antigens in Topic 5.3 and their importance in immunity in Topics 5.4 and 5.5. Now would be a good time to look through these topics again.

▼ **Table 1** *Possible genotypes of blood groups in the ABO system*

Blood group	Possible genotypes
A	$I^A I^A$ or $I^A I^O$
B	$I^B I^B$ or $I^B I^O$
AB	$I^A I^B$
O	$I^O I^O$

Multiple alleles

A gene may have more than two alleles, that is, it has multiple alleles. The inheritance of the human ABO blood groups is an example. There are three alleles associated with the gene I (immunoglobulin gene), which lead to the presence of different **antigens** on the cell-surface membrane of red blood cells:

- allele I^A, which leads to the production of antigen A
- allele I^B which leads to the production of antigen B
- allele I^O, which does not lead to the production of either antigen.

Although there are three alleles, only two can be present in an individual at any one time, as there are only two homologous chromosomes and therefore only two gene loci. The alleles I^A and I^B are codominant, whereas the allele I^O is recessive to both. The possible genotypes for the four blood groups are shown in Table 1. There are obviously many different possible crosses between different blood groups, but two of the most interesting are:

1   A cross between an individual of blood group O and one of blood group AB, rather than producing individuals of either of the parental blood groups, produces only individuals of the other two groups, A and B (Figure 4).
2   When certain individuals of blood group A are crossed with certain individuals of blood group B, their children may have any of the four blood groups (Figure 5).

Study tip

If in a genetics question you are given a ratio of offspring that is different to the 3.1 (monohybrid) or 9.3.3.1 (dihybrid) ratio, begin by considering the possibility that codominance might be the explanation.

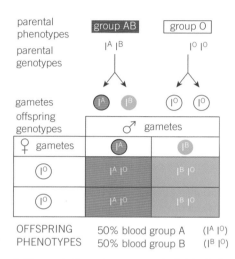

OFFSPRING   50% blood group A   ($I^A I^O$)
PHENOTYPES   50% blood group B   ($I^B I^O$)

▲ **Figure 4** *Cross between an individual of blood group AB and one of blood group O*

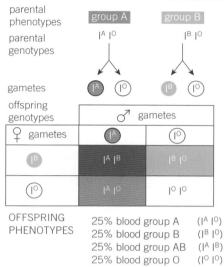

OFFSPRING   25% blood group A   ($I^A I^O$)
PHENOTYPES   25% blood group B   ($I^B I^O$)
              25% blood group AB   ($I^A I^B$)
              25% blood group O   ($I^O I^O$)

▲ **Figure 5** *Cross between an individual of blood group A and one of blood group B*

Summary questions

1   A man claims not to be the father of a child. The man is blood group O while the mother of the child is blood group A and the child is blood group AB. State, with your reasons, whether you think the man could be the father of the child.

2   In some breeds of domestic fowl, the gene controlling feather shape has two alleles that are codominant. When homozygous, the allele A^S produces straight feathers while the allele A^F produces frizzled feathers. The heterozygote for feather shape gives mildly frizzled feathers. Draw a genetic diagram to show the genotypes and phenotypes resulting from a cross between a mildly frizzled cockerel and a frizzled hen. The gene for feather shape is **not** sex-linked.

Maths link √x̄

MS 0.3, see Chapter 22.

Coats of many colours

In shorthorn cattle there is a gene **C** that determines coat colour. The gene has two alleles:

the allele **C^W** produces a white coat when homozygous

the allele **C^R** produces a red coat when homozygous.

In the heterozygous state the coat is light red, a colour also known as roan. The roan coat is a mixture of all white hairs and all red hairs. As each hair is either all red or all white, the **C^W** and **C^R** alleles are codominant.

▲ Figure 6

1   Draw a genetic diagram to show the possible genotypes and phenotypes of a cross between a bull with a white coat and a cow with a roan coat.
2   In each of the following crosses between shorthorn cattle what is the percentage of offspring with a roan coat?
    a   red coat x white coat
    b   red coat x roan coat
    c   white coat x roan coat
    d   roan coat x roan coat.

The phenotype of an organism often results from the influence of the environment on its genotype.

The fur on the ears, face, feet and tail of Siamese cats is darker in colour than the rest of the coat. This is due to the presence of a pigment. The production of this pigment is controlled by the action of an enzyme called tyrosinase. The action of tyrosinase is temperature-dependent.

3   Suggest why Siamese kittens are born with a completely light-coloured coat and only develop their characteristic markings some days later.

Humans have 23 pairs of chromosomes. 22 of these pairs have homologous partners that are identical in appearance, whether in a male or a female. The remaining pair are the sex chromosomes. In human females, the two sex chromosomes appear the same and are called the **X chromosomes**. In the human male there is a single **X** chromosome like that in the female, but the second one of the pair is smaller in size and shaped differently. This is the **Y chromosome**.

Sex inheritance in humans

In humans the sex-chromosomes are **X** and **Y**. This means:

- as females have two **X** chromosomes, all the gametes are the same in that they contain a single **X** chromosome
- as males have one **X** chromosome and one **Y** chromosome, they produce two different types of gamete – half have an **X** chromosome and half have a **Y** chromosome.

The inheritance of sex is shown in Figure 1.

Sex-linkage – haemophilia

Any gene that is carried on either the **X** or **Y** chromosome is said to be sex-linked. However, the **X** chromosome is much longer than the **Y** chromosome. This means that, for most of the length of the **X** chromosome, there is no equivalent homologous portion of the **Y** chromosome. Those characteristics that are controlled by recessive alleles on this non-homologous portion of the **X** chromosome will appear more frequently in the male. This is because there is no homologous portion on the **Y** chromosome that might have the dominant allele, in the presence of which the recessive allele does not express itself.

The **X** chromosome carries many genes. An **X**-linked genetic disorder is a disorder caused by a defective gene on the **X** chromosome. One example in humans is the condition called haemophilia, in which the blood clots only slowly and there may be slow and persistent internal bleeding, especially in the joints. As such it is potentially lethal if not treated. This has resulted in some selective removal of the gene from the population, making its occurrence relatively rare (about one person in 20 000 in Europe). Although haemophiliac females are known, the condition is almost entirely confined to males, in part because haemophiliac females usually died with the onset of menstruation at puberty.

One of a number of causes of haemophilia is a recessive allele with an altered sequence of DNA nucleotide bases that therefore codes for a faulty protein which does not function. This results in an individual being unable to produce a functional protein that is required in the clotting process. The production of this functional protein by genetically modified organisms means that it can now be given to haemophiliacs, allowing them to lead near-normal lives. Figure 3 shows the usual way in which a male inherits haemophilia. Note that the alleles are shown

Learning objectives

→ Explain how sex is determined genetically.

→ State what is meant by sex-linkage.

→ Explain how sex-linked diseases such as haemophilia are inherited.

Specification reference: 3.7.1

parental phenotypes	male	female
parental genotypes	XY	XX
gametes	X Y	X X

offspring genotypes

	♂ gametes	
♀ gametes	X	Y
X	XX	XY
X	XX	XY

offspring phenotypes

50% male (XY)
50% female (XX)

▲ **Figure 1** *Sex inheritance in humans*

▲ **Figure 2** *Scanning electron micrograph (SEM) of human X (left) and Y chromosomes as found in a male*

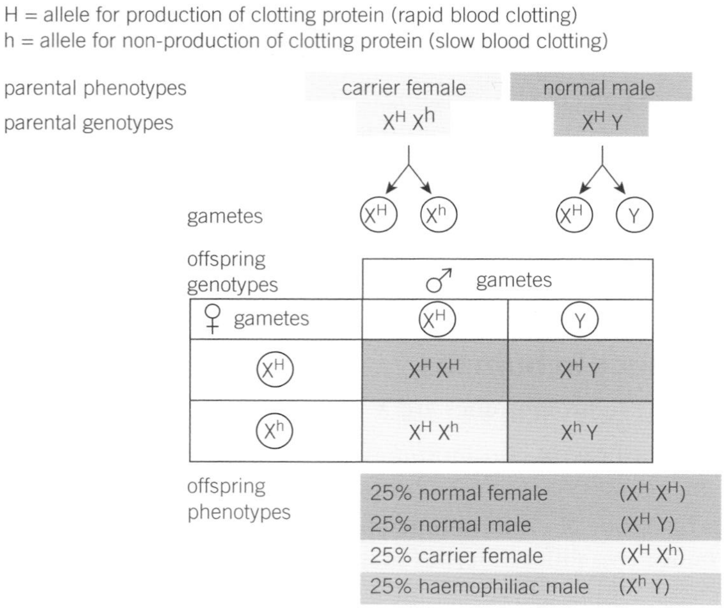

▲ **Figure 3** *Inheritance of haemophilia from a carrier female*

Hint

Remember that males have a **Y** chromosome and this could only have come from their fathers. Their **X** chromosome must therefore have come from their mothers.

Study tip

Sometimes people think that the **X** chromosome passes from father to son. If this were so, the mother must have provided the son's **Y** chromosome, in which case she would be male! Clearly this is nonsense.

in the usual way (**H** = dominant allele for production of the clotting protein, and **h** = recessive allele for the non-production of clotting protein). However, as they are linked to the **X** chromosome, they are not shown separately, but always attached to the **X** chromosome, that is, as $\mathbf{X^H}$ and $\mathbf{X^h}$ respectively. There is no equivalent allele on the **Y** chromosome as it does not carry the gene for producing clotting protein.

As males can *only* obtain their **Y** chromosome from their father, it follows that their **X** chromosome comes from their mother. As the defective allele that does not code for the clotting protein is linked to the **X** chromosome, males always inherit the disease from their mother. If their mother does not suffer from the disease, she may be **heterozygous** for the character ($\mathbf{X^H X^h}$). Such females are called carriers because they carry the allele without showing any signs of the disease in their phenotype. This is because these carriers possess one dominant **H** allele and this leads to the production of enough functional clotting protein.

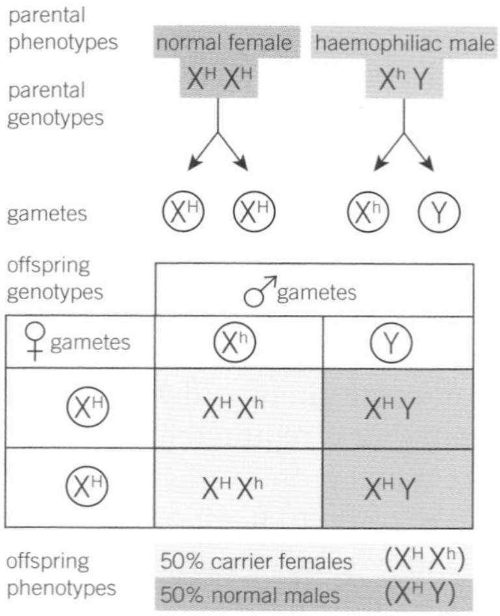

◀ **Figure 4** *Inheritance of the haemophiliac allele from a haemophiliac male*

As males pass the **Y** chromosome on to their sons, they cannot pass haemophilia to them. However, they can pass the allele to their daughters, via their **X** chromosome, who would then become carriers of the disease (Figure 4).

Pedigree charts

One useful way to trace the inheritance of sex-linked characters such as haemophilia is to use a pedigree chart. In these:

- a male is represented by a square
- a female is represented by a circle
- shading within either shape indicates the presence of a character, such as haemophilia, in the phenotype

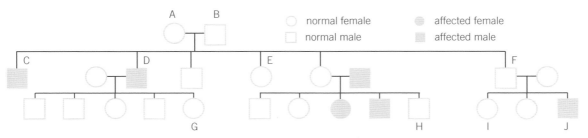

▲ Figure 5

Summary questions √x̄

Red-green colour blindness is linked to the **X** chromosome. The allele (**r**) for red-green colour blindness is recessive to the normal allele (**R**). Figure 5 shows the inheritance of this characteristic in a family.

1 State what sex chromosomes are present in individuals labelled E and F?

2 In terms of colour blindness, identify the phenotypes of each of the individuals labelled A, B and D.

3 In terms of colour blindness, identify the genotypes of each of the individuals labelled G, H, I and J.

4 √x̄ If individual C was to have children with a normal female (one who does not have any r alleles), determine the probability of any sons having colour blindness.

5 Individual J is colour blind. Assuming no history of colour blindness in either parent's family tree, suggest how this might have occurred.

Maths link √x̄

MS 1.4, see Chapter 22.

Maths link √x̄

MS 3.0, see Chapter 22.

A right royal disease

The royal families of Europe have been affected by haemophilia for many years. The origins of the disease stretch back to Queen Victoria. A pedigree chart showing the inheritance of haemophilia from Queen Victoria in members of various European royal families is shown in Figure 6.

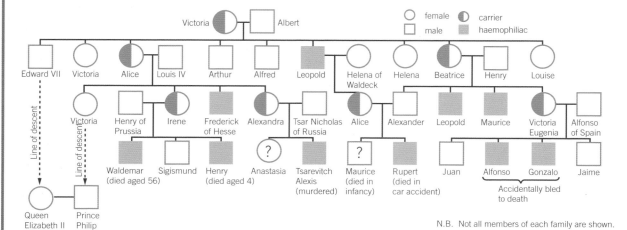

▲ **Figure 6** *Pedigree chart showing the transmission of haemophilia from Queen Victoria*

1   Suggest why haemophilia is not present in the current British royal family of Queen Elizabeth II and Prince Philip, and their children.

2   Give evidence from the chart which shows that haemophilia is:
  a   sex-linked
  b   recessive.

3   Using the symbols X^H for the chromosome carrying an allele that produces a clotting protein and X^h for a chromosome carrying an allele that does not produce a clotting protein, list the possible genotypes of the following people:
  a   Queen Elizabeth II
  b   Gonzalo
  c   Irene

4   Suppose Waldemar and Anastasia had married and produced children. Using the same symbols, list all the possible genotypes of their:
  a   sons
  b   daughters.
  Explain your answers.

In humans, just 23 pairs of chromosomes carry the genes that determine many thousands of different characteristics. It follows that each chromosome must possess many different genes. Any two genes that occur on the same chromosome are said to be **linked**. All the genes on a single chromosome form a linkage group. We saw in Topic 17.6 that genes carried on the sex chromosomes are said to be sex linked. The remaining 22 chromosomes, other than the sex chromosomes, are called **autosomes**. The name given to the situation where two or more genes are carried on the same autosome is called **autosomal linkage**.

Assuming there is no crossing over, all the linked genes remain together during meiosis and so pass into gametes, and hence the offspring, together. They do not segregate in accordance with Mendel's Law of Independent Assortment.

Figure 1 shows the different gametes that are produced if a pair of genes **A** and **B** are linked compared with if they are on separate chromosomes.

Learning objectives

→ Describe autosomal linkage.

→ Explain how autosomal linkage affects the combinations of alleles in gametes.

Specification reference: 3.7.1

Synoptic link

You learnt about crossing over in Topic 9.2, Meiosis and genetic variation.

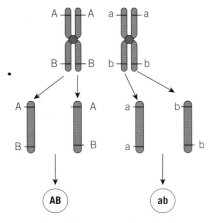

If genes A and B occur on separate chromosomes, that is, they are not linked.
If genes A and B occur on the same chromosome, that is, they are linked

Two homologous pairs are needed if all four alleles are to be present

According to Mendel's Law of Independent Assortment, any one of a pair of characters may combine with any of another pair. There are therefore four different possible types of gamete.

AB   Ab   aB   ab

If genes A and B occur on the same chromosome, that is, they are linked

Only one homologous pair is needed if all four alleles are to be present.

Possible types of gamete   AB   ab

▲ **Figure 1** *Comparison of gametes produced by an organism that is heterozygous for two genes **A** and **B** when they are linked and not linked*

We see from Figure 1 that where the two genes **A** and **B** with heterozygous alleles are on different chromosomes, there are four possible combinations of the alleles in the gametes: **AB**, **Ab**, **aB** and **ab**. However, if the two genes are linked and provided there is no crossing over, there are only two possible combinations of the alleles in the gametes: **AB** and **ab**. This makes things simpler when dealing with genetic crosses.

For example, consider two linked genes of the fruit fly *Drosophila melanogaster* – one that determines body colour and the other that determines wing size. There are two alleles for body colour. One produces a grey body and is dominant to the other which produces a black body. There are also two alleles for wing size. One that produces normal sized wings and is dominant to the other that produces vestigial wings (tiny wings that do not function). One possible cross is shown in Figure 2.

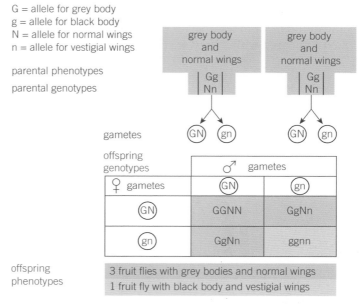

▲ Figure 2 *The offspring of a cross between two flies that are heterozygous for both characteristics*

If the alleles for body colour and wing size were not linked but on separate chromosomes, then an individual that is heterozygous for both characters would produce four different gametes (**GN**, **Gn**, **gN**, and **gn**) rather that just two (**GN**, **gn**) as shown in our cross. We saw in Topic 17.4 that when there is random fertilisation of gametes, the offspring of this dihybrid cross would be fruit flies with the following characters:

9 grey body and normal wings (**G-N-**)

3 grey body and vestigial wings (**G-nn**)

3 black body and normal wings (**ggN-**)

1 black body and vestigial wings (**ggnn**)

Summary questions

1   Distinguish between sex-linkage and autosomal linkage.

2   A variety of rabbit has an allele (**H**) for long hairs and another allele (**h**) short hairs. It also has an allele (**G**) for grey hairs and another allele (**g**) for white hairs. The gene for hair length and the gene for hair colour are on the same chromosome.

Draw a genetic diagram to show the results of a cross between a rabbit that has short, white fur and one that is heterozygous for both characters.

Tales of the unexpected

In tomato plants, flower colour is determined by a gene that has two alleles, one producing yellow flowers and the other white flowers. Another gene determines the colour of the fruit. This gene also has two alleles, one producing red fruit and the other yellow fruit.

A tomato plant with yellow flowers and red fruit is crossed with a tomato plant with white flowers and yellow fruit. All the F_1 offspring had yellow flowers and red fruit.

A plant of this F_1 generation was self-pollinated (self-fertilised by its own gametes).

1   Suggest suitable symbols for each of the alleles involved in these crosses and annotate them to show which allele each symbol represents.

2   Using the symbols you have chosen, state the genotype of a tomato plant in the F_1 generation.

3   The two genes are linked. Predict what the ratio of plants would be in the F_2 generation.

4   Deduce what the ratio of plants would be in the F_2 generation if the two genes were not linked.

5   When a plant from the F_1 generation was self-pollinated, the actual results were:

yellow flowers and red fruit   68
yellow flowers and yellow fruit   7
white flowers and red fruit   7
white flowers and yellow fruit   18

a   These results included some unexpected varieties of tomato plant. Suggest which ones these are.

b   Using your knowledge of meiosis, suggest what might have happened that would explain the appearance of these unexpected varieties.

We have looked at examples of how conditions such as linkage modify Mendelian ratios but without compromising the basic ideas. Let us consider a similar situation, epistasis.

Epistasis arises when the allele of one gene affects or masks the expression of another in the phenotype.

An example occurs in mice where several genes determine coat colour. Let us look at two such genes.

- Gene A controls the distribution of a black pigment called melanin in hairs and therefore whether they are banded or not. The dominant allele **A** of this gene leads to hairs that have black bands while the recessive allele **a** produces uniform black hairs when it is present with another recessive allele **a** (homozygous = **aa**).

- Gene B controls the colour of the coat by determining or otherwise, the expression of gene A. The dominant allele **B** leads to the production of melanin while the recessive allele **b** leads to no pigment and any hair will therefore be white when it is present with another recessive allele = **bb** – the homozygous state.

The usual (wild type) mouse has a grey-brown coat known as agouti. This is the result of having hairs with black bands. If a mouse has uniform black hairs its coat is black and if the hairs lack melanin altogether its coat is albino (white).

▲ **Figure 1** *Black, agouti, and albino mice*

If an agouti mouse with the genotype **AABB** is crossed with an albino mouse with the genotype aabb, then the offspring are all agouti. If individuals from the F_1 generation are crossed to produce the F_2 generation the following ratio is produced:

    9 agouti mice
    4 albino mice
    3 black mice.

The crosses are shown in the genetic diagram in Figure 2.

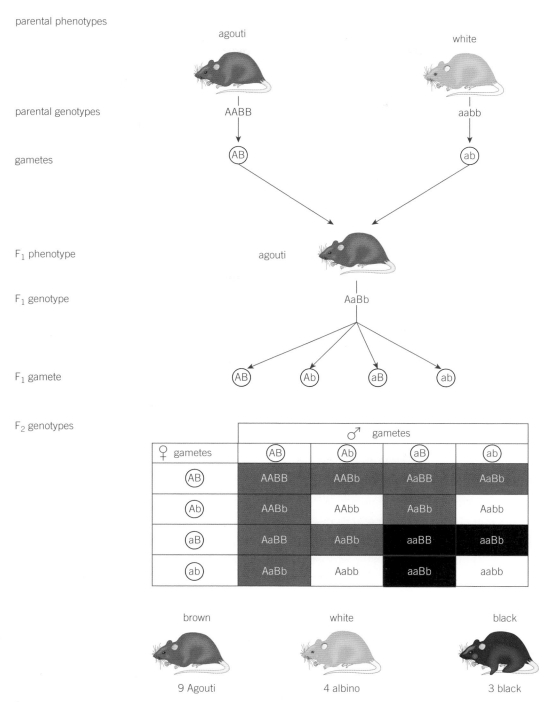

▲ **Figure 2** *Epistasis in mice*

The explanation of the results is as follows:

- The expression of gene A (black bands) is affected by the expression of gene B (production of melanin).
- If gene B is in the homozygous recessive state (**bb**) then no melanin is produced and the coat is albino.

- In the absence of melanin, gene A cannot be expressed.
- Therefore regardless of which alleles (**AA**, **Aa** or **aa**) are present, as there is no pigment, the hairs can be neither coloured nor banded.
- Where a dominant allele **B** is present, melanin is produced.
- If this allele is present with a dominant allele **A**, then banding occurs and an agouti coat results.
- Where allele **B** is present with two recessive alleles (**aa**), the hairs, and hence the coat, are uniform black.

There are other forms of epistasis. For example where genes act in sequence by determining the enzymes in a biochemical pathway. Suppose that a plant produces a red pigment in its petals using the following biochemical pathway.

$$\text{starting molecule} \xrightarrow{\text{enzyme A}} \text{intermediate molecule} \xrightarrow{\text{enzyme B}} \text{red pigment}$$

The production of enzymes A and B is coded for by genes A and B, respectively. Dominant alleles of each gene code for a functional enzyme, while recessive alleles code for a non-functional enzyme. It follows that if the alleles of either gene are both recessive, then that enzyme will be non-functional and the pathway cannot be completed. This affects the other gene in that, even if it is functional and produces its enzyme, its effects cannot be expressed because no pigment can be manufactured.

Summary questions

1 Using the example of epistasis in mice in this topic, consider a cross between mouse 1 that is heterozygous for gene A (colour distribution) and homozygous recessive for gene B (melanin production) with mouse 2 that is heterozygous for both genes.

   a State the colour of mouse 1 and the colour of mouse 2.

   b Calculate, and list, the genotypes of the offspring resulting from this cross.

   c State the ratio of different phenotypes produced by this cross.

2 Some varieties of corn, *Zea mays*, have purple seeds due to the presence of a pigment called anthocyanin in their seed coats. In the absence of the pigment, the seeds are white. The production of anthocyanin is controlled by two genes, A and B.

   One pure-breeding variety of white-seeded corn with the genotype **AAbb** was crossed with another pure-breeding variety of white-seeded corn with the genotype **aaBB**. All the offspring had purple seeds. A cross between two of the F_1 generation produced a ratio of nine purple-seeded plants to seven white-seeded plants.

   a Deduce the genotype of the F_1 generation.

   b Draw a genetic diagram showing the cross between two F_1 individuals.

   c i Analyse your table of results and deduce which nine genotypes you think might represent purple-seeded plants.

      ii State what these nine genotypes have in common.

   d Suggest how epistasis can explain the production of anthocyanin in some of the plants but not in others.

17.9 The chi-squared (χ^2) test

If you toss a coin 100 times it would be reasonable to expect it to land heads on 50 occasions and tails on 50 occasions. In practice, it would be unusual if these exact results were obtained. If it lands heads 55 times and tails only 45 times, does this mean that the coin is weighted or biased in some way, or is it purely a chance deviation from the expected result? How can we test which of these two options is correct?

What is the chi-squared test?

The chi-squared (χ^2) test is used to test the null hypothesis. The null hypothesis is used to examine the results of scientific investigations and is based on the assumption that there will be no statistically significant difference between sets of observations, any difference being due to chance alone. In our coin tossing example, the null hypothesis would be that there is no difference between the number of times it lands heads and the number of times it lands tails. The chi-squared test is a means of testing whether any deviation between the observed and the expected numbers in an investigation is significant or not. It is a simple test that can be used only if certain criteria are met:

- the sample size must be relatively large, that is, over 20
- the data must fall into discrete categories
- only raw counts and not percentages, rates, etc., can be used
- it is used to compare experimental results with theoretical ones, for example, in genetic crosses with expected Mendelian ratios

The formula is:

chi squared = sum of $\dfrac{\text{[observed numbers (O) – expected numbers (E)]}^2}{\text{expected numbers (E)}}$

summarised as:

$$\chi^2 = \Sigma \frac{(O - E)^2}{E}$$

The value obtained is then read off on a chi-squared distribution table (Table 1) to determine whether any deviation from the expected results is significant or not. To do this we need to know the number of **degrees of freedom**. This is simply the number of classes (categories) minus one, that is, if a human can have blood group A or B or AB or O, there are four classes and three degrees of freedom in this case.

Calculating chi-squared

Using our example of the coin tossed 100 times, we can calculate the chi-squared value:

Class (category)	Observed (O)	Expected (E)	O – E	(O – E)²	$\dfrac{(O-E)^2}{E}$
Heads	55	50	+5	25	0.5
Tails	45	50	−5	25	0.5
					$\Sigma = 1.0$

Learning objectives

→ Explain what the chi-squared test is.

→ Calculate values for chi-squared.

→ Demonstrate how the chi-squared test is used in genetics.

Specification reference 3.7.1

Hint

The reason for squaring the value for (O−E) in the chi-squared test is to remove negative numbers.

Maths link

MS 1.9, see Chapter 22.

Therefore the value of $\chi^2 = 1.0$.

▼ **Table 1** *Part of a χ^2 table (based on Fisher)*

Degrees of freedom	Number of classes	χ^2							
1	2	0.00	0.10	0.45	1.32	2.71	3.84	5.41	6.64
2	3	0.02	0.58	1.39	2.77	4.61	5.99	7.82	9.21
3	4	0.12	1.21	2.37	4.11	6.25	7.82	9.84	11.34
4	5	0.30	1.92	3.36	5.39	7.78	9.49	11.67	13.28
5	6	0.55	2.67	4.35	6.63	9.24	11.07	13.39	15.09
Probability that deviation is due to chance alone		0.99 (99%)	0.75 (75%)	0.50 (50%)	0.25 (25%)	0.10 (10%)	0.05 (5%)	0.02 (2%)	0.01 (1%)

← Accept null hypothesis
(Any difference is due to chance and not significant)

↑ Critical value

Reject null hypothesis and therefore accept experimental hypothesis. The difference is significant

of χ^2 at 0.05 p level as this is the smallest value accepted by statisticians for results being due to chance

Using the chi-squared table

To find out whether this value of 1.0 is significant or not we use a chi-squared table, part of which is given in Table 1. Before trying to read this table it is necessary to decide how many **classes of results** there are. In our case there are two classes of results, heads and tails. This corresponds to one **degree of freedom**, as the degrees of freedom are the number of classes minus one. We now look along the row showing two classes (i.e., one degree of freedom) for our calculated value of 1.0. This lies between the values 0.45 and 1.32. Reading along the probability row we see that our value lies between of 0.50 (50%) and 0.25 (25%). This means that the probability that chance alone could have produced the deviation is between 0.50 (50%) and 0.25 (25%). In the chi-squared test the critical value is p = 0.05. This is the attribution to chance accepted by statisticians, that is, 5% due to chance. If the probability that the deviation is due to chance is equal to or greater than 0.05 (5%), the deviation is said to be **not significant** and the null hypothesis would be accepted. If the deviation is less than 0.05 (5%), the deviation is said to be **significant**. In other words, some factor other than chance is affecting the results and the null hypothesis must be rejected. In our example the value is greater than 0.05 (5%) and so we assume the deviation is due to chance and accept the null hypothesis. Had we obtained 60 heads and 40 tails, a chi-squared value of slightly less than 0.05 (5%) would be obtained, in which case the null hypothesis would be rejected and we would assume that the coin might be weighted or biased in some way.

Chi-squared test in genetics

The chi-squared test is especially useful in genetics. To take the example of the genetic cross described in Topic 17.4 (also refer back to remind yourself how the expected results were obtained for the test), if we cross F_1 plants producing round, yellow seeds that we know are **heterozygous** we could expect a typical dihybrid F_2 ratio of:

9 round, yellow seeds   3 wrinkled, yellow seeds

3 round, green seeds   1 wrinkled, green seeds

Suppose we obtained 320 plants in the ratio 186:48:72:14. Could this variation be due to statistical chance or could some other factor be the reason for the differences? Our null hypothesis states that there is no significant difference between the observed and the expected results. Applying the chi-squared test:

Maths link $\sqrt{x}$

MS 1.9, see Chapter 22.

Class (category)	Observed (O)	Expected (E)	O – E	$(O - E)^2$	$\dfrac{(O - E)^2}{E}$
Round, yellow seeds	186	180	+6	36	0.2
Round, green seeds	48	60	−12	144	2.4
Wrinkled, yellow seeds	72	60	+12	144	2.4
Wrinkled, green seeds	14	20	−6	36	1.8
					$\Sigma = 6.8$

In this example there are four classes and therefore three degrees of freedom. Using the chi-squared table (Table 1) we see that our value falls between 6.25 and 7.82 shown on the row for three degrees of freedom and that these values correspond to between 0.1 (10%) and 0.05 (5%) probability that the deviation is due to chance alone. In this case there is between a 5% and 10% probability that the deviation is due to chance. Therefore we accept the null hypothesis and accept that the results are not significantly different from a 9:3:3:1 ratio.

Summary questions $\sqrt{x}$

In an experiment, domestic fowl with walnut combs were crossed with each other. The expected offspring ratio of comb types was 9 walnut, 3 rose, 3 pea, and 1 single. In the event, the 160 offspring produced 103 walnut combs, 20 rose combs, 33 pea combs, and 4 single combs (Figure 1).

1   Devise an appropriate null hypothesis.

2   Determine the number of degrees of freedom.

3   Calculate the value of chi-squared.

4   Determine the probability that the deviation is due to chance alone.

5   Assess whether the null hypothesis should be accepted or rejected.

▲ **Figure 1** *The different types of comb in domestic fowl are the result of dihybrid inheritance*

1 (a) In fruit flies, the genes for body colour and wing length are linked.
Explain what this means.   (*1 mark*)
A scientist investigated linkage between the genes for body colour and wing length.
He carried out crosses between fruit flies with grey bodies and long wings and fruit
flies with black bodies and short wings.
Figure 2 shows his crosses and the results.
  • **G** represents the dominant allele for grey body and **g** represents the recessive allele
    for black body.
  • **N** represents the dominant allele for long wings and **n** represents the recessive
    allele for short wings.

▼ **Figure 2**

Phenotype of parents	grey body, long wings	×	black body, short wings
Genotype of parents	GGNN		ggnn
Genotype of offspring		GgNn	
Phenotype of offspring		all grey body, long wings	

These offspring were crossed with flies homozygous for black body and short wings.
The scientist's results are shown in **Figure 3**.

▼ **Figure 3**

	GgNn	crossed with	ggnn	
	Grey body, long wings	Black body, short wings	Grey body, short wings	Black body, long wings
Number of offspring	975	963	963	194

(b) Use your knowledge of gene linkage to explain these results.   (*4 marks*)
(c) If these genes were not linked, what ratio of phenotypes would the scientist
     have expected to obtain in the offspring?   (*1 mark*)
(d) Use the chi-squared test to determine whether there is a significant difference
     between the observations recorded in Figure 3 and those expected if the
     genes are not linked.   (*2 marks*)

AQA Specimen 2014 (apart from 1 (d))

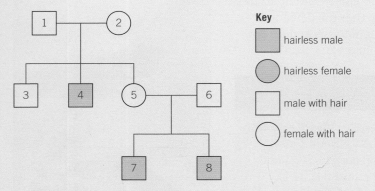

Key
  ■ hairless male
  ● hairless female
  □ male with hair
  ○ female with hair

2 A single gene controls the presence of hair
   on the skin of cattle. The gene is carried
   on the X chromosome. Its dominant allele
   causes hair to be present on the skin and
   its recessive allele causes hairlessness. The
   diagram shows the pattern of inheritance of
   these alleles in a group of cattle.
(a) Use evidence from the diagram to explain
     (i)   that hairlessness is caused by a
           recessive allele   (*2 marks*)
     (ii) that hairlessness is caused
           by a gene on the X chromosome.   (*1 mark*)

(b) What is the probability of the next calf born to animals 5 and 6 being hairless?
Copy and complete the genetic diagram to show how you arrived at your answer.
Phenotypes of parents   Female with hair   Male with hair
Genotypes of parents
Gametes
Genotypes of offspring
Phenotypes of offspring
Probability of next calf being hairless   (*4 marks*)

AQA Jan 2012

3   The fruit fly is a useful organism for studying genetic crosses. Female fruit flies are
    approximately 2.5 mm long. Males are smaller and possess a distinct black patch on their bodies.
    Females lay up to 400 eggs which develop into adults in 7 to 14 days. Fruit flies will survive and
    breed in small flasks containing a simple nutrient medium consisting mainly of sugars.
    (a) Use this information to explain two reasons why the fruit fly is a useful
         organism for studying genetic crosses.   (2 marks)
    (b) Male fruit flies have the sex chromosomes XY and the females have XX. In the
         fruit fly, a gene for eye colour is carried on the X chromosome. The allele for red
         eyes, **R**, is dominant to the allele for white eyes, **r**. The genetic diagram shows a
         cross between two fruit flies.
         (i) Copy and complete the genetic diagram for this cross.

              Phenotypes of parents   red-eyed female   ×   white-eyed male

              Genotype of parents

              Gametes   and   and

              Phenotypes of offspring   red-eyed females   and   red-eyed males

              Genotype of offspring
                                                                                      (3 marks)

         (ii) The number of red-eyed females and red-eyed males in the offspring was
              counted. The observed ratio of red-eyed females to red-eyed males was
              similar to, but not the same as, the expected ratio. Suggest **one** reason
              why observed ratios are often **not** the same as expected ratios.   (1 mark)
    (c) Male fruit flies are more likely than female fruit flies to show a phenotype
         produced by a recessive allele carried on the X chromosome. Explain why.   (2 marks)
                                                                                      AQA Jan 2013

4   A breeder crossed a black male cat with a black female cat on a number of occasions.
    The female cat produced 8 black kittens and 4 white kittens.
    (a) (i) Explain the evidence that the allele for white fur is recessive.   (1 mark)
         (ii) Predict the likely ratio of colours of kittens born to a cross between
              this black male and a white female.   (1 mark)
    (b) The gene controlling coat colour has three alleles. The
         allele **B** gives black fur, the allele **b** gives chocolate fur
         and the allele **b^i** gives cinnamon fur.

Genotype	Phenotype
Bb^i	
bb^i	
Bb	

         • Allele **B** is dominant to both allele **b** and **b^i**.
         • Allele **b** is dominant to allele **b^i**.
         (i) Copy and complete the table to show the phenotypes
              of cats with each of the genotypes shown.   (1 mark)
              A chocolate male was crossed several times with a
              black female. They produced
         • 11 black kittens
         • 2 chocolate kittens
         • 5 cinnamon kittens.
         (ii) Using the symbols given on the previous page, copy and complete the genetic
              diagram to show the results of this cross.

              Parental phenotypes   Chocolate male   Black female
              Parental genotypes
              Gametes

              Offspring genotypes
              Offspring phenotypes   Black   Chocolate   Cinnamon
                                                                                      (3 marks)

         (iii) The breeder had expected equal numbers of chocolate and cinnamon
               kittens from the cross between the chocolate male and black female.
               Explain why the actual numbers were different from those expected.   (1 mark)
         (iv) The breeder wanted to produce a population of cats that would all
              have chocolate fur. Is this possible? Explain your answer.   (2 marks)
                                                                                      AQA June 2011

We have so far looked at how genes and their alleles are passed between individuals in a population. Let us now consider the genes and alleles of an entire population.

A population is a group of organisms of the same species that occupies a particular space at a particular time and that can potentially interbreed. Any species exists as one or more populations. All the alleles of all the genes of all the individuals in a population at a given time time are known as the **gene pool**. The number of times an allele occurs within the gene pool is referred to as the **allelic frequency**.

Let us look at this more closely by considering just one gene that has two alleles, one of which is dominant and the other recessive. An example is the gene responsible for cystic fibrosis, a disease of humans in which the mucus produced by affected individuals is thicker than normal. The gene has a dominant allele (**F**) that leads to normal mucus production and a recessive allele (**f**) that leads to the production of thicker mucus and hence cystic fibrosis. Any individual human has two of these alleles in every one of their cells, one on each of the pair of homologous chromosomes on which the gene is found. As these alleles are the same in every cell of a single person, we only count one pair of alleles per gene per individual when considering a gene pool. If there are 10 000 people in a population, there will be twice as many (20 000) alleles in the gene pool *of this gene*.

The pair of alleles of the cystic fibrosis gene has three different possible combinations, namely homozygous dominant (**FF**), homozygous recessive (**ff**) and heterozygous (**Ff**). When we look at allele frequencies, however, it is important to appreciate that the heterozygous combination can be written as **Ff** or **fF**. (It is just conventional to put the dominant allele first in all cases.)

In our population of 10 000 people, if all 10 000 had the genotype **FF**, then:

- The probability of anyone being **FF** would be 1.0 and the probability of anyone being **ff** would be 0.0
- The frequency of the **F** allele would be 100% and the frequency of the **f** allele would be 0%.

If everyone in our population was heterozygous (**Ff**), then:

- The probability of anyone being **Ff** would be 1.0 and the frequency of the **F** allele would be 50%, and the frequency of the **f** allele would also be 50%.

Of course, in practice, the population is not made up of one genotype but of a mixture of all three, the proportions of which vary from population to population. How then can we work out the allele frequency of these mixed populations?

The Hardy–Weinberg principle

The Hardy–Weinberg principle provides a mathematical equation that can be used to calculate the frequencies of the alleles of a particular gene in a population. The principle makes the assumption that the proportion of dominant and recessive alleles of any gene in a population remains the same from one generation to the next this can be the case provided that five conditions are met:

- No mutations arise.
- The population is isolated, that is, there is no flow of alleles into or out of the population.
- There is no selection, that is, all alleles are equally likely to be passed to the next generation.
- The population is large.
- Mating within the population is random.

Although these conditions are probably never totally met in a natural population, the Hardy–Weinberg principle is still useful when studying gene frequencies.

To help us understand the principle let us consider a gene that has two alleles: a dominant allele (**A**) and a recessive allele (**a**).

Let the probability of allele A = p

and the probability of allele a = q

The first equation we can write is:

$$p + q = 1.0$$

because there are only two alleles and so the probability of one plus the other must be 1.0 (100%).

As there are only four possible arrangements of the two alleles, it follows that the probability of all four added together must equal 1.0. Therefore we can state that:

AA + Aa + aA + aa = 1.0   or, expressing this as a probability:

$$p^2 + 2pq + q^2 = 1.0$$

We can now use these equations to determine the probability of any allele in a population. For example, suppose that a particular characteristic is the result of the recessive allele **a**, and we know that one person in 25 000 displays the character.

- The character, being recessive, will only be observed in individuals who have two recessive alleles **aa**.
- The probability of **aa** must be $\dfrac{1}{25\,000}$ or 0.00004.
- The probability of **aa** is q^2.
- If $q^2 = 0.00004$, then $q = \sqrt{0.00004}$ or 0.00063 approximately.
- We know that the probability of both alleles **A** and **a** is $p + q$ and is equal to 1.0.
- If $p + q = 1.0$, and $q = 0.00063$ then:
- $p = 1.0 - 0.00063 = 0.9937$, that is, the probability of allele A = 0.9937.
- We can now calculate the probability of heterozygous individuals (and therefore the probability of genotypes and phenotypes) in the population.
- From the Hardy–Weinberg equation we know that the probability of the heterozygotes is $2pq$.

> **Hint**
>
> A probability of 0.0125 is the equivalent of 125 in 10 000 or 313 in our population of 25 000.

- In this case, $2pq = (2 \times 0.9937 \times 0.0063) = 0.0125$.
- Heterozygous individuals act as a reservoir of recessive alleles in the population, although they themselves do not express the allele in their phenotype.

Maths link $\sqrt{x}$

MS 2.2, see Chapter 22.

Summary questions $\sqrt{x}$

1 Define the term allelic frequency.

2 State what the Hardy–Weinberg principle predicts.

3 State the **five** conditions that need to be met for this prediction to hold true.

4 $\sqrt{x}$ The frequency p of a dominant allele is 0.942. Calculate the frequency of the heterozygous genotype in the population. Show your working and express your answer as a percentage of the population.

5 $\sqrt{x}$ The ability to roll the tongue is determined by a dominant allele. In a sample of 416 people, 26 were unable to roll their tongues. Using the Hardy–Weinberg equation, calculate how many people in the sample were homozygous dominant for this allele.

Hint

Unless an allele leads to a phenotype with an advantage or a disadvantage compared with other phenotypes, its allele frequency in a population will probably stay the same from one generation to the next.

Maths link $\sqrt{x}$

MS 2.4, see Chapter 22.

Not as black and white as it seems $\sqrt{x}$

A gene that controls wing colour in the peppered moth has two alleles. The expression of the dominant allele produces moths with light-coloured wings while the expression of the recessive allele produces moths with dark-coloured wings. Scientists sampled a population of moths by catching them in a trap and recording their sex and wing colour. The numbers in Table 1 show how many of the 2215 moths caught were of each type.

▲ **Figure 1** *Dark- and light-coloured wing forms of the peppered moth*

▼ **Table 1**

	Light-coloured wings	Dark-coloured wings
Male	836	269
Female	817	293
Total	**1653**	**562**

1 State, with your reasons, whether you think the gene for wing colour is sex-linked.

2 $\sqrt{x}$ Determine what proportion of the total sample has two recessive alleles.

3 $\sqrt{x}$ In the Hardy–Weinberg equation $(p^2 + 2pq + q^2 = 1.0)$, $p =$ the frequency of the dominant allele and $q =$ the frequency of the recessive allele. For the population of moths that were caught, use this equation to calculate:

   a the frequency (q) of the recessive allele;

   b the frequency (p) of the dominant allele;

   c the percentage of heterozygotes.

4 Some scientists wanted to estimate the size of the total moth population. Describe how they might do this.

Individuals within a population of a species often show a wide range of variation in their phenotypes. This variation is due to both genetic and environmental factors. The primary source of genetic variation is mutation, a subject that we explored in Topic 9.1. Further genetic variation results from meiosis and the random fertilisation of gametes during sexual reproduction, subjects that were covered in Topic 9.2. Although variation is the result of genetic factors and environmental influences, it is rarely entirely due to one or the other but rather a combination of both.

Variation due to genetic factors

Within a population, all members have the same genes. Genetic differences, however, occur as members of this population will have different alleles of these genes. These differences not only occur in living individuals but also change from generation to generation. Genetic variation arises as a result of:

*   **mutations**. These sudden changes to genes and chromosomes may, or may not, be passed on to the next generation. Mutations are a main source of variation.

*   **meiosis**. This special form of nuclear division produces new combinations of alleles before they are passed into the gametes, all of which are therefore different.

*   **random fertilisation of gametes**. In sexual reproduction this produces new combinations of alleles and the offspring are therefore different from parents. Which gamete fuses with which at fertilisation is a random process further adding to the variety of offspring two parents can produce.

Sexually reproducing organisms, increase variation by all three methods.

Where variation is very largely the result of genetic factors organisms fit into a few distinct forms and there are no intermediate types. In the ABO blood grouping system, for example, there are four distinct groups – A, B, AB and O (Figure 1). A character displaying this type of variation is usually controlled by a single gene. This variation can be represented on a bar chart or pie graph. Environmental factors have little influence on this type of variation.

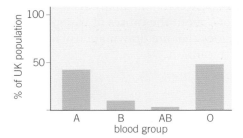

▲ **Figure 1** *Variation due to genetic factors illustrated by the percentage of the UK population with blood groups A, B, AB and O*

▲ **Figure 2** *Variation within a species (intraspecific variation) – despite the immense variety that they show, all dogs, including this Pug and Great Dane, belong to the same species –* Canis familiaris

Synoptic link

The normal distribution curve was covered in more depth in Topic 10.5, Quantitative investigations of variation and a recall of the information there would be helpful.

Summary questions

1   State **three** ways in which genetic variation can be increased in sexually reproducing organisms.

2   State how genetic variation is increased in asexually reproducing organisms

3   In the following list of statements, determine whether each refers to variation due to genetic or environmental factors.

   a   It can be represented by a line chart.

   b   It is usually controlled by a single gene.

   c   It can be represented as a bar chart.

   d   A mean can be calculated.

   e   An example is the length of the body in rats.

Variation due largely to environmental influences

The environment exerts an influence on all organisms. These influences affect the way the organism's genes are expressed. The genes set limits, but it is largely the environment that determines where, within those limits, an organism lies. In buttercups, for example, one plant may be determined by its genes to grow much taller than other plants. If, however, the seed germinated in an environment of poor light or low soil nitrate, the plant may not grow properly and it will be short. Environmental influences include climatic conditions (e.g., temperature, rainfall, and sunlight), soil conditions, pH, and food availability.

Some characteristics of organisms grade into one another, forming a continuum. In humans, two examples are height and mass. Characters that display this type of variation are not controlled by a single gene, but by many genes (polygenes). Environmental factors play a major role in determining where on the continuum an organism actually lies. For example, individuals who are genetically predetermined to be the same height actually grow to different heights due to variations in environmental factors, such as diet. This type of variation is the product of polygenes and the environment. If we measure the heights of a large population of people and plot the number of individuals against heights on a graph we will most probably obtain a bell-shaped curve known as a **normal distribution curve** (Figure 3).

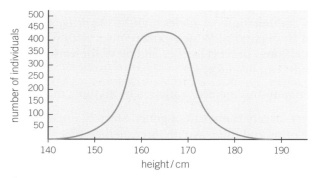

▲ **Figure 3** *Graph of frequency against height for a sample of humans*

In most cases variation is due to the combined effects of genetic differences and environmental influences. It is very hard to distinguish between the effects of the many genetic and environmental influences that combine to produce differences between individuals. As a result, it is very difficult to draw conclusions about the causes of variation in any particular case. Any conclusions that are drawn are usually tentative and should be treated with caution.

Every organism is subjected to a process of selection, based on its suitability for survival under the conditions that exist at the time. The environmental factors that limit the population of a species are called **selection pressures**. These selection pressures include predation, disease and competition. Selection pressures vary from time to time and place to place. These selection pressures determine the frequency of all alleles within the gene pool. A **gene pool** is the total number of all the alleles of all the genes of all the individuals within a particular population at a given time.

The process of evolution by means of natural selection depends upon a number of factors. These include:

- organisms produce more offspring than can be supported by the available supply of food, light, space, etc.
- there is genetic variety within the populations of all species.
- a variety of phenotypes that selection operates against.

The role of over-production of offspring in natural selection

Charles Darwin appreciated that all species have the potential to increase their numbers exponentially (Figure 1). He realised that, in nature, populations rarely, if ever, increased in size at such a rate. He rightly concluded that the death rate of even the most slow-breeding species must be extremely high. High reproductive rates have evolved in many species to ensure a sufficiently large population survives to breed and produce the next generation. This compensates for high death rates from predation, competition for food (including light in plants) and water, extremes of temperature, natural disasters such as earthquake, and fire, disease etc. Some species have evolved lower reproductive rates along with a high degree of parental care. The lower death rates that result help to maintain their population size.

Learning objectives

→ Define a gene pool.

→ Explain the role of overproduction of offspring in natural selection.

→ Explain the role of variation in natural selection.

Specification reference: 3.7.3

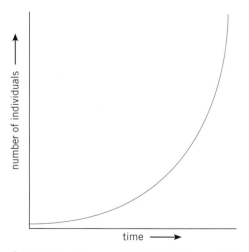

▲ **Figure 1** *The exponential rise in the population of a species whose growth is left unchecked by environmental factors such as predation, climate, disease, and competition*

The link between over-production and natural selection lies in the fact that, where there are too many offspring for the available resources, there is competition amongst individuals (**intraspecific competition**) for the limited resources available. The greater the numbers, the greater this competition and the more individuals will die in the struggle to survive. These deaths, however, are not totally random. Those individuals in a population best suited to prevailing conditions (e.g., better able to hide from or escape predators, better able to obtain light or catch prey, better able to resist disease or find a mate) will be more likely to survive than those less well adapted. These individuals will be more likely to breed and so pass on their more favourable allele combinations to the next generation, which will therefore have a different allele frequency from the previous one. The population will have evolved a combination of alleles that is better adapted to the prevailing conditions. This selection process, however, depends on individuals of a population being genetically different from one another.

The role of variation in natural selection

If an organism can survive in the conditions in which it lives, you may wonder why it doesn't always produce offspring that are identical to itself. These will, after all, be equally likely to survive in these conditions, whereas variation may produce individuals that are less suited. However, conditions change over time and having a wide range of genetically different (and therefore phenotypes) in the population means that some will have the combination of genes needed to survive in almost any new set of circumstances. Populations showing little individual genetic variation are often more vulnerable to new diseases and climate changes. It is also important that a species is capable of adapting to changes resulting from the evolution of other species.

The larger a population is, and the more genetically varied the individuals within it, the greater the chance that one or more individuals will have the combination of alleles that lead to a phenotype which is advantageous in the struggle for survival. These individuals will therefore be more likely to breed and pass their allele combinations on to future generations. Variation therefore provides the potential for a population to evolve and adapt to new circumstances.

The influence of variation on natural selection is best summarised by Darwin himself who, nearly a hundred and fifty years ago, wrote:

> How can it be doubted, from the struggle each individual has to obtain subsistence, that any minute variation in structure, habits or instinct, adapting that individual better to the new conditions, would tell upon its vigour and health? In the struggle it would have a better chance of surviving, and those of its offspring which inherited the variation, be it ever so slight, would have a better chance.

Summary questions

1 Define a gene pool.

2 State four factors that lead to differential survival and reproduction.

3 Sickle cell anaemia is a debilitating genetic disease that causes premature death but provides some resistance to the malarial parasite. Explain how selection might affect the distribution of the gene causing sickle cell anaemia in both malarial and non-malarial regions.

Maths link $\sqrt{x}$

MS 1.3 and 2.4, see Chapter 22.

How genetic variation leads to natural selection – copper tolerance in grasses

Species of grasses such as *Agrostis tenuis* and *Agrostis capillaris* have a variety of different forms with respect to copper tolerance (Figure 2). Some varieties are very tolerant and grow readily in soils with a high concentration of copper **ions**. Other varieties have a very low tolerance and do not survive, even when soil copper concentrations are very low. The majority of the varieties lie somewhere between these two extremes.

The soil around old copper workings is heavily contaminated with copper. If seeds of *Agrostis* spp. are planted in this soil, only those plants with a high tolerance to copper ions will survive. These plants are therefore the only ones to breed and so pass on their alleles to the next generation. Amongst these alleles will be those giving resistance to copper ions. Over time, the frequency of these alleles in the gene pool of the population of *Agrostis spp* increases as a result of the selection pressure from copper ions in the soil. The frequency of certain alleles in the population has changed and so the gene pool has changed and the population of *Agrostis spp* has evolved. On soils with low levels of copper, there is no selective advantage in being copper-tolerant and so the proportion

of these varieties remains low. They nevertheless survive where soil copper concentrations are low because there is no disadvantage in having tolerance to copper but there is no particular advantage either.

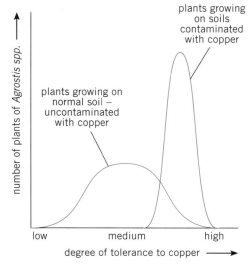

▲ **Figure 2** *Copper tolerance in* Agrostis *spp.*

1 Identify the types of curve shown in Figure 2.
2 Name the type of inheritance these curves suggest.
3 Suggest why the population of copper-tolerant varieties where the soil is heavily contaminated with copper is much larger than the population of non-tolerant varieties where the soil is less contaminated.
4 Describe how you might determine whether a plant was a variety of *Agrostis capillaris* rather than a separate species.
5 √x̄ A gene associated with copper tolerance has two alleles, **T** and **t**. Varieties of *Agrostis tenuis* that have both recessive alleles are copper tolerant, while varieties of *Agrostis tenuis* possessing one or more dominant alleles are not. A sample of plants was tested for copper tolerance and 72 plants were found to be copper tolerant while 378 were not.
   a Assuming the Hardy–Weinberg principle applies, calculate the percentage of the sample that have the genotypes **TT**, **Tt**, and **tt**.
   b Justify, giving **two** reasons, why the Hardy–Weinberg principle does not apply in the situation described here.

Learning objectives

→ Describe stabilising selection.

→ Describe directional selection.

→ Describe disruptive selection.

→ Explain the effects of each form of selection on evolution.

Specification reference: 3.7.3

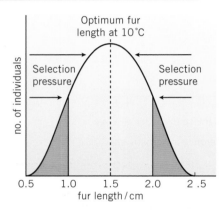

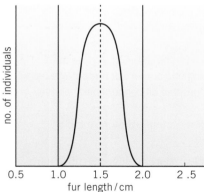

1. Initially there is a wide range of fur length about the mean of 1.5 cm. The fur lengths of less than 1.0 cm or greater than 2.0 cm in individuals are maintained by rapid breeding in years when the average temperature is much warmer or colder than normal.

2. When the average environmental temperature is consistently around 10 °C with little annual variation, individuals with very long or very short hair are eliminated from the population over a number of generations.

▲ **Figure 1** *Stabilising selection*

Environmental factors help create variation within a population. These environmental factors may be an agent for constancy or an agent for change according to the type of selection pressure they exert. Earlier we looked at two different forms of selection, stabilising and directional, and touched on the effects these forms of selection have on evolution. Let us now explore these effects more closely and investigate a third form of selection – disruptive selection.

The three main types of selection affect the characteristics of a population in the following ways:

- **stabilising selection** preserves the average phenotype (phenotypes around the mean) of a population by favouring average individuals, in other words, selection against the extreme phenotypes

- **directional selection** changes the phenotypes of a population by favouring phenotypes that vary in one direction from the mean of the population, in other words, selection for one extreme phenotype

- **disruptive selection** favours individuals with extreme phenotypes rather than those with phenotypes around the mean of the population.

Stabilising selection

Stabilising selection tends to eliminate the extremes of the **phenotype** range within a population and with it the capacity for evolutionary change. It tends to occur where the environmental conditions are constant over long periods of time. One example is fur length in a particular mammalian species. In years when the environmental temperatures are hotter than usual, the individuals with shorter fur length will be at an advantage because they can lose body heat more rapidly. In colder years the opposite is true and those with longer fur length will survive better as they are better insulated.

Therefore, if the environment fluctuates from year to year, both extremes will survive because each will have some years when it can thrive at the expense of the other. If, however, the environmental temperature is constantly 10 °C, individuals at the extremes will never be at an advantage. They will therefore be selected against in favour of those with average fur length. The mean will remain the same, but there will be fewer individuals at either extreme (Figure 1). An example of stabilising selection is the body mass of human children at birth. We saw in Topic 9.4 that babies born with a body mass greater or less than the optimum of 3.2 kg have a higher risk of dying in the few months after birth.

Directional selection

Within a population there will be a range of genetically different individuals in respect of any one phenotype. The continuous variation amongst these individuals forms a normal distribution curve. This curve has a mean that represents the optimum value for the phenotypic character under the existing conditions. If the environmental conditions

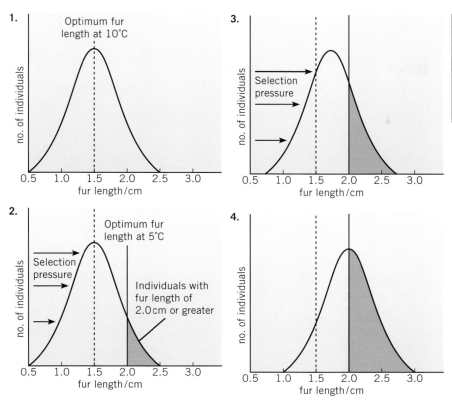

Synoptic link

Revising Topic 9.4, Types of selection, will be invaluable in helping you understand the ideas in this topic.

In a population of a particular mammal, fur length shows continuous variation.

1. When the average environmental temperature is 10 °C, the optimum fur length is 1.5 cm. This then represents the mean fur length of the population.

2. A few individuals in the population already have a fur length of 2.0 cm or greater. If the average environmental temperature falls to 5 °C, these individuals are better insulated and so are more likely to survive to breed. There is a selection pressure favouring individuals with longer fur.

3. The selection pressure causes a shift in the mean fur length towards longer fur over a number of generations. The selection pressure continues.

4. Over further generations the shift in the mean fur length continues until it reaches 2.0 cm – the optimum length for the prevailing average environmental temperature of 5 °C. The selection pressure now ceases.

▲ **Figure 2** *Directional selection*

change, so will the optimum value for survival. Some individuals, either to the left or the right of the mean, will possess a combination of alleles with the new optimum for the phenotypic character. As a result there will be a selection pressure favouring the combination of alleles that results in the mean moving to either the left or the right of its original position. Directional selection therefore results in one extreme of a range of variation being selected against in favour of the other extreme or even the average. Figure 2 illustrates a theoretical example of directional selection. A specific example is antibiotic resistance in bacteria.

Disruptive selection

Disruptive selection is the opposite of stabilising selection. It favours extreme phenotypes at the expense of the intermediate phenotypes. Although the least common form of selection, it is the most important in bringing about evolutionary change. Disruptive selection occurs when an environmental factor, such as temperature, takes two or more distinct forms. In our example this might arise if the

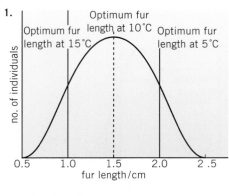

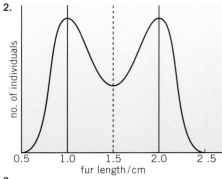

1. When there is a wide range of temperatures throughout the year, there is continuous variation in fur length around a mean of 1.5 cm.
2. Where the summer temperature is static around 15 °C and the winter temperature is static around 5 °C, individuals with two distinct fur lengths predominate – 1.0 cm types, which are active in summer, and 2.0 cm types, which are active in winter.
3. After many generations two distinct sub-populations are formed.

▲ **Figure 3** *Disruptive selection*

temperature alternated between 5 °C in winter (favouring long fur length) and 15 °C in summer (favouring short fur length). This could ultimately lead to two separate species of the mammal – one with long fur and active in winter, the other with short fur and active in summer (Figure 3). An example of disruptive selection is the coho salmon where large males and small males have a selective advantage over intermediate-sized males in passing on their alleles to the next generation. The small males are able to sneak up to the females in the spawning grounds. The large males are fierce competitors. This leaves intermediate-sized males at a disadvantage.

Selection in the peppered moth

Synoptic link

You previously learnt about the peppered moth in the context of natural selection in Topic 9.3. This is explored in more detail here.

Some species of organisms have two or more distinct forms. These different forms are genetically distinct but exist within the same interbreeding population. This situation is called polymorphism (poly = many; morph = form). One example is the peppered moth (*Biston betularia*). It existed almost entirely in its natural light form until the middle of the nineteenth century. Around this time a melanic (black) variety arose as the result of a mutation. These mutant moths had undoubtedly occurred before (one existed in a collection made before 1819) but they were highly conspicuous against the light background of the lichen-covered trees and rocks on which they normally rest. As a result, the black mutants were subjected to greater predation from insect-eating birds, for example, robins and hedge sparrows, than were the better camouflaged, light forms.

When, in 1848, a melanic form of the peppered moth was captured in Manchester, most buildings, walls and trees were blackened by the soot of 50 years of industrial development. The sulfur dioxide in smoke emissions killed the lichens that formerly covered trees and walls. Against this black background the melanic form was less, not more, conspicuous than the light form. As a result, the light form was eaten by birds more frequently than the melanic form and, by 1895, 98% of Manchester's population of the moth was of the melanic type.

This is an example of selective predation by birds favouring individuals that lie at one extreme or the other of a range of different colour types. It illustrates directional selection of different types in different populations. The melanic form is is selected for in industrial areas while the natural form is selected for in rural areas. It also shows evolution in action whereby there has been a change in the allelic frequency in populations of moths in industrial areas. However, as the two populations overlap and interbreed they are still one species. As we shall see in Topic 18.5, to become two distinct species, the two populations would need to become reproductively separated from one another.

▲ **Figure 4** *Melanism in the peppered moth (*Biston betularia*). Against a natural background the dark melanic form is far more visible and more readily predated on by birds. This natural selection leads to a predominance of the light form in rural areas. In polluted areas with blackened buildings, however, the melanic form is better camouflaged and this selective advantage leads to this form predominating (Topic 18.1)*

Summary questions

1 Consider each of the following statements and suggest which form of selection it best relates to.

a A baby with a birth weight greater than 4.0 kg or less than 2.5 kg has an increased risk of dying.

b Some species of insects have changed very little over millions of years.

c Elephants have evolved longer trunks enabling them to reach leaves higher up in trees.

d Small mammals can escape from predators by hiding in small spaces while large ones can resist attack by predators.

e It is the most important type of selection in bringing about evolutionary change.

f The mean ear length in arctic foxes has reduced over time.

g It preserves the characteristics of a population.

2 Suggest which form of the peppered moth, *Biston betularia*, is now most common in cities like Manchester and explain why in terms of selection pressure.

Learning objectives

→ Explain how selection affects allelic frequencies.

→ Explain how new species are formed.

→ Explain how populations can become geographically isolated.

→ Describe allopatric and sympatric speciation.

Specification reference: 3.7.3

Study tip

Remember that some environmental factors may influence the overall mutation rate (Topic 9.1) but that this is a general and random process rather than one that affects a specific allele in a specific way.

In topic 18.4, we examined the different forms of selection. We learnt that those organisms with phenotypes that gave them a selective advantage were more likely to produce offspring and so pass on their favourable alleles to the next generation. Now we will turn our attention to the effect that this differential reproductive success has on the allelic frequencies within a gene pool of a population.

Allelic frequencies and how selection affects them

In theory, any sexually mature individual in a population is capable of breeding with any other. This means that the alleles of any individual organism may be combined with the alleles of any other. We saw in Topic 18.3 that all the alleles of all the genes of all the individuals in a population at a given time is known as the gene pool. The number of times an allele occurs within the gene pool is referred to as the **allelic frequency**. The allelic frequency is affected by selection and, as we saw in Topic 18.4, selection is due to environmental factors. Environmental changes therefore affect the probability of an allele being passed on in a population and hence the number of times it occurs within the gene pool. It must be emphasised that environmental factors do not affect the probability of a particular mutant allele arising, they simply affect the frequency of a mutant allele that is already present in the gene pool.

Evolution by natural selection is a change in the allelic frequencies within a population.

Speciation

Speciation is the evolution of new species from existing ones. A **species** is a group of individuals that have a common ancestry and so share the same genes but different alleles and are capable of breeding with one another to produce fertile offspring. In other words, members of a species are **reproductively separated** from other species.

It is through the process of speciation that evolutionary change has taken place over millions of years. This has resulted in great diversity of forms amongst organisms, past and present.

How new species are formed

By far the most important way in which new species are formed is through reproductive separation followed by genetic change due to natural selection. Within a species there are one or more populations. Although individuals tend to breed only with others in the same population, they are nevertheless capable of breeding with individuals in other populations.

Suppose that a population becomes separated in some way from other populations and undergoes different mutations – it will become genetically different from the other populations. Each of the populations will experience different selection pressures because the environment of each will be slightly different. Natural selection will then lead to changes in the allelic frequencies of each population. The different phenotypes each combination of alleles produces will be subject to selection pressure that will lead to each population becoming adapted to its local environment. This is known as **adaptive radiation** and results in changes to the allele frequencies (evolution) of each population, in other words, each population evolves. As a result of these genetic differences it may be that, even if the populations were no longer physically separated from one another, they would be unable to interbreed successfully. Each population would now be a different species, each with its own gene pool. An example is shown in Figure 1.

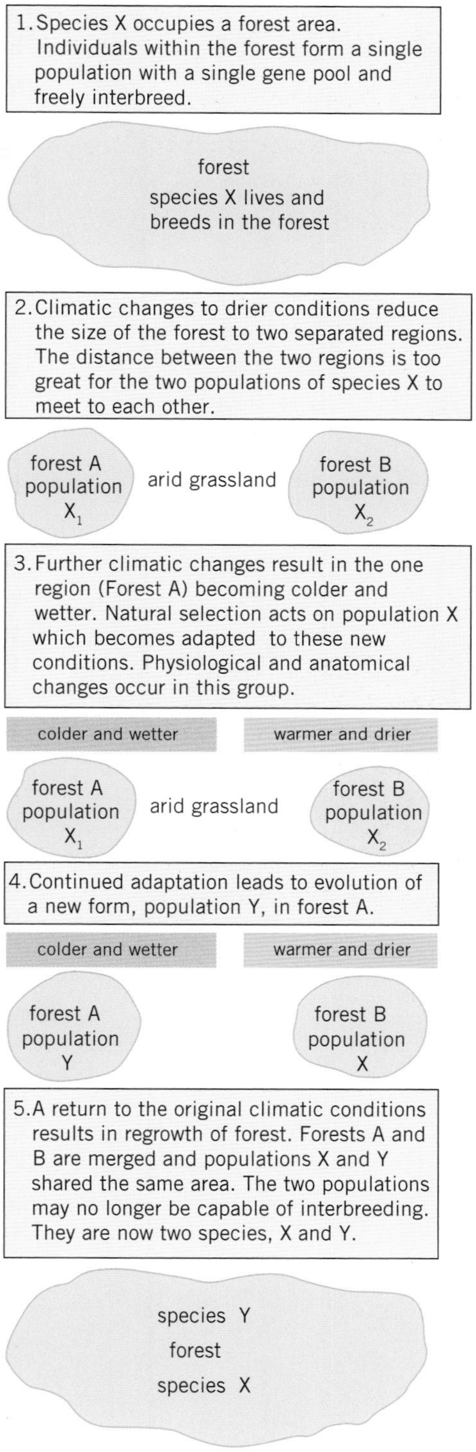

Genetic drift is something that can take place in small populations. This is because the relatively few members of a small population possess a smaller variety of alleles than the members of a large population. In other words, their genetic diversity is less. As these few individuals breed, the genetic diversity of the population is restricted to those few alleles in the original population. As there are only a small number of different alleles there is not an equal chance of each being passed on. Those that are passed on will quickly affect the whole population as their frequency is high. Any mutation to one of these alleles that is selectively favoured will also more quickly affect the whole population because its frequency will be high. The effects of genetic drift will be greater and the population will change relatively rapidly, making it more likely to develop into a separate species. In large populations the effect of a mutant allele will be diluted because its frequency is far less in the much larger gene pool. The effects of genetic drift are likely to be less, and development into a new species is likely to be slower.

Two forms of speciation are **allopatric speciation** and **sympatric speciation**.

Allopatric speciation

Allopatric means different countries and describes the form of speciation where two populations become **geographically separated**. Geographical separation may be the result of any physical barrier between two populations which prevents them interbreeding. These barriers include oceans, rivers, mountain ranges and deserts. What proves a barrier to one species may be no problem to another. While an ocean may separate populations of hedgehogs, it can be crossed by many birds and for marine fish it is their very means of getting from place to place. A tiny stream may be a barrier to snails, whereas the whole of the Pacific Ocean fails to separate populations of certain birds.

If environmental conditions either side of the barrier vary, then natural selection will influence the two populations differently and each will evolve leading to adaptations to their local conditions. These changes may take many hundreds or even thousands of generations, but ultimately may lead to reproductive separation and the formation of separate species. Figure 1 shows how speciation might occur when two populations of a forest-living species become geographically separated by a region of arid grassland.

1. Species X occupies a forest area. Individuals within the forest form a single population with a single gene pool and freely interbreed.

forest
species X lives and breeds in the forest

2. Climatic changes to drier conditions reduce the size of the forest to two separated regions. The distance between the two regions is too great for the two populations of species X to meet to each other.

forest A population X_1 — arid grassland — forest B population X_2

3. Further climatic changes result in the one region (Forest A) becoming colder and wetter. Natural selection acts on population X which becomes adapted to these new conditions. Physiological and anatomical changes occur in this group.

colder and wetter — warmer and drier

forest A population X_1 — arid grassland — forest B population X_2

4. Continued adaptation leads to evolution of a new form, population Y, in forest A.

colder and wetter — warmer and drier

forest A population Y — forest B population X

5. A return to the original climatic conditions results in regrowth of forest. Forests A and B are merged and populations X and Y shared the same area. The two populations may no longer be capable of interbreeding. They are now two species, X and Y.

species Y
forest
species X

▲ **Figure 1** *Speciation as a result of geographical separation*

An example of allopatric speciation is the Galapagos finch. A single ancestral species is thought to have colonised one of the Galapagos islands. In the absence of competition its population increased and populations became established in other habitats on the same and different islands. Each population evolved adaptations to suit its new environment, including available food resources. These adaptations included different shapes and sizes of beak to deal with different seed types. Being geographically separated from its mainland population, these changes have led to the various populations being so different that they can no longer interbreed and now form separate species.

Sympatric speciation

Sympatric means same country and describes the form of speciation that results within a population in the same area leading to them becoming reproductively separated.

A likely example of sympatric speciation taking place is the apple maggot fly. Originally this insect only laid its eggs inside the fruit of hawthorns, which are native to North America. When apples trees were introduced, the fly started to lay its eggs in apples also. Females tend to lay their eggs on the type of fruit in which they developed and males tend to look for mates on the type of fruit in which they developed. So flies raised in hawthorns usually mate with each other and flies raised in apples tend to mate with each other. While the two types of flies are not yet separate species, mutations in each population have led to the evolution of genetic differences. In time this could result in them being incapable of successfully breeding with one another and therefore being separate species.

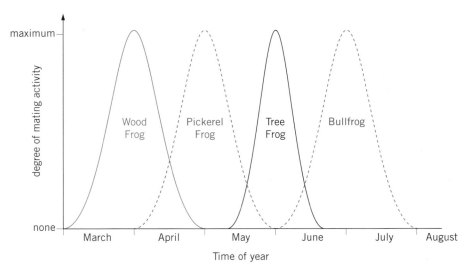

▲ **Figure 2** *Seasonal reproductive separation as illustrated by four varieties of frog*

Table 1 summarises forms of isolating mechanisms.

▼ **Table 1** *Summary of the forms of isolating mechanisms*

Type of variation	Number produced
Geographical	Populations are isolated by physical barriers such as oceans, mountain ranges, rivers, etc.
Ecological	Populations inhabit different habitats within the same area and so individuals rarely meet
Temporal	The breeding seasons of each population do not coincide and so they do not interbreed. Figure 2 illustrates this in relation to four types of frog
Behavioural	Mating is often preceded by courtship, which is stimulated by the colour or markings of the opposite sex, the call or particular actions of a mate. Any mutations which cause variations in these patterns may prevent mating, for example, if a female stickleback does not respond appropriately to the actions of the male, he ceases to court her
Mechanical	Anatomical differences may prevent mating occurring, for example, it may be physically impossible for the penis to enter the vagina in mammals
Gametic	The gametes may be prevented from meeting due to genetic or biochemical incompatibility. For instance, some pollen grains fail to germinate or grow when they land on a stigma of different genetic makeup. Some sperm are destroyed by chemicals in the female reproductive tract
Hybrid sterility	Hybrids formed from the fusion of gametes from different species are often sterile because they cannot produce viable gametes. For example, in a cross between a horse ($2n = 64$) and a donkey ($2n = 62$) the resultant mule has 63 chromosomes. It is impossible for these chromosomes to pair up appropriately during meiosis and so the gametes formed are not viable and the mule is sterile

Study tip

The examples of isolating mechanisms in Table 1 are for illustration only. You do not need to learn them.

Summary questions

1 Define a species.
2 Explain the meaning of the term speciation.
3 Describe the process of geographical separation.
4 Explain how geographical separation of two populations of a species can result in the accumulation of the differences in their gene pools.
5 Distinguish between allopatric and sympatric speciation.

City	Frequency of allele			
	White	Non-agouti	Blotched	Long-haired
Athens	0.001	0.72	0.25	0.50
Paris	0.011	0.71	0.78	0.24
London	0.004	0.76	0.81	0.33

1 (a) What does the Hardy–Weinberg principle predict? *(3 marks)*
The table shows the frequencies of some alleles in the population of cats in three cities.

(b) White cats are deaf. Would the Hardy–Weinberg principle hold true for white cats? Explain your answer. *(2 marks)*

(c) What is the evidence from the table that non-agouti and blotched are alleles of different genes? *(1 mark)*

(d) Hair length in cats is determined by a single gene with two alleles. The allele for long hair (h) is recessive. The allele for short hair (H) is dominant. Use the information in the table and the Hardy–Weinberg equation to estimate the percentage of cats in London that are heterozygous for hair length. Show your working.
(2 marks)
AQA June 2010

	United Kingdom	Sudan
Life expectancy males / years	76.5	50.5
Life expectancy females / years	81.6	52.4

2 (a) Explain what is meant by birth rate. *(1 mark)*

(b) The table shows life expectancies for babies born in the United Kingdom and in the Sudan in 2009.
(i) Describe the patterns shown by these data. *(2 marks)*
(ii) Suggest reasons for the differences in the life expectancy shown by these data.
(2 marks)
AQA June 2011

▼ Table 1

Days after release	Number of marked insects remaining in population	Number of insects captured	Number of captured insects that were marked
1	1508	524	78
2	1430	421	30
3	1400	418	18
4	1382	284	2
5	1380	232	9

3 Ecologists investigated the size of an insect population on a small island. They used a mark-release-recapture method. To mark the insects they used a fluorescent powder. This powder glows bright red when exposed to ultraviolet (UV) light.

(a) The ecologists captured insects from a number of sites on the island. Suggest how they decided where to take their samples. *(2 marks)*

(b) Give **two** assumptions made when using the mark-release-recapture method.
(2 marks)

(c) Suggest the advantage of using the fluorescent powder in this experiment. *(2 marks)*
The ecologists did **not** release any of the insects they captured 1–5 days after release of the marked insects.
Table 1 shows the ecologists' results.
Calculate the number of insects on this island 1 day after release of the marked insects. Show your working.
(2 marks)

(d) The ecologists expected to obtain the same result from their calculations of the number of insects on this island on each day during the period 1–5 days after release. In fact, their estimated number increased after day 1.
During the same period, the number of insects they caught decreased.
The method used by the ecologists might have caused these changes. Use the information provided to suggest **one** way in which the method used by the ecologists might have caused the increase in their estimates of the size of the insect population.
(2 marks)
AQA Specimen 2014

4   Snow geese fly north to the Arctic in the spring and form breeding colonies. Different colonies form at different latitudes. The greater the latitude, the further north is the colony. The further north a breeding colony forms, the colder the temperature and the greater the risk of snow.

Colony	Latitude in degrees north	Percentage of white snow geese each year			
		1930	1950	1960	1970
A	72	100		100	100
B	71		>99	>99	>99
C	66	95	85	76	
D	63	86	75	67	65
E	55		62		28

   (a) There is a positive correlation between the size of snow geese and how far north they breed. A large size results in snow geese being adapted for breeding in colder conditions. Explain how.   *(2 marks)*

Snow geese are either white or blue in colour. The table shows the percentage of white snow geese in colonies at different latitudes at different times over a 40-year period. The blank cells in the table are years for which no figures are available.

   (b) (i)   Describe how the percentage of white snow geese varies with distance north.
                                                                      (1 mark)
       (ii) The further north, the greater the risk of snow. Use this information to explain how natural selection might have accounted for the effect of latitude on the percentage of white snow geese.   *(3 marks)*

   (c) The percentage of white snow geese in these colonies changed over the period shown in the table. Use your knowledge of climate change to suggest an explanation.
                                                                      (2 marks)

   (d) Snow geese breed in large colonies. Scientists studied the nests in one colony. For each nest, they recorded the day on which the first egg hatched. They also recorded the number of young that survived from the nest. They used the data to plot a graph.

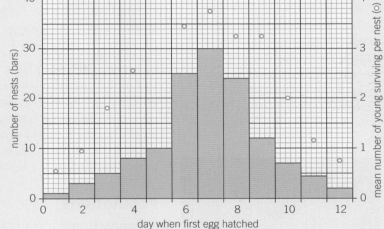

       (i)   What type of natural selection is shown in the graph?   *(1 mark)*
       (ii) Describe the evidence for your answer.   *(1 mark)*
                                                    AQA Jan 2010

5   The Amazonian forest today contains a very high diversity of bird species.
   • Over the last 2 000 000 years, long periods of dry climate caused this forest to separate into a number of smaller forests.
   • Different plant communities developed in each of these smaller forests.
   • Each time the climate became wetter again, the smaller forests grew in size and merged to reform the Amazonian forest.

   (a) Use the information provided to explain how a very high diversity of bird species has developed in the Amazonian forest.   *(5 marks)*
   (b) Speciation is far less frequent in the reformed Amazonian forest. Suggest **one** reason for this.   *(1 mark)*
                                                    AQA June 2013

19 Populations in ecosystems
19.1 Populations and ecosystems

Learning objectives

- → Define the terms environment, biotic, abiotic, and biosphere.
- → Explain what is meant by an ecosystem.
- → Explain what is meant by the terms population, community, and habitat.
- → Explain what a niche is.

Specification reference: 3.7.4

In this chapter we shall look at how living organisms form communities within ecosystems through which energy is transferred and elements are recycled.

We shall learn how populations of different species live in communities and how competition for survival arises both within and between these populations. We shall also see how populations within a single community are affected by living and non-living factors in an ecosystem.

Ecology is the study of the inter-relationships between organisms and their environment. The environment includes both non-living (**abiotic**) factors, such as temperature and rainfall, and living (**biotic**) factors, such as competition and predation.

Ecosystems

Ecosystems are dynamic systems made up of a community and all the non-living factors of its environment. Ecosystems can range in size from very small to very large. Within an ecosystem there are two major processes to consider:

- the flow of energy through the system
- the cycling of elements within the system.

An example of an ecosystem is a freshwater pond or lake. It has its own community of plants to collect the necessary sunlight energy to supply the organisms within it. Nutrients such as nitrate ions and phosphate ions are recycled within the pond or lake. There is little or no loss or gain between it and other ecosystems. Another example of an ecosystem is an oak woodland (Figure 2). Within each ecosystem, there are a number of species. Each species is made up of a group of individuals that make up a **population**.

Hint

Many students confuse the terms biotic and abiotic. Biotic refers to living things, for example, predation or competition between different species or individuals of one species. Abiotic refers to non-living factors, such as temperature, rainfall, or light intensity.

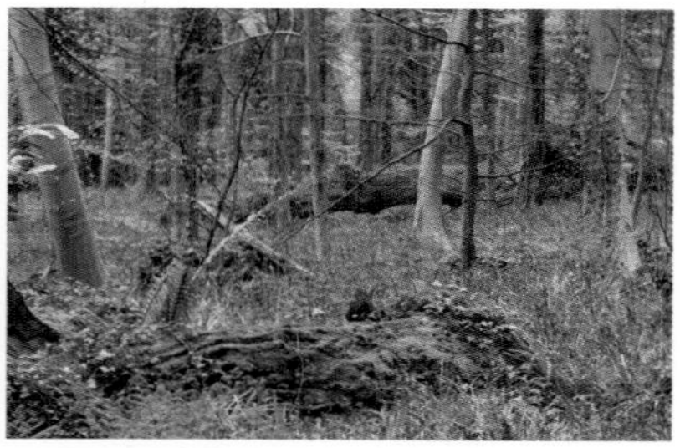

▲ **Figure 1** *Woodland ecosystem*

Populations

A **population** is a group of individuals of one species that occupy the same habitat at the same time and are potentially able to interbreed. An ecosystem supports a certain size of population of a species called the **carrying capacity**. The size of a population can vary as a result of:

- the effect of abiotic factors
- interactions between organisms, for example, intraspecific and interspecific competition and predation.

In the different habitats of an oak woodland there are populations of nettles, worms, green woodpeckers, beetles, etc. The boundaries of a population are often difficult to define. In our oak woodland, for example, all the mature green woodpeckers can breed with one another and so form a single population. Populations of different species form a **community**.

Study tip

Population is a very important concept that should be understood. It comes up in many areas of biology, including genetics and evolution.

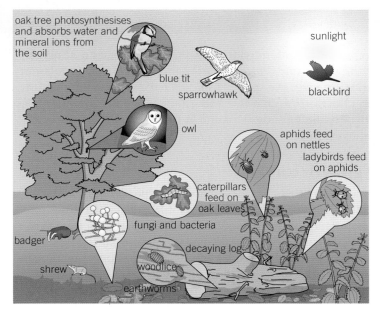

▲ **Figure 2** *Part of an oak woodland ecosystem*

Community

A **community** is defined as all the populations of different species living and interacting in a particular place at the same time. Within an oak woodland, a community may include a large range of organisms, such as oak trees, hazel shrubs, bluebells, nettles, sparrowhawks, blue tits, ladybirds, aphids, woodlice, earthworms, fungi, and bacteria (Figure 2).

Habitat

A **habitat** is the place where an organism normally lives and is characterised by physical conditions and the other types of organisms present. Within an ecosystem there are many habitats. For example, in an oak woodland, the leaf canopy of the trees may be a habitat for blue tits while a decaying log is the habitat for woodlice. A stream flowing through the woodland provides a very different habitat, within which aquatic plants and water beetles live. For a water vole, the stream and its banks are its habitat. Within each habitat there are smaller units, each with their own microclimate. These are called **microhabitats**. The mud at the bottom of the stream may be the microhabitat for a bloodworm while a crevice on the bark of an oak tree may be the microhabitat for a lichen.

Ecological niche

A **niche** describes how an organism fits into the environment. A niche refers to where an organism lives and what it does there. It includes all the biotic and abiotic conditions to which an organism is adapted in order to survive, reproduce and maintain a viable population. Some species may appear very similar, but their nesting habits or other aspects of their behaviour will be different, or they may show different levels of tolerance to environmental factors, such as a pollutant or a shortage of oxygen or nitrates. No two species occupy exactly the same niche. This is known as the competitive exclusion principle.

▲ **Figure 3** *This lake is an example of a habitat*

Summary questions

In the following passage, state the word that best replaces each of the numbers in brackets.

The study of the interrelationships between organisms and their environment is called (1). An ecosystem is a more or less self-contained functional unit made up of all the living or (2) features and non-living or (3) features in a specific area. Within each ecosystem are groups of different organisms, called a (4), which live and interact in a particular place at the same time. A group of organisms occupying the same place at the same time is called a (5), and the place where they live is known as a (6). The population size of a species that an ecosystem can support is known as (7).

▲ **Figure 1** *A population of lesser flamingos*

Hint

Humans exist in populations just like other species and therefore the rules also apply to us.

▲ **Figure 2** *The collared dove only arrived in Britain in the 1950s but its population has increased rapidly since then*

Maths link

MS 2.5, see Chapter 22.

A population is a group of individuals of the same species that occupy a habitat at the same time. The number of individuals in a population is the **population size**. We saw in Topic 19.1 that all the populations of the different organisms that live and interact together are known as a community. Populations are dynamic in that they vary in size and composition over time.

Plotting growth curves

Where a population grows in size slowly over a period of time it is possible to plot a graph of numbers in a population against time. Where the population grows rapidly over a short period of time this may not be possible. This is often the case when measuring the growth of microorganisms. Consider Table 1 which shows the increase in population size of a bacterium that initially doubles its numbers each hour. If we try to plot a graph of numbers against time using a time scale that allows us to differentiate each point, the curve runs off the graph after the point plotted at 4 hours (Figure 3).

In these cases it is necessary to use a logarithmic scale to represent the number of bacteria. The logarithms of bacterial numbers are shown in Table 1. When the graph of log bacterial numbers is plotted against time all points can be represented on the graph (Figure 4) and we can see that the rate of growth starts to slow after 8 hours.

Population size

Imagine a situation in which a single photosynthetic bacterial cell, capable of asexual reproduction, is placed in a newly created pond. It is summer and so there is plenty of light and the temperature of the water is around 12 °C – mineral nutrients have been added to the water. In these circumstances the bacterial cell divides rapidly because

▼ **Table 1**

Time/hours	Number of bacteria	Log number of bacteria
0	50	1.7
1	100	2.0
2	200	2.3
3	400	2.6
4	800	2.9
5	1 600	3.2
6	3 200	3.5
7	6 400	3.8
8	12 800	4.1
9	20 300	4.3
10	31 500	4.5

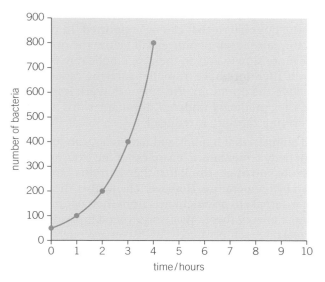

▲ **Figure 3** *Graph of number of bacteria against time*

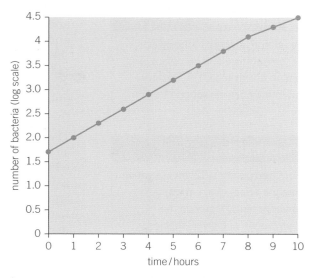

▲ **Figure 4** *Graph of log number of bacteria against time*

all the factors needed for the growth of the population are present. There are no **limiting factors**. In time, however, things change. For example:

- Mineral ions are consumed as the population becomes larger.
- The population becomes so large that the bacteria at the surface prevent light reaching those at deeper levels.
- Other species are introduced into the pond, carried by animals or the wind, and some of these species may use the bacteria as food or compete for light or minerals.
- Winter brings much lower temperatures and lower light intensity of shorter duration.

In short, the good life ends and the going gets tough. As a result the growth of the population slows, and possibly ceases altogether, and the population size may even diminish. Over the winter the population is likely to reach a relatively constant size. There are many factors, living (biotic) and non-living (abiotic), which affect this population size. Changes in these factors will influence the rate of growth and the size of the population.

In summary, no population continues to grow indefinitely because certain factors limit growth, for example, the availability of food, light, water, oxygen and shelter, and the accumulation of toxic waste, disease and predators. Each population has a certain size, the **carrying capacity**, that can be sustained over a relatively long period and this is determined by these limiting factors.

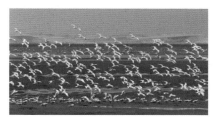

▲ **Figure 5** *A population of migrating birds, like these terns, fluctuates seasonally*

Hint

Remember that the size of any population is eventually determined by a limiting factor.

Hint

A species can only live within a certain range of abiotic factors and this range differs from species to species.

Abiotic factors

The abiotic conditions that influence the size of a population include:

- **temperature**. Each species has a different optimum temperature at which it is best able to survive. The further away from this optimum, the fewer individuals in a population are able to survive and the smaller is the population that can be supported. In plants and cold-blooded animals, as temperatures fall below the optimum, the enzymes work more slowly and so their metabolic rate is reduced. Populations therefore have a smaller carrying capacity. At temperatures above the optimum, the enzymes work less efficiently because they gradually undergo denaturation. Again the population's carrying capacity is reduced.

The warm-blooded animals, that is, birds and mammals, can maintain a relatively constant body temperature regardless of the external temperature. Therefore you might think that their carrying capacity would be unaffected by temperature. However, the further the temperature of the external environment gets from their optimum temperature, the more energy these organisms expend in trying to maintain their normal body temperature. This leaves less energy for individual growth and so they mature more slowly and their reproductive rate slows. The carrying capacity of the population is therefore reduced.

- **Light**. As the ultimate source of energy for most ecosystems, light is a basic necessity of life. The rate of photosynthesis increases as light intensity increases. The greater the rate of photosynthesis, the faster plants grow and the more spores or seeds they produce. Their carrying capacity is therefore potentially greater. In turn, the carrying capacity of animals that feed on plants is potentially larger.

- **pH**. This affects the action of enzymes. Each enzyme has an optimum pH at which it operates most effectively. A population of organisms is larger where the appropriate pH exists and smaller, or non-existent, where the pH is different from the optimum.

- **Water and humidity**. Where water is scarce, populations are small and consist only of species that are well adapted to living in dry conditions. Humidity affects the transpiration rates in plants and the evaporation of water from the bodies of animals. Again, in dry air conditions, the populations of species adapted to tolerate low humidity will be larger than those with no such adaptations.

In general terms, when any abiotic factor is below the optimum for a population, fewer individuals are able to survive because their adaptations are not suited to the conditions. If no individuals have adaptations that allow survival, the population becomes extinct.

▲ **Figure 6** *This cactus is adapted to survive in conditions where water is scarce. Its population in dry regions is therefore relatively large as there is little competition from other species, most of which are not adapted to survive in such conditions*

The influence of abiotic factors on plant populations

Species X and species Y are two species of flowering plants. Each is able to tolerate different temperatures and different pHs. In this way they avoid direct competition by occupying different niches. This is called niche separation. The chart (Figure 7) below illustrates the way each species is able to tolerate each of these two abiotic factors.

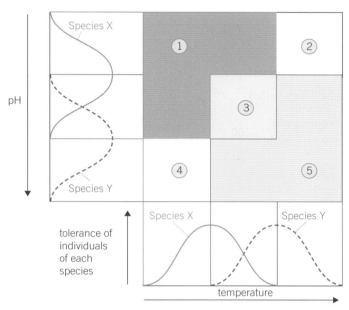

▲ Figure 7

1 State the numbered box that best fits each of the descriptions below.
  a Only a population of species X is found.

  b Both temperature and pH allow a population of both species to exist.

  c The temperature is too high for a population of species X and the pH is too low for a population of species Y.

  d There is competition between species X and species Y.

2 Explain why there is no population of either species in box 4.

Summary questions ⓥ𝑥̄

1 Explain why populations never grow indefinitely.

2 Distinguish between biotic and abiotic factors.

3 Suggest the level and type of abiotic factor that is most likely to limit the population size of the organisms and their habitats given below.

  a Ground plants on a forest floor

  b Hares in a sandy desert

  c Bacteria on the summit of a high mountain.

4 ⓥ𝑥̄ Table 2 shows the estimated world population over the past 12 000 years.

▼ Table 2

Time / years before present (BP)	Estimated human population / billions
0	600
2 000	200
4 000	35
6 000	20
8 000	10
10 000	5
12 000	1

  a Explain the benefits of using a logarithmic scale for population numbers when plotting a graph of these data.

  b Calculate to three significant places the log values for the human population in each case.

  c Plot a suitable graph to show the growth of the human population over the past 12 000 years.

Maths link ⓥ𝑥̄

MS 0.3 and 3.1, see Chapter 22.

The growth and size of human populations

The human population has doubled in less than 50 years and now totals over 7 billion. The basic factors that affect the growth and size of human populations are the **birth rate** and the **death rate**. It is the balance between these two factors that determines whether a human population increases, decreases or remains the same.

Individual populations are further affected by **migration**, which occurs when individuals move from one population to another. There are two types of migration:

- **immigration**, where individuals join a population from outside
- **emigration**, where individuals leave a population.

Again it is a balance between these two components that affects population size.

> population growth = (births + immigration) − (deaths + emigration)

$$\frac{\text{percentage population growth rate (in a given period)}}{} = \frac{\text{population change during the period}}{\text{population at the start of the period}} \times 100$$

> **1** The figures below show some population statistics for a country.
>
> Total population at the start of 2007 = 1 000 000
>
> Birth rate in 2007 = 25 per 1000 of population
>
> Death rate in 2007 = 20 per 1000 of population
>
> Calculate the percentage population growth for this country in 2007. Show your working.

As the future size of a human population depends upon the number of females of child-bearing age, it is useful to know the age and gender profile of a population. This is displayed graphically by a series of stacked bars representing the percentages of males and females in each age group. These graphs, called age **population pyramids**, give useful information on the future trends of different populations. Three typical types of population are represented in the age population pyramids in Figure 9. These are:

- **stable population** (Figure 9a), where the birth rate and death rate are in balance and so there is no increase or decrease in the population size.
- **increasing population** (Figure 9b), where there is a high birth rate, giving a wider base to the population pyramid (compared to a stable population) and fewer older people, giving a narrower apex to the pyramid. This type of population is typical of economically less developed countries.
- **decreasing population** (Figure 9c), where there is a lower birth rate (narrower base of the population pyramid) and a lower mortality rate leading to more elderly people (wider apex to pyramid). This type of population occurs in certain economically more developed countries, such as Japan.

As countries have developed economically their human populations have, so far, displayed a pattern of growth known as demographic transition. This pattern can be divided into four stages depending on the birth rate, death rate and total population size. The relationship between these four stages and the birth rates, death rates and total population are illustrated in Figure 10.

▲ **Figure 8** *The human population now exceeds 7 billion*

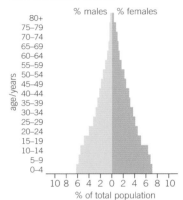

a *Population pyramid for a stable population*

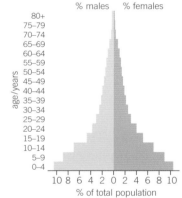

b *Population pyramid for an increasing population*

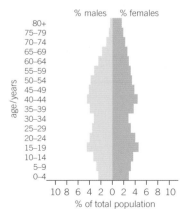

c *Population pyramid for a decreasing population*

▲ **Figure 9** *Age population pyramids*

2 Using Figure 10, suggest which of the four stages (1, 2, 3 or 4) best applies to each of the descriptions below.

a A country that has a rapidly falling birth rate and a relatively low death rate.

b A country in which there is a high birth rate but much starvation and periodic epidemic disease.

c A country where there have been many years of improved nutrition, far less infectious disease and a large number of children.

d Britain 20 000 years ago when famine and disease led to regular population crashes.

e Britain today.

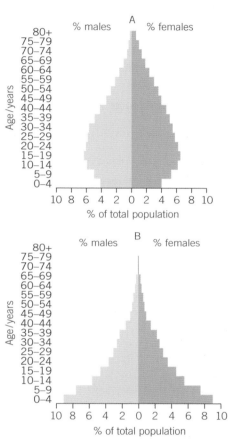

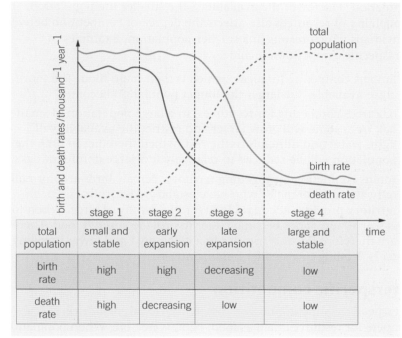

total population	small and stable	early expansion	late expansion	large and stable
birth rate	high	high	decreasing	low
death rate	high	decreasing	low	low

▲ **Figure 10** *Demographic transition*

▲ **Figure 11** *Age population pyramids A and B*

3 Figure 11 shows two age population pyramids: A and B. Suggest which stage of the demographic transition model shown in Figure 10 each pyramid represents. Give reasons for your answer in each case.

19.3 Competition

Learning objectives

→ Describe what is meant by intraspecific competition.

→ Summarise the factors that different species compete for.

→ Describe interspecific competition.

→ Explain how interspecific competition influences population size.

Specification reference: 3.7.4

Synoptic link

To help you understand competition and predation, revise the information on energy and ecosystems in Topics 13.1 and 13.2.

Hint

Which of two species in competition has the competitive advantage depends on the conditions at any point in time. If one species can tolerate a higher temperature than another, a rise in environmental temperature will favour it. If however there is a fall in environmental temperature, the other species is more likely to become dominant.

Where two or more individuals share any resource (e.g., light, food, space, oxygen) that is insufficient to satisfy all their requirements fully, then competition results. Where such competition arises between members of the same species it is called **intraspecific competition**. Where it arises between members of different species it is termed **interspecific competition**.

Intraspecific competition

Intraspecific competition occurs when individuals of the *same* species compete with one another for resources such as food, water, breeding sites, etc. It is the availability of such resources that determines the size of a population. The greater the availability, the larger the population. The lower the availability, the smaller the population. Availability of resources also affects the degree of competition between individuals which results in a smaller population. Examples of intraspecific competition include:

- limpets competing for algae, which is their main food. The more algae available, the larger the limpet population becomes.
- oak trees competing for resources. In a large population of small oak trees some will grow larger and restrict the availability of light, water and minerals to the rest, which then die. In time the population will be reduced to relatively few large dominant oaks.
- robins competing for breeding territory. Female birds are normally only attracted to males who have established territories. Each territory provides adequate food for one family of birds. When food is scarce, territories become larger to provide enough food. There are therefore fewer territories in a given area and fewer breeding pairs, leading to a smaller population size.

Interspecific competition

Interspecific competition occurs when individuals of *different* species compete for resources such as food, light, water, etc. When populations of two species are in competition one will normally have a competitive advantage over the other. The population of this species will gradually increase in size while the population of the other will diminish. If conditions remain the same, this will lead to the complete removal of one species. This is known as the competitive exclusion principle.

This principle states that where two species are competing for limited resources, the one that uses these resources most effectively will ultimately eliminate the other. In other words, no two species can occupy the same niche indefinitely when resources are limiting. Two species of sea birds, shags and cormorants, appear to occupy the same niche, living and nesting on the same type of cliff face and eating fish from the sea. Analysis of their food, however, shows that shags feed largely on sand eels and herring, whereas cormorants eat mostly flat fish, gobies, and shrimps. They therefore occupy different niches.

To show how a factor influences the size of a population it is necessary to link it to the birth rate and death rate of individuals in a population. For example, an increase in food supply does not necessarily mean there will be more individuals - it could just result in bigger individuals. It is therefore important to show how a factor, such as a change in food supply, affects the number of individuals in a population. For example, a decrease in food supply could lead to individuals dying of starvation and directly reduce the size of a population. An increase in food supply means that more individuals are likely to survive and so there is an increased probability that they will produce offspring and the population will increase. This effect therefore takes longer to influence population size.

The effects of interspecific competition on population size

The red squirrel is native to the British Isles and exclusively occupied a particular niche until around 130 years ago, when the grey squirrel was introduced from North America. Since then the two species have been competing for food and territory. There are now an estimated 2.5 million grey squirrels and just 160 000 red squirrels in the British Isles. The red squirrel population lives mostly in Wales and Scotland, with smaller groups in north eastern England and on islands such as Anglesey and the Isle of Wight. Figure 1 illustrates the changes in red and grey squirrel populations in Wales and Scotland between 1970 and 1990.

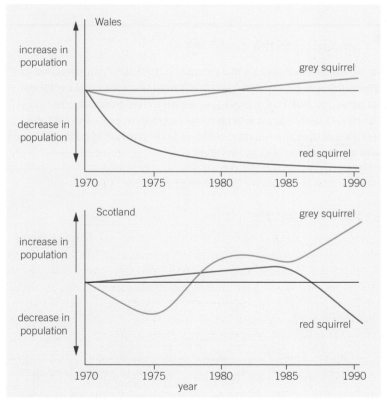

▲ **Figure 1** *Changes in red and grey squirrel populations in Wales and Scotland between 1970 and 1990. The lines show changes in comparison with the 1970 population*

In many cases we suspect that competition is the reason for variations in population. In practice it is difficult to prove for a number of reasons:

● There are many other factors that influence population size, such as abiotic factors.
● A causal link has to be established to show that competition is the cause of an observed correlation.

Summary questions

1 Distinguish between intraspecific competition and interspecific competition.
2 Name any two resources that species compete for.

Maths link

MS 3.1, see Chapter 22.

▲ **Figure 2** *Red squirrel*

▲ **Figure 3** *Grey squirrel*

Maths link $\sqrt{x}$

MS 3.1, see Chapter 22.

- There is a time lag in many cases of competition and so a population change may be due to competition that took place many years earlier.
- Data on natural population sizes are hard to obtain and not always reliable.

Study Figure 1 and answer the following questions.

1   State one piece of evidence from the graph for Scotland which indicates that changes in the red squirrel population are due to competition from the grey squirrel.
2   In Wales the populations of both grey and red squirrels declined between 1970 and 1975. Suggest a possible reason for this.
3   Both types of squirrels eat nuts, seeds and fruit as part of their diet. Grey squirrels spend more time foraging on the forest floor than red squirrels. Suggest how this behaviour might give the grey squirrel a competitive advantage over the red squirrel.
4   Suggest an explanation why islands such as Anglesey and the Isle of Wight still have significant red squirrel populations while they have disappeared from much of the rest of England and Wales.

 Competing to the death

In an experiment, two species of a genus of unicellular organism called *Paramecium* were grown separately in different test tubes that contained yeast as a source of food. The two species were then grown together in the same test tube – again with yeast as a food source. In each case the populations of both species were measured over a period of 20 days. The results are shown in the graph in Figure 4.

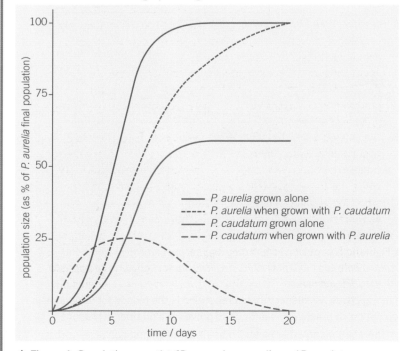

▲ **Figure 4** *Population growth of* Paramecium aurelia *and* P. caudatum *grown separately and together*

1. Describe the population growth curve of *P. caudatum* when grown alone over the 20-day period.
2. Compare the population growth curve of *P. caudatum* when grown with *P. aurelia* to the curve when *P. caudatum* is grown alone.
3. Suggest an explanation for the difference in the final population size of *P. caudatum* when grown with *P. aurelia* compared with when it is grown alone.
4. Suggest why the growth rate of *P. aurelia* is slower in the presence of *P. caudatum* than when grown alone.
5. Suggest why, after 20 days, the population size of *P. aurelia* grown with *P. caudatum* is the same as that when *P. aurelia* is grown alone.

Hint

Although the population of one species may increase as another decreases, this does not prove that this is due to direct competition between them. To be certain, it is necessary to establish a causal link for the observed correlation.

 ### Effects of abiotic and biotic factors on population size

Oak trees produce acorns in the autumn. Deer mice feed on acorns. Table 1 shows the dry mass of acorns produced per hectare (ha) from 1992 to 1997 in an area of woodland. It also shows the estimated population size of deer mice per hectare of the same area of woodland in the spring of each year from 1993 to 1998.

Maths link

MS 1.2 and 1.3, see Chapter 22.

1. Suggest a method by which the population of deer mice might be estimated.
2. Calculate the mean annual growth rate in deer mice population over the period 1993 to 1995. Show your working.
3. With reference to the data in the table, describe the relationship between acorn production in autumn and the deer mice population the following spring.
4. Acorn seeds begin to form in spring. It has been suggested that the higher the temperature in spring, the more acorns are produced the following autumn. From the table, state which year probably had the coldest spring.
5. The caterpillars of the gypsy moth feed on oak leaves. When the population of gypsy moth caterpillars is large, the damage they cause to oak trees reduces acorn production. Suggest how and why a rise in the population of gypsy moth caterpillars might affect the population of deer mice.
6. As well as acorns, deer mice also eat the pupae of gypsy moths.
   a Explain how a warm spring might result in a fall in the gypsy moth population the following year.
   b Owls are natural predators of deer mice. Suggest the possible effect of an increase in the owl population on the production of acorns. Explain your answer.

▼ Table 1

Year	Dry mass of acorns/kg ha⁻¹ produced in autumn	Estimated deer mice population/ number ha⁻¹ in spring
1992	28	–
1993	131	260
1994	318	550
1995	211	1320
1996	726	990
1997	39	3440
1998	–	340

213

19.4 Predation

Learning objectives

→ Explain what is meant by predation.

→ Explain how the predator–prey relationship affects the population size of the predator and prey.

Specification reference: 3.7.4

In Topic 19.3 we looked at interspecific competition. We shall now turn our attention to one type of interspecific relationship, the predator–prey relationship. A **predator** is an organism that feeds on another organism, known as their **prey**.

As predators have evolved they have become better adapted for capturing their prey - faster movement, more effective camouflage, better means of detecting prey. Prey have equally become more adept at avoiding predators - better camouflage, more protective features such as spines, concealment behaviour. In other words the predator and the prey have evolved alongside each other. If either of them had not matched the adaptations of the other, it would most probably have become extinct.

Predation

Predation occurs when one organism is consumed by another. When a population of a predator and a population of its prey are brought together in a laboratory, the prey is usually exterminated by the predator. This is largely because the range and variety of the habitat provided is normally limited to the confines of the laboratory. In nature the situation is different. The area over which the population can travel is far greater and the variety of the environment is much more diverse. In particular, there are many more potential refuges. In these circumstances some of the prey can escape predation because the fewer there are the harder they are to find and catch. Therefore, although the prey population falls to a low level, it rarely becomes extinct.

Evidence collected on predator and prey populations in a laboratory does not necessarily reflect what happens in the wild. At the same time, it is difficult to obtain reliable data on natural populations because it is not possible to count all the individuals in a natural population. Its size can only be estimated from sampling and surveys. These are only as good as the techniques used, none of which guarantee complete accuracy. We must therefore treat all data produced in this way with caution.

Effect of predator–prey relationship on population size

The relationship between predators and their prey and its effect on population size can be summarised as follows:

- Predators eat their prey, thereby reducing the population of prey.
- With fewer prey available the predators are in greater competition with each other for the prey that are left.
- The predator population is reduced as some individuals are unable to obtain enough prey for their survival or to reproduce.
- With fewer predators left, fewer prey are eaten and so more survive and are able to reproduce.
- The prey population therefore increases.
- With more prey now available as food, the predator population in turn increases.

This general predator–prey relationship is illustrated in Figure 1. In natural ecosystems, however, organisms eat a range of foods and therefore the fluctuations in population size shown in the graph are often less severe.

Study tip

Herbivores are sometimes considered as predators on plants.

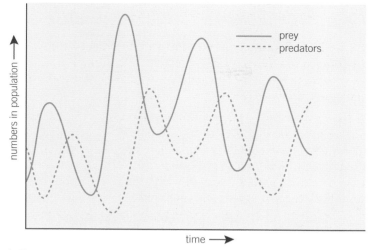

▲ **Figure 1** *Relationships between prey and predator populations*

Although predator–prey relationships are significant reasons for cyclic fluctuations in populations, they are not the only reasons, disease and climatic factors also play a part. These periodic population crashes are important in evolution as there is a **selection pressure** which means that those individuals who are able to escape predators, or withstand disease or an adverse climate, are more likely to survive to reproduce. The population therefore evolves to be better adapted to the prevailing conditions.

 The Canadian lynx and the snowshoe hare

The long-term study of the predator–prey relationship of the Canadian lynx and the snowshoe hare was made possible because records exist of the number of furs traded by companies such as the Hudson Bay Company in Canada over 200 years. By analysing these records the relative population size of the Canadian lynx and the snowshoe hare can be determined. The data collected are shown as a graph in Figure 3.

1 State what assumption is being made if we use the number of each type of fur traded as a measure of the population size of each species.
2 Describe the changes that occur in the populations of Canadian lynx and snowshoe hare.
3 Explain the changes that you have described.

It has long been observed that the population of snowshoe hares fluctuates in cycles. The question is whether these fluctuations are due mostly to predation by the lynx, mostly to changes in the food supply or mostly to a combination of both. To find out, ecologists fenced off 1 km² areas of coniferous forest in Canada where the hares lived. Separate areas were treated in four different ways:

● In the first set of areas, the hares were given extra food.
● In the second set of areas, lynx were excluded.
● In the third set of areas, the hares were given extra food and lynx were excluded.
● In the fourth set of areas, conditions were left unaltered as a control.

The results of the experiment are shown in Figure 4.

Study tip

When asked to describe predator–prey relationships from a graph, use names to describe precisely the changes taking place.

Summary questions

1 Explain why a predator population often exterminates its prey population in a laboratory but rarely does so in natural habitats.

2 Explain how a fall in the population of a predator can lead to a rise in its prey population.

3 A species of mite (A) is fed on oranges in a laboratory tank until its population is stable. A second mite species (B), that preys on species A, is introduced into the tank. Sketch a graph of the likely cycle of population change that the two species will undergo. Explain the changes that the graph illustrates.

▲ **Figure 2** *Canadian lynx catching a snowshoe hare*

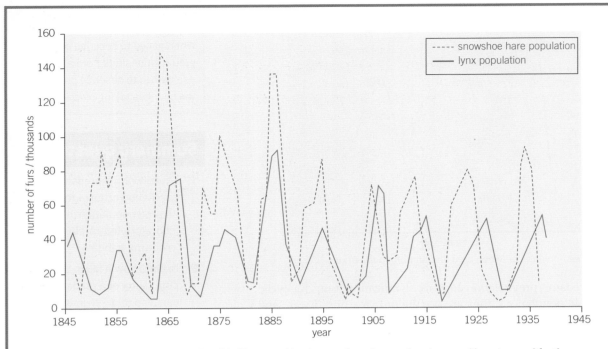

▲ **Figure 3** *The predator–prey relationship illustrated by the number of snowshoe hare and lynx trapped for the Hudson Bay Company between 1845 and 1940*

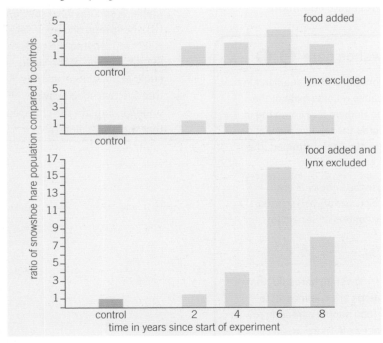

▲ **Figure 4** *Snowshoe hare population experiment*

4   ✓x̄ Calculate by how many times the addition of food increased the population after six years compared with the control.
5   Deduce which had the greater influence on the population of hares— the addition of food or the exclusion of the lynx. Explain your answer.
6   Deduce what conclusions can be drawn from this experiment.

To study a **habitat**, it is often necessary to count the number of individuals of a species in a given space. This is known as **abundance**. It is virtually impossible to identify and count every organism. To do so would be time-consuming and would almost certainly cause damage to the habitat being studied. For this reason only small samples of the habitat are usually studied in detail. As long as these samples are representative of the habitat as a whole, any conclusion drawn from the findings will be reliable. There are a number of sampling techniques used in the study of habitats. These include:

- random sampling using frame quadrats or point quadrats
- systematic sampling along a belt transect.

Quadrats

Two types of quadrat frequently used are:

A point quadrat which consists of a horizontal bar supported by two legs. At set intervals along the horizontal bar are ten holes, through each of which a long pin may be dropped (Figure 1). Each species that the pin touches is then recorded.

A frame quadrat which is a square frame divided by string or wire into equally sized subdivisions (Figure 2). It is often designed so that it can be folded to make it more compact for storage and transport. The quadrat is placed in different locations within the area being studied. The abundance of each species within the quadrat is then recorded.

Learning objectives

→ Name the factors to be considered when using a quadrat.

→ Explain how a transect is used to obtain quantitative data about changes in communities along a line.

→ Describe how the abundance of different species is measured.

→ Explain how the mark-release-recapture method can be used to measure the abundance of motile species.

Specification reference: 3.7.4

Synoptic link

Random sampling was considered in Topic 10.5, Quantitative investigations of variation, and provides further information on the subject.

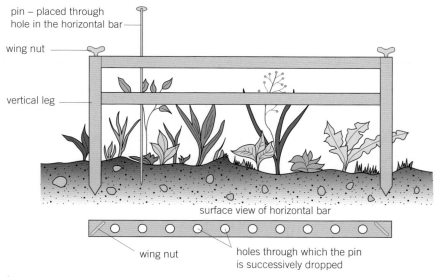

▲ **Figure 1** *A point quadrat*

There are three factors to consider when using quadrats:

- **The size of quadrat to use**. This will depend on the size of the plants or animals being counted and how they are distributed within the area. Larger species require larger quadrats. Where a population of species is not evenly distributed throughout the area, a large number of small quadrats will give more representative results than a small number of large ones.

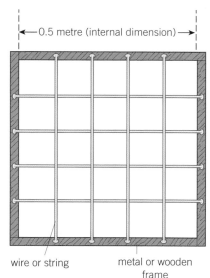

▲ **Figure 2** *A frame quadrat*

▲ **Figure 3** *Students carrying out fieldwork*

- **The number of sample quadrats to record within the study area**. The larger the number of sample quadrats the more reliable the results will be. As the recording of species within a quadrat is a time-consuming task a balance needs to be struck between the reliability of the results and the time available. The greater the number of different species present in the area being studied, the greater the number of quadrats required to produce reliable results for a valid conclusion.

- **The position of each quadrat within the study area**. To produce statistically significant results a technique known as random sampling must be used.

Sampling at random

In Topic 10.5 we introduced the idea that sampling at random is important to avoid any bias in collecting data. Avoiding bias ensures that the data obtained are reliable.

Suppose we wish to investigate the effects of grazing animals on the species of plants growing in a field. We begin by choosing two fields as close together as possible in order to minimise soil, climatic, and other abiotic differences. One field is regularly grazed by animals such as sheep, whereas the other has not been grazed for many years. We then take samples at many random sites in each field by placing the quadrat on the ground and recording the names and numbers of every species found within the area of the quadrat.

But how do we get a truly random sample? We could simply stand in one of the fields and throw the quadrat over our shoulder. A better method of sampling at random is to:

1 Lay out two long tape measures at right angles, along two sides of the study area.
2 Obtain a series of coordinates by using random numbers taken from a table or generated by a computer.
3 Place a quadrat at the intersection of each pair of coordinates and record the species within it.

Systematic sampling along belt transects

It is sometimes more informative to measure the abundance and distribution of a species in a systematic rather than a random manner. This is particularly important where some form of gradual change (transition) in the communities of plants and animals takes place. For example, the distribution of organisms along a line of succession, such as, through sand dunes by the edge of the sea and inland up into woodland. The stages of succession are especially well shown using transects. A belt transect can be made by stretching a string or tape across the ground in a straight line. A frame quadrat is laid down alongside the line and the species within it recorded. It is then moved its own length along the line and the process repeated. This gives a record of species in a continuous belt.

Measuring abundance

Random sampling with quadrats and counting along transects are used to obtain measures of **abundance**. Abundance is the number of individuals of a species within a given area. For species that don't move around, it can be measured in several ways, depending on the size of the species being counted and the habitat. Examples include:

- **frequency**, which is the likelihood of a particular species occurring in a quadrat. If, for example, a species occurs in 15 out of 30 quadrats, the frequency of its occurrence is 50%. This method is useful where a species, such as grass, is hard to count. It gives a quick idea of the species present and their general distribution within an area. However, it does not provide information on the density and detailed distribution of a species.

- **percentage cover**, which is an estimate of the area within a quadrat that a particular plant species covers. It is useful where a species is particularly abundant or is difficult to count. The advantages in these situations are that data can be collected rapidly and individual plants do not need to be counted. It is less useful where organisms occur in several overlapping layers (more probably plants).

To obtain reliable results, it is necessary to ensure that the sample size is large, that is, many quadrats are used and the mean of all the samples is obtained. The larger the number of samples, the more representative of the community as a whole will be the results.

Mark-release-recapture techniques

The methods of measuring abundance described above work well with plant species and non-motile (sessile) or very slow moving animal species that remain in one place but not with motile organisms. Motile animals move away when approached. They are often hidden and are therefore difficult to find and identify. To estimate the abundance of most animals requires an altogether different technique.

A known number of animals are caught, marked in some way, and then released back into the community. Some time later, a given number of individuals is collected randomly and the number of marked individuals is recorded. The size of the population is then calculated as follows:

$$\text{estimated population size} = \frac{\text{total number of individuals in the first sample} \times \text{total number of individuals in the second sample}}{\text{number of marked individuals recaptured}}$$

This technique relies on a number of assumptions:

- The proportion of marked to unmarked individuals in the second sample is the same as the proportion of marked to unmarked individuals in the population as a whole.
- The marked individuals released from the first sample distribute themselves evenly amongst the remainder of the population and have sufficient time to do so.
- The population has a definite boundary so that there is no immigration into or emigration out of the population.
- There are few, if any, deaths and births within the population.
- The method of marking is not toxic to the individual nor does it make the individual more conspicuous and therefore more liable to predation.
- The mark or label is not lost or rubbed off during the investigation.

Summary questions

1 An ecologist was estimating the population of sandhoppers on a beach. One hundred sandhoppers were collected, marked and released again. A week later 80 sandhoppers were collected, of which five were marked. Calculate the estimated size of the sandhopper population on the beach. Show your working.

2 When using the mark-release-recapture technique, explain how each of the following might affect the final estimate of a population.

a The marks put on the individuals captured in the first sample make them more easily seen by predators and so proportionately more are eaten than unmarked individuals.

b Between the release of marked individuals and the collection of a second sample an increased birth rate leads to a very large increase in the population.

c Between the release of marked individuals and the collection of a second sample, disease kills large numbers of all types of individual.

3 In a mark-release-recapture exercise, a sample of 120 woodlice were marked. After five days a second sample of 120 woodlice were collected. The population size was found to be 960. Calculate the number of marked woodlice that there were in the second sample.

Learning objectives

→ Describe changes that occur in the variety of species that occupy an area over time.

→ Define the terms succession and climax community.

→ Explain how managing succession can help to conserve habitats.

Specification reference: 3.7.4

We have seen that ecosystems are made up of all the interacting biotic and abiotic factors in a particular area within which there are a number of communities of organisms. As we look around at natural ecosystems, such as moorland or forest, we may get the impression that they have been there forever. This is far from the case. Ecosystems are dynamic. This means that they change day to day as populations fluctuate, sometimes slowly and sometimes very rapidly. **Succession** is the term used to describe these changes, over time, in the species that occupy a particular area.

One example of succession is when bare rock or other barren land is first colonised. Barren land may arise as a result of:

- a glacier retreating and depositing rock, sand being piled into dunes by wind or sea, volcanoes erupting and depositing lava, lakes or ponds being created by land subsiding, and silt and mud being deposited at river estuaries.

Stages of succession

Succession takes place in a series of stages. At each stage new species colonise the area and these may change the environment. These species may alter the environment in a way that makes it:

- less suitable for the existing species. As a result the new species may out-compete the existing one and so take over a given area.

- more suitable for other species with different adaptations. As a result this species may be out-competed by the better adapted new species.

In this way there is a series of successional changes which alter the abiotic environment. These alterations can result in a less hostile environment that makes it easier for other species to survive. As a consequence new communities are formed and biodiversity may be changed and/or increased.

The first stage of this type of succession is the colonisation of an inhospitable environment by organisms called **pioneer species**. Pioneer species make up a pioneer community and often have features that suit them to colonisation. These may include:

- asexual reproduction so that a single organism can rapidly multiply to build up a population.

- the production of vast quantities of wind-dispersed seeds or spores, so they can easily reach isolated situations such as volcanic islands

- rapid germination of seeds on arrival as they do not require a period of dormancy

- the ability to photosynthesise, as light is normally available but other food is not. They are therefore not dependent on animal species

- the ability to fix nitrogen from the atmosphere because, even if there is soil, it has few or no nutrients

- tolerance to extreme conditions.

Imagine an area of bare rock. One of the few kinds of organism capable of surviving on such an inhospitable area is lichens. Lichens are therefore pioneer species. Lichens can survive considerable drying out.

Hint

Pioneer communities put some organic material into the soil when they die. This allows recycling to start and increases mineral ions in the soil allowing other species of plants to grow.

In time, weathering of the base rock by the action of the lichens produces sand or soil, although this in itself cannot support other plants. However, as the lichens die and decompose they release sufficient nutrients to support a community of small plants. In this way the lichens change the abiotic environment by creating soil and nutrients for the organisms that follow. Mosses are typically the next stage in succession, followed by ferns. With the continuing erosion of the rock and the increasing amount of organic matter available from the death of these plants, a thicker layer of soil is built up. The organic material holds water making it easier for other plants to grow. Again these species change the abiotic environment, making it less hostile and so more suitable for the organisms that follow, for example, small flowering plants such as grasses and, in turn, shrubs and trees. These species provide more sources of food, leading to more food chains that develop into complex food webs and lead to more stable communities. In the UK the ultimate community is most likely to be **deciduous** oak woodland. This stable state comprises a balanced equilibrium of species with few, if any, new species replacing those that have become established. In this state, many species flourish and there is much biodiversity. This is called the **climax community** which remains more or less stable over a long period of time. This community consists of animals as well as plants.

The animals have undergone a similar series of successional changes, which have been largely determined by the plant types available for food and as **habitats**. The dead lichens provide food for animals such as detritus-feeding mites. The growth of mosses and grasses provides food and habitats for insects, millipedes, and worms. These are followed in turn by secondary consumers, such as centipedes, which feed on these organisms. The development of flowering plants, including trees, helps to support communities of butterflies and moths as well as larger organisms, such as reptiles, mammals, and birds.

During any succession there are a number of common features that emerge:

- **the non-living (abiotic) environment becomes less hostile**, for example, soil forms (which helps retain water) nutrients are more plentiful, and plants provide shelter from the wind. This leads to:
- **a greater number and variety of habitats and niches** that in turn produce:
- **increased** biodiversity as different species occupy these habitats. This is especially evident in the early stages, reaching a peak in mid-succession, but decreasing as the climax community is reached. The decrease is due to dominant species out-competing pioneer and other species, leading to their elimination from the community. With increased biodiversity comes:
- **more complex food webs**, leading to:
- **increased** biomass, especially during mid-succession.

Climax communities are in a stable equilibrium with the prevailing climate. It is abiotic factors such as climate that determine the dominant species of the community. In the lowlands of the UK, the climax community is deciduous woodland. In other climates of the world it may be tundra, steppe, or rain forest.

Another type of succession occurs when land that has already sustained life is suddenly altered. This may be the result of land clearance for agriculture or a forest fire. The process by which the ecosystem returns

▲ **Figure 1** *Lichens, with their ability to withstand dry conditions and to colonise bare rock, are frequently the first pioneer species on barren terrain*

Synoptic link

To appreciate successional change it would help to look again at Topics 13.1 Food chains and energy transfer, 13.2 Energy transfer and productivity and 13.3 Nutrient cycles

Hint

The climax community is determined by the limiting abiotic factor. For example, trees may not develop on very high mountains because it is too cold, too windy, or the soil layer is too thin (especially at the start of a succession).

▲ **Figure 2** *Deciduous woodland is normally the climax community in lowland Britain*

to its climax community is the same as described above, except that it normally occurs more rapidly. This is because soil already exists in which spores and seeds often remain alive in the soil, and there is an influx of animals and plants through dispersal and migration from the surrounding area. This type of succession is called **secondary succession**. Because the land has been altered in some way, for example, by fire, some of the species in the climax community will be different.

Figure 3 summarises the events of ecological succession on land.

Maths link $\sqrt{x}$

MS 3.1, see Chapter 22.

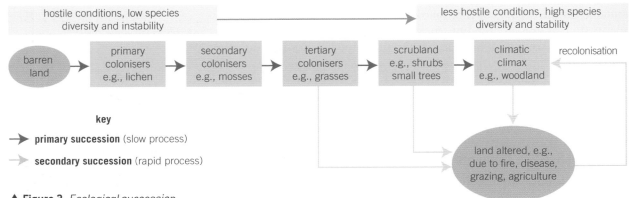

▲ **Figure 3** *Ecological succession*

Summary questions

1   State the general name given to the first organisms to colonise bare land.

2   Describe how changes in the environment lead to increased biodiversity during succession.

3   State the name that is given to the stable, final stage of any succession.

▲ **Figure 4** *The grassland in the foreground is grazed by sheep and so is prevented from reaching its natural climax. The land behind the fence has not been grazed for many years and has reverted to the climax community of woodland. This is therefore an example of secondary succession*

 ## Warming to succession

Many glaciers in the northern hemisphere have been melting over the past 200 years. This retreat is, in part, the result of the additional global warming that has taken place since the industrial revolution and the burning of fossil fuels that has accompanied it. When glaciers melt and retreat they leave behind gravel deposits known as moraines. The retreat of the glaciers in Glacier Bay, Alaska, has been measured since 1794 and so the age of the moraines in this region is recorded.

Although no ecologist has been present to watch the succession that has taken place on these moraines, they can infer the changes that have occurred by examining the plant and animal communities on the moraines of different ages. The youngest moraines (those nearest the retreating glacier) have the earliest colonisers (pioneer species), whereas those successively further away from the glacier show a time sequence of later communities.

Each stage of a succession has its own distinctive community of plants and animals that alters the environment in a way that allows the next stage and its community to develop. The stages that follow the retreat of an Arctic glacier are:

* **pioneer stage**. In the early years after the ice has retreated, photosynthetic bacteria and lichens colonise patches of land. Both of these pioneers fix nitrogen This is essential because nitrogen is virtually absent from glacial moraines. They also form tough mats that help to stabilise the loose surface of the moraines. When these pioneer species die, they decompose to form humus. Humus provides the nutrients that enable mosses to colonise. The pioneer stage occurs when the land has been ice free for 10–20 years.

- *Dryas* **stage**. Some 30 years after the ice has retreated, the ground is an almost continuous mat of the herbaceous plant *Dryas*. Its roots stabilise the thin and fragile soil layer formed from the erosion of the rocks that make up the moraine. Bacteria in root nodules of *Dryas* also fixes nitrogen, further adding nitrogenous nutrients to this poor-quality soil. Other plants found at this stage are the Arctic poppy and moss campion.

- **alder stage**. This arises about 60 years after the ice has retreated. Alder is a tree that has nitrogen-fixing nodules on its roots, enabling it to grow on nitrogen-poor soil. Alder sheds its leaves, which decompose into nitrogen-rich humus that further enriches the soil. The alder stage occurs some 50–70 years after the retreat of the glacial ice.

- **spruce stage**. About 100 years after the ice has retreated, spruce trees develop amongst the alder. A period of transition takes place and during the next 50 years or so the taller spruce out-competes the alder and ultimately displaces it altogether.

Figure 5 summarises changes in soil nitrogen, plant diversity and biomass following the retreat of an Arctic glacier.

▲ **Figure 6** *Dryas (mountain avens) is the most common pioneer species in Glacier Bay, Alaska. It is able to fix nitrogen and forms dense mats and therefore enriches and stabilises the thin fragile soil*

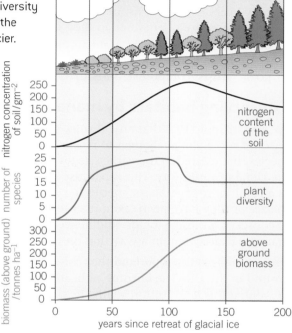

▲ **Figure 5**

▲ **Figure 7** *Arctic poppy (yellow flower) and moss campion (pink flowers) are early flowering pioneer species on Arctic moraines*

▲ **Figure 8** *Spruce trees are the final succession stage following the retreat of glacial ice in the Arctic. They begin to grow around 100 years after the ice has retreated and persist as the dominant vegetation for centuries*

1 Using the information on the graphs, describe and explain the changes in above-ground biomass over the 200-year period.

2 a Using your knowledge of the nitrogen cycle, explain how nitrogen from the atmosphere becomes incorporated into the soil, causing its level to increase during the first 100 years after the glacier retreats.

   b Suggest two reasons for the fall in soil nitrogen levels after 150 years.

3 Suggest a reason for:

   a the rapid increase in plant species during the first 30 years after the retreat of the glacier.

   b the fall in the number of plant species 100 years after the retreat of the glacier.

4 Explain why it would be more appropriate to use a transect rather than quadrats placed at random when investigating this succession.

▲ **Figure 1** *Moorland is an example of the conservation of a habitat by managing succession. Burning of heather and grazing by sheep has prevented shrubs and trees from developing*

What is conservation?

Conservation is the management of the Earth's natural resources by humans in such a way that maximum use of them can be made in the future. This involves active intervention by humans to maintain ecosystems and biodiversity. It is therefore a dynamic process that entails careful management of existing resources and reclamation of those already damaged by human activities. The main reasons for conservation are:

* **personal** to maintain our planet and therefore our life support system.
* **ethical**. Other species have occupied the Earth far longer than we have and should be allowed to coexist with us. Respect for living things is preferable to disregard for them.
* **economic**. Living organisms contain a gigantic pool of genes with the capacity to make millions of substances, many of which may prove valuable in the future. Long-term productivity is greater if ecosystems are maintained in their natural balanced state.
* **cultural and aesthetic**. Habitats and organisms enrich our lives. Their variety adds interest to everyday life and inspires writers, poets, artists, composers, and others who entertain and fulfill us.

Conserving habitats by managing succession

We saw in Topic 19.6 that any climax community has undergone a series of successional changes to reach its current state. Many of the species that existed in the earlier stages are no longer present as part of the climax community. This is because their habitats have disappeared as a result of succession, or species have been out-competed by other species or they have been taken over for human activities. One way of conserving these habitats, and hence the species they contain, is by managing succession in a way that prevents a change to the next stage.

One example is the moorland that exists over much of the higher ground in the UK. The burning of heather and grazing by sheep has prevented this land from reaching its climax community. The burning and grazing destroy the young tree saplings and so prevent the natural succession into deciduous woodland.

Around 4000 years ago, much of lowland UK was a climax community of oak woodland, but most of this forest was cleared to allow grazing and cultivation. The many heaths and grasslands that we now refer to as natural are the result of this clearance and subsequent grazing by animals. An example is chalk downland which was cleared of forest and where sheep and rabbits now eat any new saplings preventing these saplings from developing into full grown trees.

If the factor that is preventing further succession is removed, then the ecosystem develops naturally into its climatic climax (secondary succession). For example, if grasslands are no longer grazed or mowed, or if farmland is abandoned, shrubs initially take over, followed by deciduous woodland. Sand dunes can be managed to prevent succession to woodland leaving wet areas where species like natterjack toads can thrive.

Conflicting interests

One challenging conservation issue in the UK is the conflict between the conservation of hen harriers and the commercial hunting of red grouse. One scientific survey investigated the effect of predation by hen harriers on the breeding success of red grouse on managed moorland in Scotland. Some of the results included:

- On moorland where hen harriers were present there were, on average, 17% fewer young grouse than on moorlands without hen harriers.
- Over a three-year period grouse nests were intensively observed during the six weeks following the hatching of chicks. In this period, predation by harriers accounted for 91% of grouse chick losses.
- Prey remains found around harrier nests were examined. Of the 300 items identified, 32% were grouse chicks.

1 √x̄ Calculate how many of the items of prey identified around harrier nests were grouse chicks.

2 Harriers also feed on voles and meadow pipits. Explain how a rise in the population of these organisms might affect the population of grouse.

Moorland is considered one of the most attractive landscapes in the UK. Many of the national parks are made up of moorland and are visited by millions of people each year. To rear grouse, moorland has to be carefully managed. Controlled grazing by sheep and the periodic burning of vegetation are used to maintain low-growing plant populations of heather, bilberry, and crowberry that grouse feed on and nest within. The money to support this management comes largely from charges made to those who shoot grouse.

3 Explain what might happen to moorland if sheep-grazing and burning of the vegetation ceased.

The population of grouse in the UK is in decline due mainly to disease. Currently there are around 250 000 breeding pairs. The hen harrier was persecuted to such an extent that, by 1900, it was only found on a few Scottish islands. It recolonised the UK mainland in the 1970s and there are now around 750 breeding pairs. Both harriers and grouse normally produce one clutch of eggs each year. Hen harriers are protected by law and it is illegal to kill them, collect their eggs or destroy their nests. Conservationists want to retain this protection so that the population of hen harriers can increase. Grouse managers want to be allowed to control hen harrier populations to prevent them threatening the declining grouse populations.

4 Outline the arguments for and against continued protection of hen harrier populations.

To try to help resolve this conflict, scientists are currently conducting experiments to test whether hen harrier populations can be increased at the same time as reducing their negative impact on grouse populations. The information can then be used to inform decisions about how best to conserve grouse, harriers and moorland habitats.

The experiment will be carried out in two large areas where harriers are currently rare. Within these areas, the results of two strategies on the size of harrier and grouse populations will be measured:

- Killing hen harrier chicks, or moving them to a different location, when the harrier population size reaches an agreed ceiling.
- Providing alternative sources of food for hen harriers.

▲ **Figure 2** *Hen harrier*   ▲ **Figure 3** *Red grouse*

5 In each of the following, suggest a reason why:
   a The experiment will take at least five years to produce any findings.
   b An independent body, acceptable to both conservation groups and grouse managers will be needed to oversee the experiment.
   c The sites chosen for the experiment are ones where harriers can be expected to colonise relatively quickly.
   d Each experimental area will contain a number of different moorland sites managed by different individuals.
   e Some people are concerned about the long-term implications of a suspension, however temporary, to the legal protection of harriers that would be required during the experiment.
6 Explain how scientific experiments such as this one help to inform decision-making.

1   The young of frogs and toads are called tadpoles. Ecologists investigated the effect of predation on three species of tadpole. They set up four artificial pond communities. Each community contained
    • 200 spadefoot toad tadpoles
    • 300 spring peeper frog tadpoles
    • 300 southern toad tadpoles.
    The ecologists then added a different number of newts to each pond. Newts are predators.
    Figure 1 shows the effect of increasing the number of newts on the percentage survival of the tadpoles of each species.

    (a) (i)   Describe the effect of an increase in the number of newts on the percentage survival of the tadpoles of each of the toad species.   (*2 marks*)
        (ii) Suggest an explanation for the effect of an increase in the number of newts on the percentage survival of the tadpoles of spring peeper frogs.   (*2 marks*)

    Figure 2 shows how the masses of the tadpoles were affected in each pond during the investigation.

    (b) Using the information provided in **Figure 1** explain the results obtained in **Figure 2**.   (*2 marks*)
                                                                                   AQA Jan 2011

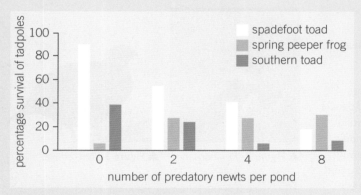

▲ Figure 1

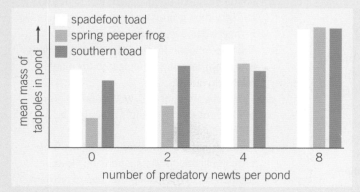

▲ Figure 2

2 Algae are photosynthesising organisms. Some algae grow on rocky shores. A scientist investigated succession involving different species of algae. He placed concrete blocks on a rocky shore. At regular intervals over 2 years, he recorded the percentage cover of algal species on the blocks. His results are shown in the graph.

(a) Name the pioneer species. *(1 mark)*

(b) **(i)** The scientist used percentage cover rather than frequency to record the abundance of algae present. Suggest why. *(1 mark)*

(ii) Some scientists reviewing this investigation were concerned about the validity of the results because of the use of concrete blocks. Suggest one reason why these scientists were concerned about using concrete blocks for the growth of algae.

(1 mark)

(c) Use the results of this investigation to describe and explain the process of succession.

(4 marks)

AQA June 2013

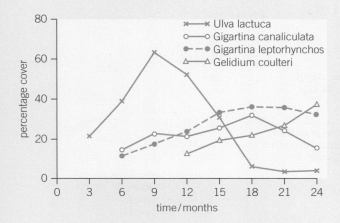

3 The diagram shows the dominant plants in communities formed during a succession from bare soil to pine forest.

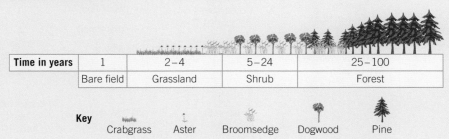

Time in years	1	2–4	5–24	25–100
	Bare field	Grassland	Shrub	Forest

Key   Crabgrass   Aster   Broomsedge   Dogwood   Pine

(a) Name the pioneer species shown in the diagram. *(1 mark)*

(b) The species that are present change during succession. Explain why. *(2 marks)*

(c) The pine trees in the forest have leaves all year. Explain how this results in a low species diversity of plants in the forest.

(1 mark)

AQA June 2012

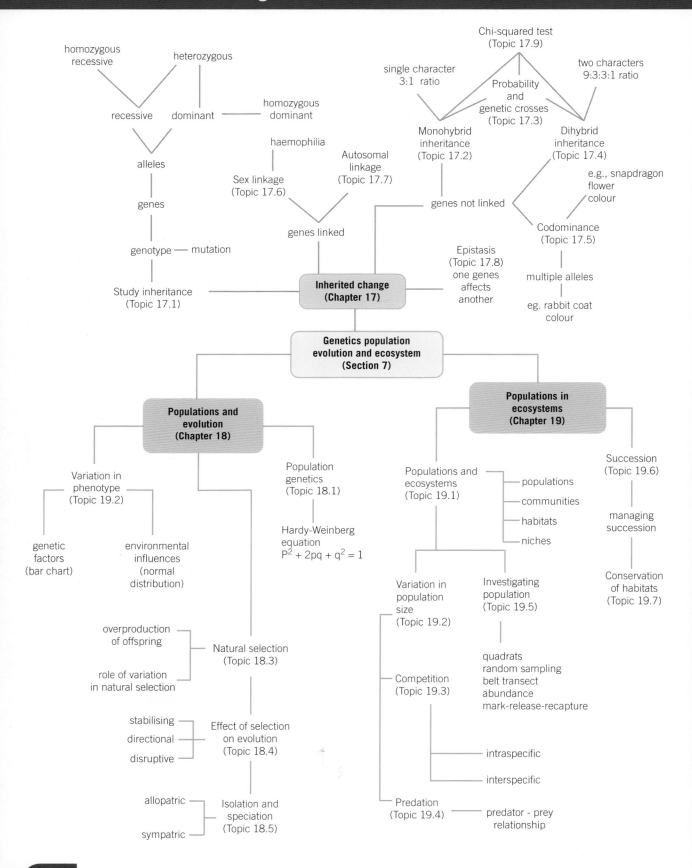

Practical skills

In this section you have met the following practical skills:

- How to plot growth curves using a logarithmic scale.
- How to carry out random sampling.
- Investigate the distribution of organisms in a habitat using randomly placed framed quadrats or a belt transect.
- Use the mark-release-recapture method to investigate the abundance of a motile species.

Extension task

Using your local newspaper, regional television news or local community websites in your area, identify a scheme in your region designed to conserve a habitat. Find out the purpose of this scheme and the organisations involved in it.

Research the various sources of funds that are available to support conservation projects like the one you have identified.

Draft a letter to one source of funds applying for a specified sum of money to support the aims of your project. Include in your letter a justification for the conservation project and the benefits it will bring to the community. Explain how the money will be used, how it will further the aims of the project and how you will evaluate whether it has been well spent.

Maths skills

In this section you have met the following maths skills:

- Calculating ratios and percentages of the offspring of genetic crosses.
- Understanding and calculating the probability associated with genetic inheritance.
- Using the chi-squared test to test the significance of the difference between observed and expected results of genetic crosses.
- Solving, and changing the subject in, algebraic equations such as the Hardy-Weinberg equation.
- Using a logarithmic scale in relation to quantities that range over several orders of magnitude.
- Using the logarithmic function on a calculator.
- Plotting two variables from experimental data provided.
- Finding arithmetical means

Section 7 Practice questions

1. A student investigated an area of moorland where succession was occurring. She used quadrats to measure the percentage cover of plant species, bare ground, and surface water every 10 metres along a transect. She also recorded the depth of soil at each quadrat. Her results are shown in the table.

	Percentage cover in each quadrat A to E				
	A	B	C	D	E
Bog moss	55	40	10	–	–
Bell heather	–	–	–	15	10
Sundew	10	5	–	–	–
Ling	–	–	–	15	20
Bilberry	–	–	–	15	25
Heath grass	–	–	30	10	5
Soft rush	–	30	20	5	5
Sheep's fescue	–	–	25	35	30
Bare ground	20	15	10	5	5
Surface water	15	10	5	–	–
Soil depth / cm	3.2	4.7	8.2	11.5	14.8

– Indicates zero percentage cover

   (a) Explain how these data suggest that succession has occurred from points **A** to **E** along the transect. *(3 marks)*
   (b) The diversity of animal species is higher at **E** than **A**. Explain why. *(2 marks)*
   (c) The student used the mark-release-recapture technique to estimate the size of the population of sand lizards on an area of moorland. She collected 17 lizards and marked them before releasing them back into the same area. Later, she collected 20 lizards, 10 of which were marked.
      (i) Give **two** conditions for results from mark-release-recapture investigations to be valid. *(2 marks)*
      (ii) Calculate the number of sand lizards on this area of moorland. Show your working. *(2 marks)*

AQA Jan 2013

2 In a species of snail, shell colour is controlled by a gene with three alleles. The shell may be brown, pink, or yellow. The allele for brown C^B, is dominant to the other two alleles. The allele for pink, C^P, is dominant to the allele for yellow, C^Y.

(a) Explain what is meant by dominant allele. *(1 mark)*

(b) Give **all** the genotypes which could result in a brown-shelled snail. *(1 mark)*

(c) A cross between two pink shelled snails produced only pink-shelled and yellow-shelled snails. Use a genetic diagram to explain why. *(3 marks)*

(d) The shells of this snail may be unbanded or banded. The absence or presence of bands is controlled by a single gene with two alleles. The allele for unbanded, **B**, is dominant to the allele for banded, **b**.
A population of snails contained 51% of unbanded snails. Use the Hardy-Weinberg equation to calculate the percentage of this population that you would expect to be heterozygous for this gene. Show your working. *(3 marks)*

AQA June 2012

3 (a) Insect pests of crop plants can be controlled by chemical pesticides or biological agents. Give **two** advantages of using biological agents. *(2 marks)*
Two-spotted mites are pests of strawberry plants. Ecologists investigated the use of predatory mites to control two-spotted mites. They then recorded the percentage of strawberry leaves occupied by two-spotted mites and by predatory mites over a 16-week period. The results are shown in Figure 1.

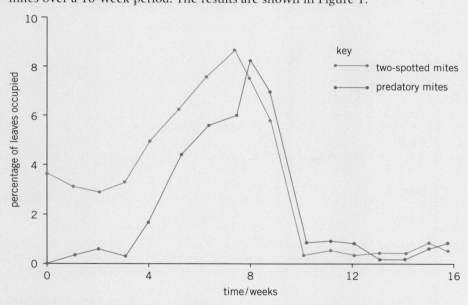

▲ Figure 1

(b) Describe how the percentage of leaves occupied by predatory mites changed during the period of this investigation. *(2 marks)*

(c) The ecologists concluded that in this investigation the control of the two-spotted mite by a biological agent was effective. Explain how the results support this conclusion. *(2 marks)*

(d) Farmers who grow strawberry plants and read about this investigation might decide **not** to use these predatory mites. Suggest **two** reasons why. *(2 marks)*

(e) The ecologists repeated the investigation but sprayed chemical pesticide on the strawberry plants after 10 weeks. After 16 weeks no predatory mites were found but the population of two-spotted mites had risen significantly. Suggest an explanation for the rise in the two-spotted mite population. *(2 marks)*

AQA June 2012

4   In birds, males are XX and females are XY.
    (a) Use this information to explain why recessive, sex-linked characteristics are more
         common in female birds than in male birds.   (*1 mark*)
    (b) In chickens, a gene on the X chromosome controls the rate of feather production.
         The allele for slow feather production, F, is dominant to the allele for rapid feather
         production, f. Figure 3 shows the results produced from crosses carried out by
         a farmer.   (*2 marks*)

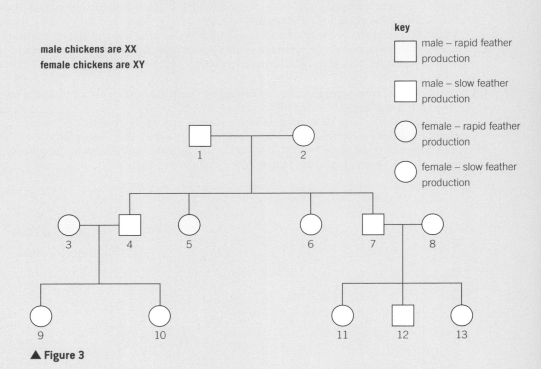

male chickens are XX

female chickens are XY

key

☐ male – rapid feather production

☐ male – slow feather production

○ female – rapid feather production

○ female – slow feather production

▲ Figure 3

    (i)   Explain one piece of evidence from Figure 3 which shows that the allele
          for rapid feather production is recessive.   (*2 marks*)
    (ii) Give all the possible genotypes of chicken 5 and chicken 7 from Figure 3.
                                                                                       (*2 marks*)
    (iii) A cross between two chickens produced four offspring. Two of these were
          males with rapid feather production and two were females with slow feather
          production. Give the genotypes of the parents.   (*1 mark*)
    (c) Feather colour in one species of chicken is controlled by a pair of codominant alleles
         which are not sex linked. The allele C^B codes for black feathers and the allele C^W
         codes for white feathers. Heterozygous chickens are blue-feathered.
         On a farm, 4% of the chickens were black-feathered. Use the Hardy-Weinberg
         equation to calculate the percentage of this population that you would expect
         to be blue-feathered. Show your working.   (*3 marks*)
                                                                                  AQA June 2014

5 The graph shows the effects of light intensity on the rate of photosynthesis of three species of tree, **X**, **Y** and **Z**. Each of these species occurs at a different stage in succession.

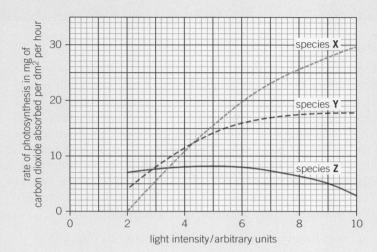

(a) Species **X** is the first tree to become established in the succession. Use the graph to explain why it is likely to become established earlier in the succession than **Y** or **Z**. *(3 marks)*

(b) Species **X** may change the environment so that it becomes more suitable for species **Z**. Use the graph to explain why. *(2 marks)*

AQA Jan 2010

Section 8
The control of gene expression

Introduction

At a cellular level, control of metabolic activities is achieved by regulating which genes of the genome are transcribed and translated, and when this takes place. Although the cells within an organism carry the same genes they translate only part of them. In multicellular organisms, this control of translation enables cells to have specialised functions and to form specific tissues and organs. Cells formed from the zygote are initially able to differentiate into any type of cell – they are totipotent. As these cells become specialised they lose the ability to become a different type of cell. In mature mammals, only a few cells retain the ability to differentiate into other cells. These are called stem cells.

It has long been known that many factors control the expression of genes and, thus, the phenotype of organisms. Some of these factors are external, environmental factors, others are internal factors. What was not generally disputed was the idea that these environmental factors are never inherited by the following generation. Only those processes such as mutations, which caused changes to the nucleotide base sequence in a DNA molecule could be inherited. This view has now been challenged by the discovery that environmental factors can cause heritable changes in gene function without any change to the base sequence of DNA. This so called epigenetic regulation of transcription is being recognised as important.

We are increasingly able to control the expression of genes by altering the epigenome. This allows us to alter an organism's genomes and the proteins they produce (proteomes). Along with our ability to manipulate the transcription and translation of genes, this has opened up many medical and technological applications. The use of DNA technology allows us to clone genes for use in medical techniques such as gene therapy. Other aspects include the use of DNA probes and DNA hybridisation in the diagnosis and treatment of human diseases, as well as the use of genetic fingerprinting for medical, forensic, and breeding purposes.

Working scientifically

In studying this unit there will be opportunities to perform practical exercises and so develop practical skills.

In performing these exercises you will have the chance to develop practical skills such as:

- separating biological compounds using electrophoresis
- using microbiological aseptic techniques.

What you already know

The material in this unit is intended to be self-explanatory, but there is certain information from GCSE that will be useful to your appreciation of this section. This information includes:

- ◯ Different genes control the development of different characteristics of an organism.

- ◯ Differences in the characteristics of different individuals of the same kind may be due to differences in:
  - the genes they have inherited (genetic causes)
  - the conditions in which they have developed (environmental causes)
  - a combination of both of the above.

- ◯ The differences between Darwin's theory of evolution and conflicting theories, such as that of Lamark.

- ◯ In genetic engineering, genes from the chromosomes of humans and other organisms can be 'cut out' using enzymes and transferred to cells of other organisms.

- ◯ Each person (apart from identical twins) has unique DNA. This can be used to identify individuals in a process known as DNA fingerprinting.

- ◯ Embryos can be screened for the alleles that cause genetic disorders.

- ◯ Genes can also be transferred to the cells of animals, plants or microorganisms at an early stage in their development so that they develop with desired characteristics.

- ◯ New genes can be transferred to crop plants and crops that have had their genes modified in this way are called genetically modified crops (GM crops). Examples of genetically modified crops include ones that are resistant to insect attack or to herbicides

- ◯ Genetically modified crops generally show increased yields.

- ◯ Concerns about GM crops include the effect on populations of wild flowers and insects, and uncertainty about the effects of eating GM crops on human health.

- ◯ Interpreting information about cloning techniques and genetic engineering techniques.

- ◯ Making informed judgements about the economic, social and ethical issues concerning cloning and genetic engineering, including genetically modified (GM) crops.

20 Gene expression
20.1 Gene mutations

Learning objectives

→ Describe the types of gene mutation.

→ Explain how the different types of gene mutation result in different amino acid sequences in polypeptides.

→ Explain why some mutations do not result in a changed amino acid sequence.

→ Discuss the causes of gene mutations.

Specification reference: 3.8.1

Synoptic link

Throughout this topic reference is made to the effects of changes to polypeptide structure as a result of mutations. A review of Topic 1.6, as well as Topic 9.1 make essential background reading.

Hint

Consider the following sentence, which consists only of three-letter words: THE RED HEN ATE HER TEA. If we delete the first letter T but continue to divide the sentence into three-letter words, it becomes HER EDH ENA TEH ERT EA and is incomprehensible. If we delete the final T, this leaves the strange but mostly readable sentence THE RED HEN ATE HER EA.

Study tip

For simplicity, the effect of a mutation caused by a change to a single base is often used as an example. It must be remembered that in practice it is often more than one base that is involved.

In Topic 9.1, we saw that any change to the quantity or the structure of the DNA of an organism is known as a **mutation** and any change to one or more nucleotide bases, or any rearrangement of the bases, in DNA is known as a **gene mutation**. These gene mutations might arise during the replication of DNA.

We also learnt that any changes to one or more bases in the DNA triplets could result in a change in the amino acid sequence of the polypeptide. Let us now consider gene mutations in more detail by looking at the different types.

Substitution of bases

The type of gene mutation in which a nucleotide in a section of a DNA molecule is replaced by another nucleotide that has a different base is known as a substitution. Depending on which new base is substituted for the original base, there are three possible consequences:

- The formation of one of the three stop codons that mark the end of a polypeptide chain. As a result the production of the polypeptide coded for by the section of DNA would be stopped prematurely. The final protein would almost certainly be significantly different and the protein could not perform its normal function.

- The formation of a codon for a different amino acid, meaning that the structure of the polypeptide produced would differ in a single amino acid. The protein of which this polypeptide is a part may differ in shape and not function properly. For example, if it is an enzyme, its active site may no longer fit the substrate and it will not catalyse the reaction. An example of this form of substitution mutation causes a condition called sickle cell anaemia.

- The formation of a different codon but one that produces a codon for the same amino acid as before. This is because the genetic code is degenerate and so most amino acids have more than one codon. The mutation therefore has no effect on the polypeptide produced and so the mutation will have no effect.

Deletion of bases

We saw in Topic 9.1 that the loss of a nucleotide base from a DNA sequence is called a deletion. Minor though the loss of a single base might seem, the impact on the phenotype can be enormous. The one deleted base creates what is known as a **frame shift** because the reading frame that contains each three letters of the code has been shifted to the left by one letter. The gene is now read in the wrong three-base groups and the coded information is altered. Most triplets will then be different, as will the amino acids they code for. The polypeptides will be different and lead to the production of a non-functional protein that could considerably alter the phenotype. One deleted base at the very start of a sequence could alter every triplet in the sequence. A deleted base near the end of the sequence is likely to have a smaller impact but can still have consequences (see Hint). An example of the effect of a deletion mutation is shown in Figure 1.

▲ **Figure 1** *Effects of the deletion of a DNA base on the amino acid sequence in the final polypeptide*

Other types of gene mutation

There are a number of other ways in which the base sequence of DNA may be changed. These include:

- **Addition of bases** – an extra base becomes inserted in the sequence. This usually has a similar effect to a base deletion in that there is usually a frame shift and the whole sequence of triplets becomes altered. The frame shift is to the right not to the left as it is when a base is deleted. If three extra bases are added, or any multiple of three bases, there will not be a frame shift. The resulting polypeptide will be different from the one produced from a non-mutant gene, but not to the same extent as if there was a frame shift.

- **Duplication of bases** – one or more bases are repeated. This produces a frame shift to the right.

- **Inversion of bases** – a group of bases become separated from the DNA sequence and rejoin at the same position but in the inverse order (back to front). The base sequence of this portion is therefore reversed and effects the amino acid sequence that results.

- **Translocation of bases** – a group of bases become separated from the DNA sequence on one chromosome and become inserted into the DNA sequence of a different chromosome. Translocations often have significant effects on gene expression leading to an abnormal phenotype. These effects include the development of certain forms of cancer and also reduced fertility.

Causes of mutations

Gene mutations can arise spontaneously during DNA replication. Spontaneous mutations are permanent changes in DNA that occur without any outside influence. Despite being random occurrences, mutations occur with a predictable frequency. The natural mutation rate varies from species to species, but is typically around one or two mutations per 100 000 genes per generation. This basic mutation rate can be increased by outside factors known as **mutagenic agents** or mutagens. These include the following:

- **High energy ionising radiation**, for example, α and β particles as well as short wavelength radiation such as X-rays and ultra violet light. These forms of radiation can disrupt the structure of DNA.

- **Chemicals** such as nitrogen dioxide may directly alter the structure of DNA or interfere with transcription. Benzopyrene, a consitituent of tobacco smoke, is a powerful mutagen that inactivates a tumour-suppressor gene TP53 leading to a cancer. We will learn more about this in Topic 20.5.

▲ **Figure 2** *This albino hedgehog is the result of a mutation that prevents the production of the pigment melanin*

Mutations have both costs and benefits. On the one hand they produce the genetic diversity necessary for natural selection and speciation (see Topics 18.4 and 18.5). On the other hand they are almost always harmful and produce an organism that is less well suited to its environment. Additionally, mutations that occur in body cells rather than in gametes leading to disruption of normal cellular activities, such as cell division, for example, cancer.

Summary questions

1 A translocation mutation is, in effect, a combination of two other different types of gene mutation. Deduce which two types of mutation these are and explain your answer.

2 A section of DNA has the following sequence – AGT TCT GAT CGC TG. State the type of mutation that has taken place in each of the following variants of this DNA.

   a AGT TCT GAT CCT G

   b AGT TCT TAG CGC TG

   c AGT TCT GAG CGC TG

   d AGT TCT GAT CGT CTG

3 Explain why the effects of a single additional base in a sequence of DNA bases may have:

   a a considerable effect on the polypeptide produced.

   b little effect on the polypeptide produced.

4 A mutation causes three bases in the DNA of a gene to become duplicated. Explain how the effects of this mutation might differ if the duplicated bases are consecutive rather than in three separate locations on the DNA molecule.

5 Suggest **two** reasons why the addition of a single base into a DNA sequence may not alter the amino acid sequence in the resultant polypeptide.

Mutagenic agents

Mutations can be induced by external influences called mutagenic agents. These cause damage in a number of ways.

- **Certain chemicals can remove groups from nucleotide bases.** Nitrous acid can remove an $-NH_2$ group from cytosine in DNA, changing it into uracil.

> 1 Suggest what the result of this change might be on the codons on a mRNA molecule that is transcribed from a section of DNA with the triplets GCA CTC ATC.

- **Other chemicals can add groups to nucleotides.** Benzopyrene is a chemical found in tobacco smoke. It adds a large group to guanine that makes it unable to pair with cytosine. When DNA polymerase reaches the affected guanine it inserts any of the other bases.

> 2 What type of mutation is caused by benzopyrene?

- **Ionising radiation**, such as X-rays, can produce highly reactive agents, called free radicals, in cells. These free radicals can alter the shape of bases in DNA so that DNA polymerase can no longer act on them.

> 3 Explain why DNA polymerase cannot act on DNA that has been damaged by X-rays.
> 4 State *one* genetic effect of DNA polymerase being unable to act on DNA.

Ultraviolet radiation from the Sun or tanning lamps affects thymine in DNA, causing it to form bonds with the nucleotides on either side of it. This seriously disrupts DNA replication.

Scientific research and experimentation has enabled us to identify potentially dangerous mutagenic agents. The effects of such agents are complex and the amount of harm they cause is often a matter of debate. Commercial organisations, such as the tobacco industry, manufacturers of sunbeds, and producers and retailers of sun-block lotions all have an interest in the research that is undertaken. They are more likely to fund research that may benefit their business than research that may harm it. It is therefore important that the results of any research are subjected to the scrutiny of other scientists from a wide variety of backgrounds, views, interests and organisations, in a process known as peer review.

This is usually achieved by publishing research findings in reputable scientific journals that have an extensive global readership. The conclusions and claims made by researchers and their sponsors can then be debated and the scientific community at large can test the claims by further experimentation. These claims then become accepted, modified or rejected, depending on the outcome of this further research.

Armed with all this scientific information, decision-makers such as governments and heads of business can take appropriate action that benefits society. Governments, for example, can introduce legislation that controls cigarette sales and smoking, and the use of sunbeds and sets a minimum age at which cigarettes or tanning treatments can be bought. The decisions are often not clear-cut however. X-rays, for example, can be harmful on one hand but are an invaluable diagnostic tool, with countless health benefits, on the other.

> 5 Leaders in business and government have to make decisions about the use of scientific discoveries. Who else, apart from research scientists, might influence the advice that these leaders give to the public on the use of sunbeds?

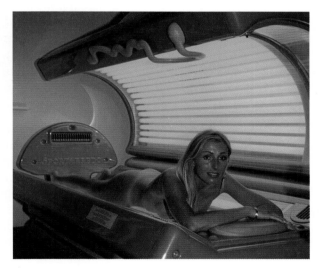

▲ **Figure 3** *Ultraviolet radiation from sunbeds has the potential to disrupt DNA replication*

Specification reference: 3.8.2.1

Learning objectives

→ State what totipotent cells are.

→ Explain how cells lose their totipotency and become specialised.

→ Describe cell differentiation and cell specialisation.

→ Describe the origins and types of stem cells.

→ Explain how pluripotent stem cells can be used to treat human disorders.

In multicellular organisms, cells are specialised to perform specific functions. The process by which each cell develops into a specialised structure suited to the role that it will carry out is known as **cell differentiation**. Let us investigate the process in more detail.

Cell differentiation and specialisation

Single-celled organisms perform all essential life functions inside the boundaries of a single cell. Although they perform all functions adequately, they cannot be totally efficient at all of them, because each function requires a different type of cellular structure, enzymes, and other proteins. One activity may be best carried out by a long, thin cell, while another might suit a spherically shaped cell. No one cell can provide the best conditions for all functions. The cells of multicellular organisms are each adapted in different ways to perform a particular role. In early development, an organism is made up of a tiny ball of cells, all of which are identical. As it matures, each cell takes on its own individual characteristics that adapt it to the function that it will perform when it is mature.

All the cells in an organism, such as a human, are derived by mitotic divisions of the fertilised egg (zygote). It follows that they all contain exactly the same genes. Every cell is therefore capable of making everything that the body can produce. A cell in the lining of the small intestine has the gene coding for insulin just as a β cell of the pancreas has the gene coding for maltase. So why do the cells of the small intestine produce maltase rather than insulin and β cells of the pancreas produce insulin rather than maltase? The answer is that, although all cells contain all genes, only certain genes are expressed (switched on) in any one cell at any one time.

Some genes are permanently expressed (switched on) in all cells. For example, the genes that code for essential chemicals, such as the enzymes involved in respiration, are expressed in all cells. Other genes permanently switched on in all cells include those coding for enzymes and other proteins involved in essential processes like transcription, translation, membrane synthesis, ribosomes and tRNA synthesis. Other genes are permanently not expressed (switched off), for example, the gene for insulin in cells lining the small intestine. Further genes are switched on and off as and when they are needed. In this chapter we shall look at how the expression of genes is controlled.

Differentiated cells differ from each other, often visibly so. This is mainly because they each produce different proteins. The proteins that a cell produces are coded for by the genes it possesses or, more accurately, by the genes that are expressed (switched on).

Totipotency

An organism develops from a single fertilised egg. A fertilised egg clearly has the ability to give rise to all types of cells. Cells such as fertilised eggs, which can mature into any body cell, are known as **totipotent cells.** The early cells that are derived from the fertilised egg are also totipotent. These later differentiate and become specialised for a particular function. For example, mesophyll cells become specialised for photosynthesis and muscle cells become specialised for contraction. This is because, during the process of cell specialisation, only some of the genes are expressed. This means that only part of the DNA of a cell is translated into proteins. The cell therefore only makes those proteins that it requires to carry out its specialised function. These proteins include those required for essential processes like respiration and membrane synthesis.. Although it is still capable of making all the other proteins, these are not needed and so it would be wasteful to produce them. In order to conserve energy and resources, a variety of stimuli (controlling factors) ensure the genes for these other proteins are not expressed. The ways in which genes are prevented from expressing themselves include:

- preventing transcription and so preventing the production of mRNA
- preventing translation.

Stem cells

If specialised cells still retain all the genes of the organism, can they still develop into any other cell? The answer, depends – there are no hard and fast rules. Xylem vessels, which transport water in plants, and red blood cells, which carry oxygen in animals, are so specialised that they lose their nuclei once they are mature. As the nucleus contains the genes, then clearly these cells cannot develop into other cells. In fact, specialisation is irreversible in most animal cells. Once cells have matured and specialised they can no longer develop into other cells. In mature mammals, only a few cells retain the ability to differentiate into other cells. These are called **stem cells**.

Stem cells are undifferentiated dividing cells that occur in adult animal tissues and need to be constantly replaced. They therefore have the ability to divide to form an identical copy of themselves in a process called self-renewal.

Stem cells originate from various sources in mammals:

- **Embryonic stem cells** come from embryos in the early stages of development. They can differentiate into any type of cell in the initial stages of development.
- **Umbilical cord blood stem cells** are derived from umbilical cord blood and are similar to adult stem cells.
- **Placental stem cells** are found in the placenta and develop into specific types of cells.
- **Adult stem cells**, despite their name, are found in the body tissues of the fetus through to the adult. They are specific to a particular tissue or organ within which they produce the cells to maintain and repair tissues throughout an organism's life.

Types of stem cells

There are number of different stem cells which are classified according to their ability to differentiate.

- **Totipotent stem cells** are found in the early embryo and can differentiate into any type of cell. Since all body cells are formed from a zygote, it follows that the zygote is totipotent. As the zygote

> **Hint**
>
> Differentiation results from differential gene expression.

▲ **Figure 1** *SEM of a three-day-old human embryo at the 16-cell stage on the tip of a pin. These cells are totipotent*

divides and matures, its cells develop into slightly more specialised cells called pluripotent stem cells.

- **Pluripotent stem cells** are found in embryos and can differentiate into almost any type of cell. Examples of pluripotent stem cells are embryonic stem cells and fetal stem cells.
- **Multipotent stem cells** are found in adults and can differentiate into a limited number of specialised cells. They usually develop into cells of a particular type, for example, stem cells in the bone marrow can produce any type of blood cell. Examples of multipotent cells are adult stem cells and umbilical cord blood stem cells.
- **Unipotent stem cells** can only differentiate into a single type of cell. They are derived from multipotent stem cells and are made in adult tissue.

Induced pluripotent stem cells

Induced pluripotent stem cells (iPS cells) are a type of pluripotent cell that is produced from unipotent stem cells. The unipotent cell may be almost any body cell. These body cells are then genetically altered in a laboratory to make them acquire the characteristics of embryonic stem cells which, as we have seen, are a type of pluripotent cell. To make the unipotent cell acquire the new characters involves inducing genes and transcriptional factors (see Topic 20.3) within the cell to express themselves. In other words to turn on genes that were otherwise turned off. The fact that these genes are capable of being reactivated shows that adult cells retain the same genetic information that was present in the embryo.

The iPS cells are very similar to embryonic stem cells in form and function. However, although they express some of the same genes that are usually expressed in embryonic stem cells, they are not exact duplicates of them. One feature of particular interest is that they are capable of self-renewal. This means that they can potentially divide indefinitely to provide a limitless supply. As such they could replace embryonic stem cells in medical research and treatment and so overcome many of the ethical issues surrounding the use of embryos in stem cell research.

Pluripotent cells in treating human disorders

There are many possible uses of pluripotent cells. The cells can be used to regrow tissues that have been damaged in some way, either by accident (e.g., skin grafts for serious burn damage) or as a result of disease (e.g., neuro-degenerative diseases, such as Parkinson's disease). Table 1 lists some of the potential uses of human cells produced from stem cells.

▼ **Table 1** *Potential uses of human cells produced from stem cells*

Type of cell	Disease that could be treated
Heart muscle cells	Heart damage, for example, as a result of a heart attack
Skeletal muscle cells	Muscular dystrophy
β cells of the pancreas	Type 1 diabetes
Nerve cells	Parkinson's disease, multiple sclerosis, strokes, Alzheimer's disease, paralysis due to spinal injury
Blood cells	Leukaemia, inherited blood diseases
Skin cells	Burns and wounds
Bone cells	Osteoporosis
Cartilage cells	Osteoarthritis
Retina cells of the eye	Macular degeneration

Human embryonic stem cells and the treatment of disease

Although there are a number of types of stem cell in the human body, it is the first few cells from the division of the fertilised egg that have the greatest potential to treat human diseases. As they come from the early stages of an embryo, they are called human embryonic stem cells. These cells can be grown *in vitro* and then induced to develop into a wide range of different human tissues. The process is illustrated in Figure 2.

At present, embryonic stem cell research is only allowed in the UK under licenced and specified conditions. These conditions include its use as a means of increasing knowledge about embryo development and serious diseases, including their treatment. Embryos used in this type of research are obtained from *in vitro* fertilisation. The process nevertheless presents a number of ethical issues.

One issue surrounds the argument as to whether a human embryo less than 14 days old should be afforded the same respect as a fetus or an adult person. Some people feel that using embryos in this way undermines our respect for human life and could progress to the use of fetuses, and even newborn babies, for research or the treatment of disease. They feel that it is a further move towards reproductive cloning and, even if this remains illegal in the UK, the information gained could be used to clone humans elsewhere. Others disagree, arguing that an embryo at such an early stage of development is just a ball of identical, undifferentiated cells, bearing no resemblance to a human being. They feel that the laws prohibiting cloning, in the UK and elsewhere, provide sufficient protection.

Supporters of human embryonic stem cell research contend that it is wrong to allow human suffering to continue when there is a possibility of alleviating it. They further argue that, since embryos are produced for other purposes, for example, fertility treatments, it makes no sense to destroy spare embryos that could be used in research. Opponents of embryonic stem cell research contend that it is wrong to use humans, including human embryos, as a means to an end, even if that end is the to alleviate human suffering.

However, human embryos are not the only source of stem cells. For example, they can be obtained from the bone marrow of adult humans. As long as a person gives consent, this source of stem cells raises no real ethical issues. At present, these cells have far more restricted medical applications but scientists hope, in time, to be able to make them behave more like embryonic stem cells.

> 1 Write *two* accounts, each of around 200 words, evaluating the case for and against the continued use of embryos for stem cell research.

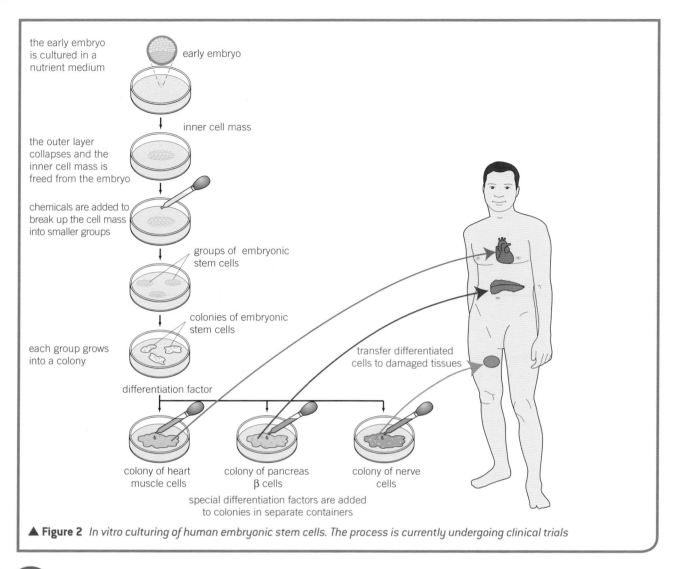

▲ **Figure 2** *In vitro culturing of human embryonic stem cells. The process is currently undergoing clinical trials*

Within Figure 2:
- the early embryo is cultured in a nutrient medium — early embryo
- inner cell mass
- the outer layer collapses and the inner cell mass is freed from the embryo
- chemicals are added to break up the cell mass into smaller groups
- groups of embryonic stem cells
- colonies of embryonic stem cells
- each group grows into a colony
- differentiation factor
- transfer differentiated cells to damaged tissues
- colony of heart muscle cells
- colony of pancreas β cells
- colony of nerve cells
- special differentiation factors are added to colonies in separate containers

✚ Growth of plant tissue cultures 🧪

Mature plants have many totipotent cells. Under the right conditions, many plant cells can develop into any other cell. For example, if we take a cell from the root of a carrot, place it in a suitable nutrient medium and give it certain chemical stimuli at the right time, we can develop a complete new carrot plant.

There are many factors that influence the growth of plant tissue cultures from totipotent cells. One group of factors consists of plant growth factors, which are chemicals involved in the growth and development of plant tissues. Plant growth factors have a number of features:

- They have a wide range of effects on plant tissues.
- The effects on a particular tissue depend upon the concentration of the growth factor.

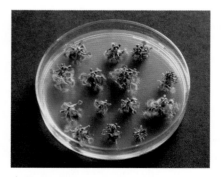

▲ **Figure 3** *Plants growing from tissue cultures in a Petri dish*

- The same concentration affects different tissues in different ways.

- The effect of one growth factor can be modified by the presence of another.

An experiment was carried out to investigate the effects on the development of a plant tissue culture of three growth factors: cytokinin, IAA, and 2,4-D. Samples of totipotent plant cells were grown on a basic growth medium in a series of test tubes. Each test tube contained a mixture of the three growth factors in different concentrations. After two weeks, the tubes were observed to see the effects of the growth factor mixtures on shoot and root growth. The results are shown in Table 2.

▼ **Table 2** *Effect of growth factors on shoot and root development*

Tube no.	Relative concentration of growth factors			Shoot development	Root development
	Cytokinin	IAA	2,4-D		
1	None	Low	None	Moderate	Little
2	Low	High	None	Extensive	Little
3	High	Low	None	Little	Moderate
4	None	High	High	Extensive	Extensive
5	None	None	None	Very little	Very little

1 Name the process by which the totipotent cells of the plant tissue culture change in appearance and develop into shoot or root cells.
2 State the general term used to describe growing living cultures like plants in a laboratory.
3 Plant tissues grown in culture often originate from a single initial cell and are therefore genetically identical.
   a State the name given to this group of genetically identical cells.
   b Name the process by which these genetically identical cells formed.
4 From Table 2, state which *two* growth factors together produced the greatest development of both shoots and roots.
5 Describe one piece of evidence from Table 2 that supports the view that the effects of one growth factor can be modified by another.

20.3 Regulation of transcription and translation

Learning objectives

→ Explain how oestrogen affects gene transcription.

→ State what small interfering RNA is.

→ Explain how small interfering RNA affects gene expression.

Specification reference: 3.8.8.2

Hint

Only target cells have the oestrogen receptor and so only these cells respond to the stimulus of oestrogen.

In Topic 20.2, we saw how cell specialisation is the result of the selective expression of certain genes out of the full complement found in every cell. Let us now investigate some ways in which cells control which genes are expressed.

The effect of oestrogen on gene transcription

In Topic 16.3 we learned how hormones, such as adrenaline, operate by using a second messenger. Here we will examine another mechanism, which is used by steroid hormones such as oestrogen. Before looking at how oestrogen operates, let us consider the general principles involved in controlling the expression of a gene by controlling transcription.

- For transcription to begin the gene is switched on by specific molecules that move from the cytoplasm into the nucleus. These molecules are called **transcriptional factors**.
- Each transcriptional factor has a site that binds to a specific base sequence of the DNA in the nucleus.
- When it binds, it causes this region of DNA to begin the process of transcription.
- Messenger RNA (mRNA) is produced and the information it carries is then translated into a polypeptide.
- When a gene is not being expressed (i.e., is switched off), the site on the transcriptional factor that binds to DNA is not active.
- As the site on the transcriptional factor binding to DNA is inactive it cannot cause transcription and polypeptide synthesis.

Hormones like oestrogen can switch on a gene and thus start transcription by combining with a receptor site on the transcriptional factor. This activates the DNA binding site by causing it to change shape. The process is illustrated in Figure 1 and operates as follows:

- Oestrogen is a lipid-soluble molecule and therefore diffuses easily through the phospholipid portion of cell-surface membranes (Figure 1, stage 1).
- Once inside the cytoplasm of a cell, oestrogen binds with a site on a receptor molecule of the transcriptional factor. The shape of this site and the shape of the oestrogen molecule complement one another (Figure 1, stage 2).
- By binding with the site, the oestrogen changes the shape of the DNA binding site on the transcriptional factor, which can now bind to DNA (it is activated) (Figure 1, stage 3).
- The transcriptional factor can now enter the nucleus through a nuclear pore and bind to specific base sequences on DNA (Figure 1, stage 4).
- The combination of the transcriptional factor with DNA stimulates transcription of the gene that makes up the portion of DNA (Figure 1, stage 4).

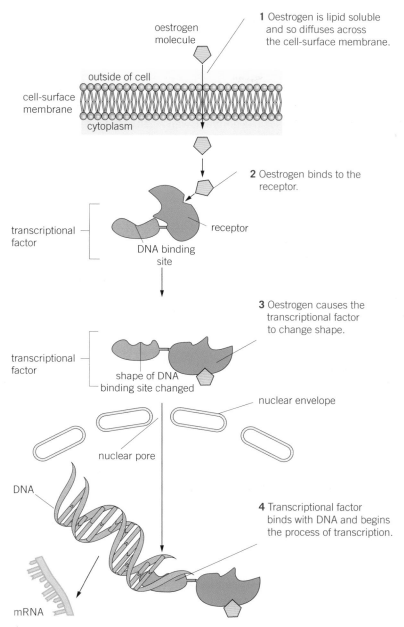

1 Oestrogen is lipid soluble and so diffuses across the cell-surface membrane.

oestrogen molecule

outside of cell

cell-surface membrane

cytoplasm

2 Oestrogen binds to the receptor.

receptor

transcriptional factor

DNA binding site

3 Oestrogen causes the transcriptional factor to change shape.

transcriptional factor

shape of DNA binding site changed

nuclear envelope

nuclear pore

DNA

4 Transcriptional factor binds with DNA and begins the process of transcription.

mRNA

▲ **Figure 1** *The effect of oestrogen on gene transcription*

Synoptic link

The attachment of oestrogen to a receptor causes changes in the shape of the receptor in the same way as the attachment of a non-competitive inhibitor to an enzyme molecule changes its shape and also its active site. This involves the same basic mechanism, which is described in Topic 1.9, Enzyme inhibition, which would be worthwhile revising.

Summary questions

1 What is the role of a transcriptional factor?
2 Describe how oestrogen stimulates the expression of a gene.

Gene expression in haemoglobin

A haemoglobin molecule is made up of four polypeptide chains each known as a globulin. In adult humans two of the polypeptides in a haemoglobin molecule are alpha-globulin and two are beta-globulin. In other words, 50% of the total globulin in all haemoglobin is alpha and 50% is beta. In a human fetus, however, the haemoglobin is different, with much of the beta-globulin being replaced by a third type, gamma-globulin. Fetal haemoglobin has a greater affinity for oxygen than adult haemoglobin. The changes in the production of the three types of globulin during early human development are shown in Figure 2.

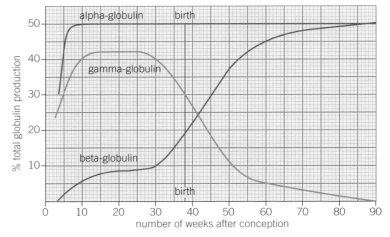

▲ **Figure 2** *Percentage total globulin production during early human development*

Humans have genes that code for the production of all three types of globulin. The production of the different haemoglobins depends upon which gene is expressed. The expression of these genes changes at different times during development.

1   Suggest an advantage of fetal haemoglobin having a greater affinity for oxygen than adult haemoglobin.
2   At birth, what percentage of the total globulin production is of each globulin type?
3   Describe the changes in gene expression that occur at 30 weeks.
4   Outline *two* possible explanations for the change in the expression of the gene for gamma-globulin after 30 weeks.
5   Sickle cell disease is the result of a mutant form of haemoglobin. In Saudia Arabia and India, some individuals have high levels of fetal haemoglobin in their blood, even as adults. Where these individuals have sickle cell disease, their symptoms are much reduced. Suggest how controlling the expression of the genes for globulin might provide a therapy for sickle cell disease.

20.4 Epigenetic control of gene expression

Ever since James Watson and Francis Crick proposed the double-helix model in 1953, it has been taken as fact that DNA possesses the instructions for making all parts of an organism. In recent years, however, we have come to realise that DNA is only part of the story of heredity. It is accepted that while genes determine the features of an organism, the environment can influence the expression of these genes (Topic 18.2). However, the changes they cause to the phenotype were thought not to be inherited by the offspring. We now believe that environmental factors can cause heritable changes in gene function without changing the base sequence of DNA. This process is known as **epigenetics**.

Epigenetics

Epigenetics is a relatively new scientific field that provides explanations as to how environmental influences such as diet, stress, toxins, etc. can subtly alter the genetic inheritance of an organism's offspring. It is helping to explain, and maybe cure, illnesses ranging from autism to cancer. It is even causing scientists to look again at previously discredited theories of evolution that suggested characteristics acquired during an organism's life could be passed on to future generations (Lamarckism).

The epigenome

We learned in Topic 8.2 that DNA is wrapped around proteins called histones. We now know that both the DNA and histones are covered in chemicals, sometimes called tags. These chemical tags form a second layer known as the **epigenome**. The epigenome determines the shape of the DNA-histone complex. For example it keeps genes that are inactive in a tightly packed arrangement and therefore ensures that they cannot be read (it keeps them switched off). This is known as epigenetic silencing. By contrast, it unwraps active genes so that the DNA is exposed and can easily be transcribed (switches them on).

We know that the DNA code is fixed. The epigenome, however, is flexible. This is because its chemical tags respond to environmental changes. Factors like diet and stress can cause the chemical tags to adjust the wrapping and unwrapping of the DNA and so switch genes on and off.

The epigenome of a cell is the accumulation of the signals it has received during its lifetime and it therefore acts like a cellular memory. In early development, the signals come from within the cells of the fetus and the nutrition provided by the mother is important in shaping the epigenome at this stage. After birth, and throughout life, environmental factors affect the epigenome, although signals from within the body, for example, hormones, also influence it. These factors cause the epigenome to activate or inhibit specific sets of genes.

Learning objectives

→ State what is meant by epigenetics.

→ Describe the nature of the epigenome.

→ Explain the effect of epigenetic factors on DNA and histones.

→ Explain the effects of decreased acetylation of histones.

→ Explain the effects of increased methylation of DNA.

Specification reference 3.8.2.2

▲ **Figure 1** *What we eat not only affects us, it may affect our children too*

The environmental signal stimulates proteins to carry its message inside the cell from where it is passed by a series of other proteins into the nucleus. Here the message passes to a specific protein which can be attached to a specific sequence of bases on the DNA. Once attached the protein has two possible effects. It can change:

- acetylation of histones leading to the activation or inhibition a gene
- methylation of DNA by attracting enzymes that can add or remove methyl groups.

Before we look in a little more detail at how each process works it will be helpful to look more closely at the DNA–histone complex.

The DNA–histone complex (chromatin)

Where the association of histones with DNA is weak, the DNA–histone complex is less condensed (loosely packed). In this condition the DNA is accessible by transcription factors, which can initiate production of mRNA, that is, can switch the gene on.

Where this association is stronger, the DNA–histone complex is more condensed (tightly packed). In this condition the DNA is not accessible by transcription factors, which therefore cannot initiate production of mRNA, that is, the gene is switched off.

Condensation of the DNA–histone complex therefore inhibits transcription. It can be brought about by decreased acetylation of the histones or by methylation of DNA. Let us turn our attention to these two processes and how they inhibit transcription.

Decreased acetylation of associated histones

Acetylation is the process whereby an acetyl group is transferred to a molecule. In this case the group donating the acetyl group is acetylcoenzyme A which you may remember from the link reaction in respiration (Topic 12.2). Deacetylation is the reverse reaction where an acetyl group is removed from a molecule.

Decreased acetylation increases the positive charges on histones and therefore increases their attraction to the phosphate groups of DNA. The association between DNA and histones is stronger and the DNA is not accessible to transcription factors. These transcription factors cannot initiate mRNA production from DNA. In other words, the gene is switched off.

Increased methylation of DNA

Methylation is the addition of a methyl group (CH_3) to a molecule. In this case the methyl group is added to the cytosine bases of DNA. Methylation normally inhibits the transcription of genes in two ways:

- preventing the binding of transcriptional factors to the DNA
- attracting proteins that condense the DNA–histone complex (by inducing deacetylation of the histones) making the DNA inaccessible to transcription factors.

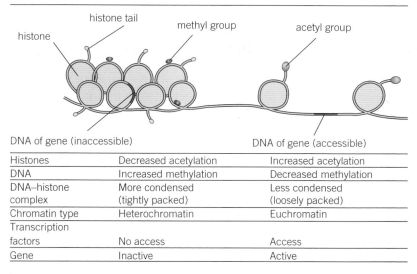

Histones	Decreased acetylation	Increased acetylation
DNA	Increased methylation	Decreased methylation
DNA–histone complex	More condensed (tightly packed)	Less condensed (loosely packed)
Chromatin type	Heterochromatin	Euchromatin
Transcription factors	No access	Access
Gene	Inactive	Active

▲ **Figure 2** *Effects of epigenetic factors such as methyl and acetyl groups on the DNA–histone complex*

Epigenetics and inheritance

Unexpected though it might be, there is now little doubt that epigenetic inheritance takes place. Experiments on rats have shown that female offspring who received good care when young, respond better to stress in later life and themselves nurture their offspring better. Female offspring receiving low-quality care, nurture their offspring less well. Good maternal behaviour in rats transmits epigenetic information onto their offspring's DNA without passing through an egg or sperm.

In humans, when a mother has a condition known as gestational diabetes, the fetus is exposed to high concentrations of glucose. These high glucose concentrations cause epigenetic changes in the daughter's DNA, increasing the likelihood that she will develop gestational diabetes herself.

It is thought that in sperm and eggs during the earliest stages of development a specialised cellular mechanism searches the genome and erases its epigenetic tags in order to return the cells to a genetic 'clean slate'. However, a few epigenetic tags escape this process and pass unchanged from parent to offspring.

Epigenetics and disease

Epigenetic changes are part of normal development and health but they can also be responsible for certain diseases. Altering any of the epigenetic processes can cause abnormal activation or silencing of genes. Such alterations have been associated with a number of diseases including cancer. In some cases the activation of a normally inactive gene can cause cancer, in other cases it is the inactivation of a normally active gene that gives rise to the disease.

In 1983, researchers found that diseased tissue taken from patients with colorectal cancer had less DNA methylation than normal tissue from the same patients. As we saw earlier, increased DNA methylation normally inhibits transcription (switches off genes). This means that these patients with less DNA methylation would have higher than normal gene activity – more genes were turned on.

It is known that there are specific sections of DNA (ones near regions called promoter regions) that have no methylation in normal cells. However, in cancer cells these regions become highly methylated causing genes that should be active to switch off. This abnormality happens early in the development of cancer.

We have seen that epigenetic changes do not alter the sequence of bases in DNA. They can, however, increase the incidence of mutations. Some active genes normally help repair DNA and so prevent cancers. In people with various types of inherited cancer, it is found that increased methylation of these genes has led to these protective genes being switched off. As a result, damaged base sequences in DNA are not repaired and so can lead to cancer.

Treating diseases with epigenetic therapy

As we have seen, many diseases, such as cancer, are triggered by of epigenetic changes that cause certain genes to be activated or silenced. It is therefore logical to try to use epigenetic treatments to counteract these changes. These treatments use drugs to inhibit certain enzymes involved in either histone acetylation or DNA methylation. For example, drugs that inhibit enzymes that cause DNA methylation can reactivate genes that have been silenced. Epigenetic therapy must be specifically targeted on cancer cells. If the drugs were to affect normal cells they could activate gene transcription and make them cancerous, so causing the very disorder they were designed to cure.

Another use of epigenetics in disease treatment has been the development of diagnostic tests that help to detect the early stages of diseases such as cancer, brain disorders and arthritis. These tests can identify the level of DNA methylation and histone acetylation at an early stage of disease. This allows those with these diseases to seek early treatment and so have a better chance of cure.

> **Hint**
>
> Double-stranded RNA can be made by in vitro transcription of a DNA template using the polymerase chain reaction. (Topic 21.3)

The effect of RNA interference on gene expression

In **eukaryotes** and some **prokaryotes** the translation of mRNA produced by a gene can be inhibited by breaking mRNA down before its coded information can be translated into a polypeptide. One type of small RNA molecule that may be involved is small interfering RNA (siRNA). The mechanism involving small double-stranded sections of siRNA operates as follows.

- An enzyme cuts large double-stranded molecules of RNA into smaller sections called small interfering RNA (siRNA) (Figure 3, stage 1).
- One of the two siRNA strands combines with an enzyme (Figure 3, stage 2).

- The siRNA molecule guides the enzyme to a messenger RNA molecule by pairing up its bases with the complementary ones on a section of the mRNA molecule (Figure 3, stage 3).
- Once in position, the enzyme cuts the mRNA into smaller sections (Figure 3, stage 4).
- The mRNA is no longer capable of being translated into a polypeptide.
- This means that the gene has not been expressed, that is, it has been blocked.

Summary questions

1 Explain what is meant by epigenetics.

2 Name two mechanisms by which changes in the environment can inhibit transcription.

3 One of the two strands of siRNA combines with an enzyme and guides it to an mRNA molecule which it then cuts. Explain why the mRNA is unlikely to be cut if the other siRNA strand combines with the enzyme.

4 Suggest how siRNA could be used to:
   a identify the role of genes in a biological pathway
   b to prevent certain diseases.

5 The enzyme histone deacetylase (HDAC) removes acetyl groups from histones. Suggest what the effect of this enzyme would be on:
   a the arrangement of chromatin (DNA–histone complex)
   b transcription.

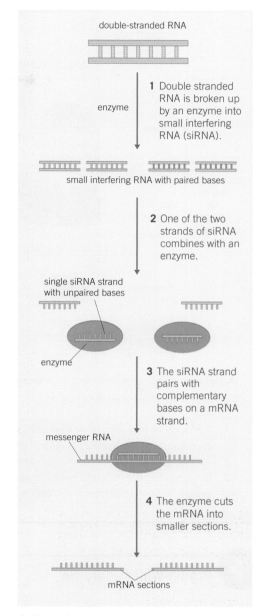

▲ **Figure 3** *The effect of siRNA on gene expression*

 Nature versus nurture

A small Californian plant, *Potentilla glandulosa*, has a number of genetic forms, each adapted to growing at different altitudes. Experiments were carried out as follows:

- plants of *Potentilla* were collected from three altitudes – high, medium, and low

- one plant from each location was split into three cuttings, each of which therefore had an identical genotype

- one of the cuttings from each location was grown at each altitude (high, medium, and low).

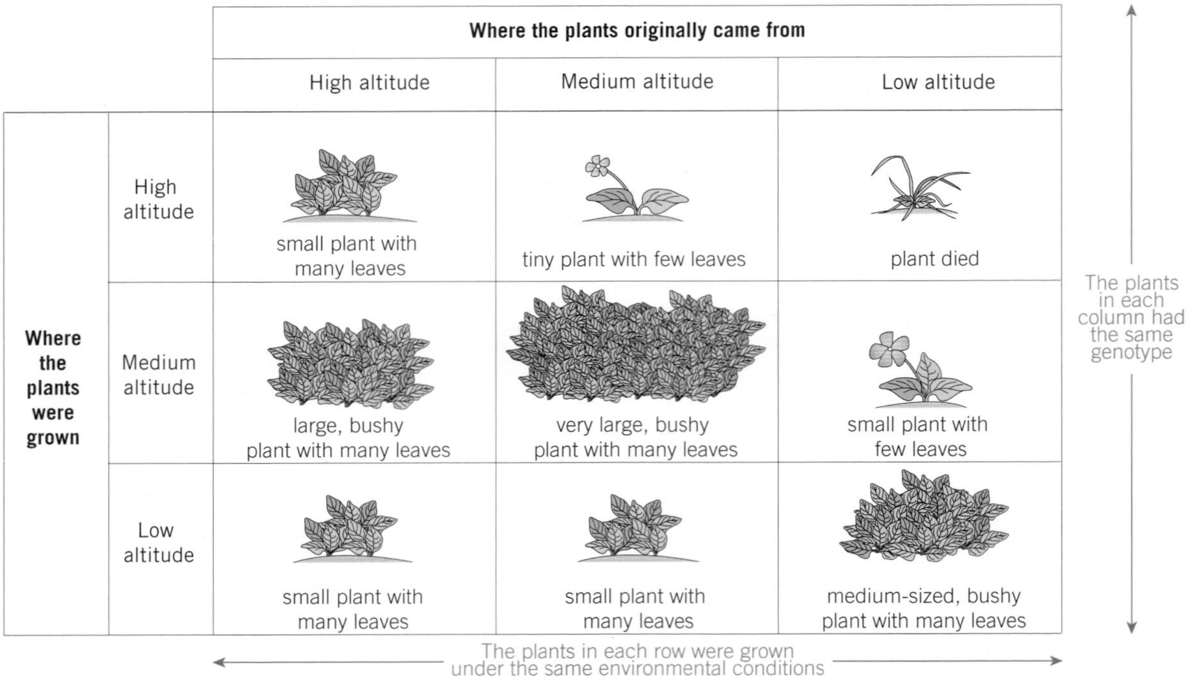

		Where the plants originally came from		
		High altitude	Medium altitude	Low altitude
Where the plants were grown	High altitude	small plant with many leaves	tiny plant with few leaves	plant died
	Medium altitude	large, bushy plant with many leaves	very large, bushy plant with many leaves	small plant with few leaves
	Low altitude	small plant with many leaves	small plant with many leaves	medium-sized, bushy plant with many leaves

The plants in each column had the same genotype

The plants in each row were grown under the same environmental conditions

▲ **Figure 4** *Effect of environment on phenotype – growing genetically identical* Potentilla glandulosa *at different altitudes*

The results are illustrated in Figure 4.

1 Deduce whether genetic or environmental factors have the greatest influence on the phenotype of *Potentilla glandulosa* as illustrated in Figure 4.
2 Justify your answer to question 1 using evidence from Figure 4.
3 Suggest why the differences in phenotype between the three genetically identical plants from low altitude are greater than the differences between the three genetically identical plants from high altitude.

Prader-Willi syndrome

Epigenetic inheritance is thought to be involved in a rare genetic disease called Prader-Willi syndrome. It is the result of seven genes on chromosome 15 being deleted.

In most people, only one copy of the genes (usually from the father) is expressed while the other copy of the genes (from the mother) is silenced through epigenetic inheritance. This means that most people have one working and one epigenetically-silenced set of these genes. However, if a mutation on chromosome 15 of the father deletes the relevant seven genes, any offspring produced will have one set of non-working genes and one set of epigenetically-silenced genes. These individuals will inherit Prader-Willi syndrome.

1 Explain why most offspring do not develop Prader-Willi syndrome despite inheriting epigenetically-silenced genes from their mother.
2 People with Prader-Willi syndrome are often infertile. Suggest how a deletion mutation to chromosome 15 inherited from the father might result in infertility in a person with Prader-Willi syndrome.

The word cancer (Latin for crab), was first used by Hippocrates 2400 years ago. He saw a similarity between the swollen veins radiating from a breast tumour and the legs of a crab. Cancer is a group of diseases caused by damage to the genes that regulate mitosis and the cell cycle (Topic 3.8). This leads to unrestrained growth of cells. As a consequence, a group of abnormal cells, called a tumour, develops and constantly expands in size. Cancer is a common and destructive disease. That being said, cancer is to some extent avoidable and, if diagnosed early enough, successfully treatable.

Types of tumour

Not all tumours are cancerous. Those that are cancerous are called **malignant** while those that are non-cancerous are called **benign**. The main characteristics of benign and malignant tumours are compared in Table 1.

▼ **Table 1** *A comparison of benign and malignant tumours*

Benign tumours	Malignant tumours
Can grow to a large size	Can also grow to a large size
Grow very slowly	Grow rapidly
The cell nucleus has a relatively normal appearance	The cell nucleus is often larger and appears darker due to an abundance of DNA
Cells are often well differentiated (specialised)	Cells become de-differentiated (unspecialised)
Cells produce adhesion molecules that make them stick together and so they remain within the tissue from which they arise = primary tumours	Cells do not produce adhesion molecules and so they tend to spread to other regions of the body, a process called **metastasis**, forming secondary tumours
Tumours are surrounded by a capsule of dense tissue and so remain as a compact structure	Tumours are not surrounded by a capsule and so can grow finger-like projections into the surrounding tissue
Much less likely to be life-threatening but can disrupt functioning of a vital organ	More likely to be life-threatening, as abnormal tumour tissue replaces normal tissue
Tend to have localised effects on the body	Often have systemic (whole body) effects such as weight loss and fatigue
Can usually be removed by surgery alone	Removal usually involves radiotherapy and/or chemotherapy as well as surgery
Rarely reoccur after treatment	More frequently reoccur after treatment

Learning objectives

→ Distinguish between benign and malignant tumours.

→ Explain the role of oncogenes and tumour suppressor genes in the development of tumours.

→ Explain the effects of abnormal methylation of tumour suppressor genes and oncogenes.

→ Explain how increased oestrogen levels can cause breast cancer.

Specification reference 3.8.2.3

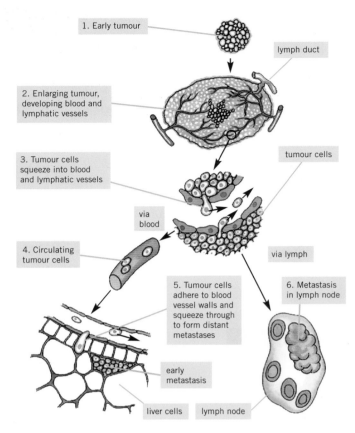

1. Early tumour

lymph duct

2. Enlarging tumour, developing blood and lymphatic vessels

tumour cells

3. Tumour cells squeeze into blood and lymphatic vessels

via blood

4. Circulating tumour cells

via lymph

5. Tumour cells adhere to blood vessel walls and squeeze through to form distant metastases

6. Metastasis in lymph node

early metastasis

liver cells   lymph node

▲ **Figure 1** *A primary tumour and its development and spread into a secondary tumour*

The development of secondary tumours from a primary tumour is illustrated in Figure 1.

Cancer and the genetic control of cell division

DNA analysis of tumours has shown that, in general, cancer cells are derived from a single mutant cell. The initial mutation causes uncontrolled mitosis in this cell. Later, a further mutation in one of the descendant cells leads to other changes that cause subsequent cells to be different from normal in growth and appearance. The two main types of genes that play a role in cancer are tumour suppressor genes and oncogenes.

Oncogenes

Most oncogenes are mutations of proto-oncogenes. As we discovered in Topic 3.8, proto-oncogenes stimulate a cell to divide when growth factors attach to a protein receptor on its cell-surface membrane. This then activates genes that cause DNA to replicate and the cell to divide. If a proto-oncogene mutates into an oncogene it can become permanently activated (switched on) for two reasons:

- The receptor protein on the cell-surface membrane can be permanently activated, so that cell division is switched on even in the absence of growth factors.
- The oncogene may code for a growth factor that is then produced in excessive amounts, again stimulating excessive cell division.

The result is that cells divide too rapidly and out of control, and a tumour or cancer, develops. A few cancers are caused by inherited mutations of proto-oncogenes that cause the oncogene to be activated but most cancer-causing mutations involving oncogenes are acquired, not inherited.

Tumour suppressor genes

Tumour suppressor genes slow down cell division, repair mistakes in DNA, and 'tell' cells when to die – a process called apoptosis (programmed cell death). They therefore have the opposite role from proto-oncogenes. As its name suggests, a normal tumour suppressor gene maintains normal rates of cell division and so prevents the formation of tumours. If a tumour suppressor gene becomes mutated it is inactivated (switched off). As a result, it stops inhibiting cell division and cells can grow out of control. The mutated cells that are formed are usually structurally and functionally different from normal cells. While most of these die, those that survive can make clones of themselves and form tumours. There are a number of forms of tumour suppressor genes including *TP53*, *BRCA1* and *BRCA2*.

Some cancers are caused by inherited mutations of tumour suppressor genes but most are acquired, not inherited. For example, more than half of human cancers display abnormalities of the *TP53* gene (which codes for the p53 protein). Acquired mutations of the *TP53* gene occur in many cancers, including lung and breast cancer. The p53 protein is involved in the process of apoptosis (programmed cell death). This process is activated when a cell is unable to repair DNA. If the gene for p53 is not functioning correctly, cells with damaged DNA continue to divide leading to cancer.

An important difference between oncogenes and tumour suppressor genes is that while oncogenes cause cancer as a result of the activation of proto-oncogenes, tumour suppressor genes cause cancer when they are inactivated.

Hint

A tumour suppressor gene's role in cell division is like the brake pedal on a car – it prevents it from going too quickly. When it mutates, it is as if the brake pedal doesn't work and cell division takes place more rapidly.

Abnormal methylation of tumour suppressor genes

In Topic 20.4 we learnt about the significance of methylation of DNA. It is now known that abnormal DNA methylation is common in the development of a variety of tumours. The most common abnormality is hypermethylation (increased methylation). The process by which hypermethylation may lead to cancer is as follows:

* Hypermethylation occurs in a specific region (promoter region) of tumour suppressor genes.
* This leads to the tumour suppressor gene being inactivated.
* As a result, transcription of the promoter regions of tumour suppressor genes is inhibited.
* The tumour suppressor gene is therefore silenced (switched off).
* As the tumour suppressor gene normally slows the rate of cell division, its inactivation leads to increased cell division and the formation of a tumour.

Abnormal methylation of this type is thought to occur in a tumour suppressor gene known as BRCA1 and leads to the development of breast cancer.

Another form of abnormal methylation is hypomethylation (reduced methylation). This has been found to occur in oncogenes where it leads to their activation and hence the formation of tumours.

Oestrogen concentrations and breast cancer

Oestrogens play a central role in regulating the menstrual cycle in women. It is known that after the menopause, a woman's risk of developing breast cancer increases. This is thought to be due to increased oestrogen concentrations. At first this seemed paradoxical because the production of oestrogens from the ovaries diminishes after the menopause. However, the fat cells of the breasts tend to produce more oestrogens after the menopause. These locally produced oestrogens appear to trigger breast cancer in postmenopausal women. Once a tumour has developed, it further increases oestrogen concentration which therefore leads to increased development of the tumour. It also appears that white blood cells that are drawn to the tumour increase oestrogen production. This leads to even greater development of the tumour.

How then can oestrogen cause a tumour to develop? We saw in Topic 20.3 the mechanism by which oestrogen effectively activates a gene by releasing an inhibitor molecule that prevents transcription. If the gene that oestrogen acts on is one that controls cell division and growth, then it will be activated and its continued division could produce a tumour. It is known, for example, that oestrogen causes proto-oncogenes of cells in breast tissue to develop into oncogenes. This leads to the development of a tumour (breast cancer).

Summary questions

1 Describe a process by which oestrogen might cause breast cancer in post-menopausal women.

2 Explain why the activation of a proto-oncogene can cause the development of a tumour while it requires deactivation of a tumour suppressor gene to do so.

3 Suggest two reasons why the surgical removal of a benign tumour is usually sufficient treatment to prevent the tumour growing again.

4 Suggest why the surgical removal of a malignant tumour requires follow-up treatments such as chemotherapy and radiotherapy.

5 The enzyme histone deacetylase (HDAC) removes acetyl groups from histones. Phenylbutyric acid is an inhibitor of the enzyme HDAC. Suggest how phenylbutyric acid might be used to treat cancer. Explain your answer.

Risk factors and cancer

Cancer is not a single disease and, likewise, does not have a single cause. Some causal factors are beyond our individual control, for example age and genetic factors. Others are lifestyle factors and therefore within our power to change.

We can do nothing about our genes or our age but our lifestyle can expose us to environmental and **carcinogenic** factors that put us at risk of contracting cancer. It is thought that about half the people who are diagnosed with cancer in the UK could have avoided getting the disease if they had changed their lifestyle. The specific lifestyle factors that contribute to cancer include:

- **smoking.** Not only smokers are in danger, those who passively breathe tobacco smoke also have an increased risk of getting cancer.

- **diet.** What we eat and drink affects our risk of contracting cancer. There is strong evidence that a low-fat, high-fibre diet, rich in fruit and vegetables, reduces the risk.

- **obesity.** Being overweight increases the risk of cancer.

- **physical activity.** People who take regular exercise are at lower risk from some cancers than those who take little or no exercise.

- **sunlight.** The more someone is exposed to sunlight or light from sunbeds, the greater is the risk of skin cancer.

Smoking and cancer

Tobacco was first introduced to Britain in the 16th century. Initially, only men smoked, but women took up smoking in the 1920s. By 1945, the equivalent of 12 cigarettes a day for every British male was being smoked. At the time the public regarded smoking as a harmless pleasure. Doctors, however, were alarmed by a phenomenal increase in deaths from lung cancer. At a 1947 conference, a number of scientists suggested tobacco smoke as a possible cause of the increase.

1 Scientists need to look at all possible explanations for the correlations that they have recognised. Suggest another possible cause of lung cancer, other than smoking, that they might have investigated.

Epidemiologists collect data on diseases and then look for correlations between these diseases and various factors in the lives of those who have them. The world's longest-running survey of smoking began in the UK in 1951. This survey, and others elsewhere in the world, has revealed a number of statistically significant correlations about smokers.

- A regular smoker is three times more likely to die prematurely than a non-smoker.
- The more cigarettes smoked per day, the earlier, on average, a smoker dies.
- Smokers who give up the habit improve their life expectancy compared with those who continue to smoke.
- Long-term smokers are more likely to die early as a result of smoking.
- The incidence of pulmonary disease increases with the number of cigarettes smoked.
- Smokers make up 98% of emphysema sufferers.

Data like those in Figure 2 were used to help establish a correlation between disease and smoking.

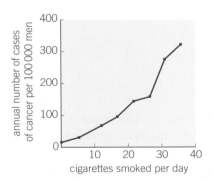

▲ **Figure 2** *Annual number of cases of lung cancer per 100 000 men in the USA correlated to daily consumption of cigarettes*

2 Describe the correlation shown by the data in Figure 2.

Epidemiological statistics show correlations between lung cancer and smoking. These include:

- A man smoking 25 cigarettes a day is 25 times more likely to die of lung cancer than a non-smoker.
- The longer a person smokes, the greater the risk of developing lung cancer. Smoking 20 cigarettes a day for 40 years increases the risk of lung cancer eight times more than smoking 40 cigarettes a day for 20 years.
- When a person stops smoking, the risk of developing lung cancer decreases and approaches that of a non-smoker after around 10–15 years (depending on age and amount of tobacco consumed).
- The death rate from lung cancer is 18 times greater in a smoker than in a non-smoker.

Cigarette manufacturers and some smokers argued that these epidemiological correlations were coincidental.

3 Many of the data linking smoking to lung cancer were collected from very large samples of the population. Suggest why this weakens the argument that the link is coincidental.

4 State whether the data provide evidence of a causal link between lung cancer and smoking. Explain your answer.

Experimental evidence linking smoking to disease

Scientists carried out experiments in the 1960s in which dogs were made to inhale cigarette smoke. The smoke was either inhaled directly or first passed through a filter tip. Those dogs that inhaled the filtered smoke remained generally healthy. Those inhaling unfiltered smoke developed pulmonary disease and early signs of lung cancer. Scientists then carried out a further series of experiments that allowed them to formulate a new hypothesis from each result, which they could then test experimentally.

- Machines were used to simulate the action of smoking and to collect the harmful constituents that accumulated in the filters.
- These were then analysed chemically and each constituent was tested in the laboratory for its ability to damage epithelial cells and mutate the genes they contain. This was done by adding tar to the skin of mice or to cells that had been grown in culture.
- As a result of such tests it was shown that the tar found in cigarette smoke contained **carcinogens**.
- The constituent chemicals of the tar were each tested and one, benzopyrene (BP), was shown to mutate DNA.
- The scientists still had to demonstrate precisely *how* it caused cancer. They carried out experiments which

showed that BP is absorbed by epithelial cells and converted to a derivative. This then binds with a **gene** and mutates it.

- Another experiment showed that this **mutation** led to uncontrolled cell division of epithelial cells and hence the growth of a **tumour**.
- Even this was not proof. In further experiments, scientists showed that the mutations of the gene in a cancer cell occurred at three specific points on the DNA. When the derivative of BP from tobacco smoke was used to mutate the gene, it caused changes to the DNA at precisely the same points.

> 5 Identify the key evidence that smoking is a cause of lung cancer?

The evidence was now conclusive. Smoking tobacco could cause lung cancer. This is not to say that it always does, but simply that there is an increased risk — it is about probabilities not certainties.

These experiments convinced the public of the health risks of smoking and led to reduced use of tobacco in the UK. This changed view in turn persuaded the government to take measures designed to reduce smoking. These included — progressively raising taxes on tobacco, banning tobacco advertising, placing health warnings on tobacco products, banning smoking in work and public places, including bars, pubs and clubs.

> 6 'My father smoked 30 cigarettes a day and lived to be 95.' This type of argument is sometimes used to suggest that smoking is not harmful. Explain why scientists do not accept this reasoning.

▲ **Figure 3** *Smoking these 20 cigarettes would, on average, reduce your life expectancy by $3\frac{1}{2}$ hours*

Cancer – the 'two hit' hypothesis

We have learnt that tumours can develop as a result of a mutation of proto-oncogenes that causes cells to divide more rapidly than normal. Tumours can also develop by a mutation of tumour suppressor genes that prevents them from inhibiting cell division.

It only takes a single mutated allele to activate proto-oncogenes but it takes a mutation of both alleles to inactivate tumour suppressor genes (two-hits). As natural mutation rates are slow, it takes a considerable time for both tumour suppressor alleles to mutate. This explains why the risk of many cancers increases as one gets older. It is thought that some people are born with one mutated allele. These people are at greater risk of cancer as they need only one further mutation, rather than two, to develop the disease. This explains why certain cancers carry an inherited increased risk.

1 Explain why a doctor may enquire about a patient's family medical history before deciding on using X-ray analysis for a condition other than cancer.
2 Suggest a reason why a single mutant allele of a proto-oncogene can cause cancer, but it requires two mutant alleles of the tumour suppressor gene to do so.
3 One experimental treatment for cancer involves introducing tumour suppressor genes into rapidly dividing cells in order to arrest tumour growth. Explain how this treatment might work.
4 Another experimental treatment is the development of an antibiotic drug that will destroy certain protein receptors on membranes of cancer cells. Explain how this treatment might be effective.

Projects to determine the entire DNA nucleotide base sequence of a wide range of organisms, including humans, have taken place over the past few decades. The idea has been to map the DNA base sequences that make up the genes of the organism and then to map these genes on the individual chromosomes of that organism. In this way a complete map of all the genetic material in an organism (the **genome**) is obtained.

Sequencing genomes

When you consider that the human genome consists of over 3 billion base pairs organised into around 20 000 genes, sequencing every one of those bases is a mammoth task and yet it took just 13 years to complete. This would have been impossible without the use of bioinformatics. Bioinformatics is the science of collecting and analysing complex biological data such as genetic codes. It uses computers to read, store, and organise biological data at a much faster rate than previously. It also utilises algorithms (mathematical formulae) to analyse and interpret biological data.

DNA sequencing

Determining the complete DNA base sequence of an organism uses the technique of whole-genome shotgun (WGS) sequencing. This involves researchers cutting the DNA into many small, easily sequenced sections and then using computer algorithms to align overlapping segments to assemble the entire genome. Sequencing methods such as these are continuously updated which, along with the increased automation of the processes involved, have led to extremely rapid sequencing of whole genomes.

The medical advances that have been made as a result of sequencing the human genome are many. For example, over 1.4 million single nucleotide polymorphisms (SNPs) have been found in the human genome. SNPs are single-base variations in the genome that are associated with disease and other disorders. Figure 1 shows some of the diseases that have been mapped on the human X chromosome. Medical screening of individuals has allowed quick identification of potential medical problems and for early intervention to treat them. As we saw in Topic 10.4, sequencing the DNA of different organisms has also made it possible to establish the evolutionary links between species.

The proteome

Of greater practical importance to humans is not the genes themselves, but the nature of the proteins these genes code for. These proteins are known as the **proteome**. A general definition of the proteome is all the proteins produced by the genome. However, as a protein is only produced when a gene is switched on, and genes are not switched on all the time, a more specific definition is all the proteins produced

Learning objectives

→ Outline the importance of genome sequencing projects.

→ Describe the nature of the proteome.

→ Describe how to determine the genome and proteome of simple organisms.

→ Describe how to determine the genome and proteome of complex organisms.

Specification reference 3.8.3

Synoptic link

Aspects of this topic require some knowledge and understanding of investigating diversity (Topic 10.4, Investigating diversity) and humoral immunity (Topic 5.4, B lymphocytes and humoral immunity). An initial read through these topics would provide helpful background information.

Hint

A primer is essential to start DNA synthesis because it makes a double strand of DNA, and DNA polymerase only works on double-stranded DNA.

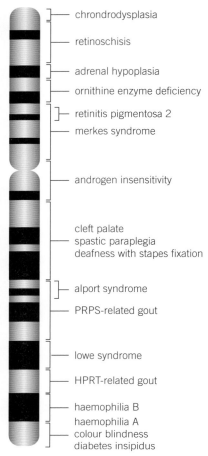

chrondrodysplasia

retinoschisis

adrenal hypoplasia

ornithine enzyme deficiency

retinitis pigmentosa 2

merkes syndrome

androgen insensitivity

cleft palate
spastic paraplegia
deafness with stapes fixation

alport syndrome

PRPS-related gout

lowe syndrome

HPRT-related gout

haemophilia B
haemophilia A
colour blindness
diabetes insipidus

▲ **Figure 1** *Some of the 60 diseases that have been mapped on the human X chromosome . This is shown for illustration only and the chromosome map does not have to be learnt*

in a given type of cell (cellular proteome) or organism (complete proteome), at a given time, under specified conditions. There are differences in the ease with which we can determine the genomes and proteomes of simple and complex organisms.

Determining the genome and proteome of simpler organisms

The first bacterium to have its genome fully sequenced was *Haemophilus influenza* in 1995. *H. influenza* contains 1700 genes comprising 1.8 million bases. The genomes of thousands of prokaryotic and single-celled eukaryotic organisms are currently being sequenced as part of the Human Microbiome Project. It is hoped that the information gained will help cure disease and provide knowledge of genes that can be usefully exploited. For example, ones from organisms that can withstand extreme or toxic environmental conditions and so have potential uses in cleaning up pollutants or in manufacturing biofuels.

Determining the proteome of prokaryotic organisms like bacteria is relatively easy because:

- the vast majority of prokaryotes have just one, circular piece of DNA that is not associated with histones
- there are none of the non-coding portions of DNA which are typical of eukaryotic cells.

Knowledge of the proteome of organisms like bacteria has a number of applications. Of particular interest is the identification of those proteins that act as antigens on the surfaces of human pathogens. These antigens can be used in vaccines against diseases caused by these pathogens. In the case of vaccines, the antigens can be manufactured and then administered to people in appropriate doses. In response to the antigen, memory cells are produced which trigger a secondary response when the antigen is encountered on a second occasion (Topic 5.4).

One example is sequencing of the DNA of *Plasmodium falciparum* which causes malaria. All 5300 genes on *Plasmodium's* 14 chromosomes have been sequenced giving us an insight into its metabolism and knowledge of the proteins it produces. All this will be invaluable in helping us to develop the elusive vaccine against this globally important disease.

Determining the genome and proteome of complex organisms

The success in mapping the human genome in 2003 is a testimony to what can be achieved in mapping DNA sequences of complex organisms. There are around 20 000 genes in the human genome although this number is constantly being revised down as our techniques for identifying genes improves. The problem in complex organisms is translating knowledge of the genome into the proteome. This is because the genome of complex organisms contains many non-coding genes as well as others that have a role in regulating other genes. In humans, it is thought that as few as 1.5% of genes may code for proteins. There is a human proteome project currently underway to identify all the proteins produced by humans. There is also the question of whose DNA is used for mapping. All individuals, except identical twins, have different base sequences on their DNA. The DNA mapped will differ, if only slightly, from everyone else's DNA.

Summary questions

1 Distinguish between a genome and a proteome.

2 Explain why determining the proteome of simple organisms like bacteria is easier than determining the proteome of complex ones like humans.

3 Explain how knowledge of the proteome of a pathogen might help to control the disease it causes.

1 SCID is a severe inherited disease. People who are affected have no immunity. Doctors carried out a trial using gene therapy to treat children with SCID. The doctors who carried out the trial obtained stem cells from each child's umbilical cord.

    (a) Give two characteristic features of stem cells. *(2 marks)*

        The doctors mixed the stem cells with viruses. The viruses had been genetically modified to contain alleles of a gene producing full immunity. The doctors then injected this mixture into the child's bone marrow. The viruses that the doctors used had RNA as their genetic material. When these viruses infect cells, they pass their RNA and two viral enzymes into the host cells.

    (b) One of the viral enzymes makes a DNA copy of the virus RNA. Name this enzyme. *(1 mark)*

        The other viral enzyme is called integrase. Integrase inserts the DNA copy anywhere in the DNA of the host cell. It may even insert the DNA copy in one of the host cell's genes.

    (c) (i) The insertion of the DNA copy in one of the host cell's genes may cause the cell to make a non-functional protein. Explain how. *(2 marks)*

        (ii) Some of the children in the trial developed cancer. How might the insertion of the DNA have caused cancer? *(2 marks)*

    (d) Five out of the 20 children in the trial developed cancer. Although the cancer was treated successfully, the doctors decided to stop the trial in its early stages. They then reviewed the situation and decided to continue. Do you agree with their decision to continue? Explain your answer. *(2 marks)*

AQA June 2010

2 Figure 1 shows part of a gene that is being transcribed.

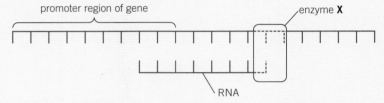

▲ **Figure 1**

    (a) Name enzyme **X**. *(1 mark)*

    (b) (i) Oestrogen is a hormone that affects transcription. It forms a complex with a receptor in the cytoplasm of target cells. Explain how an activated oestrogen receptor affects the target cell. *(2 marks)*

        (ii) Oestrogen only affects target cells. Explain why oestrogen does not affect other cells in the body. *(1 mark)*

    (c) Some breast tumours are stimulated to grow by oestrogen. Tamoxifen is used to treat these breast tumours. In the liver, tamoxifen is converted into an active substance called endoxifen. **Figure 2** shows a molecule of oestrogen and a molecule of endoxifen.

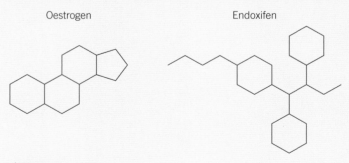

▲ **Figure 2**

Use **Figure 2** to suggest how endoxifen reduces the growth rate of these breast tumours.

(*2 marks*)

AQA June 2010

3 **(a)** Explain how the methylation of tumour suppressor genes can lead to cancer. (*3 marks*)

Scientists investigated a possible relationship between the percentage of fat in the diet and the death rate from breast cancer in women from 10 countries. Their data are shown in **Table 3**.

▼ **Table 3**

Percentage of fat in diet of population	Death rate of women from breast cancer per 100 000 women
9.5	1.5
15.0	7.0
20.0	12.0
25.0	9.0
32.0	15.0
35.0	8.0
35.0	20.0
40.5	18.0
43.0	24.0
45.0	26.0

(b) Describe how you would plot a suitable graph of these data. Explain your choice of type of graph. (*3 marks*)

(c) Use the data to calculate the correlation coefficient, *r*, to test for the strength of any correlation between fat percentage in the diet and the death rate from breast cancer. (*4 marks*)

AQA Specimen 2014 (apart from 3 (c))

$$r = \frac{\sum (x - \bar{x}) \times (y - \bar{y})}{\sqrt{\sum (x - \bar{x})^2 \times \sum (y - \bar{y})^2}}$$

NB you will only be required to carry out this form of calculation as part of your practical work.

Perhaps the most significant scientific advance in recent years has been the development of recombinant DNA technology that allows genes to be manipulated, altered and transferred from organism to organism – even to transform DNA itself. These techniques have enabled us to understand better how organisms work and to design new industrial processes and medical applications.

A number of human diseases result from individuals being unable to produce for themselves various metabolic chemicals. Many of these chemicals are proteins, such as **insulin**. They are therefore the product of a specific length of DNA, that is, the product of a gene. Treatment of such deficiencies previously involved extracting the chemical from a human or animal donor and introducing it into the patient. This presents problems such as rejection by the immune system and risk of infection. The cost is also considerable.

It follows that there are advantages in producing large quantities of 'pure' proteins from other sources. As a result, techniques have been developed to isolate genes, **clone** them and transfer them into microorganisms. These microorganisms are then grown to provide a 'factory' for the continuous production of a desired protein. The DNA of two different organisms that has been combined in this way is called **recombinant DNA**. The resulting organism is known as a **transgenic** or **genetically modified organism** (**GMO**).

How then is it possible that the DNA of one organism is not only accepted by a different species but also functions normally when it is transferred? The answer lies in the fact that the genetic code is the same in all organisms. In other words it is universal and can be used by all living organisms. This explains why the coded information on the transferred DNA can be interpreted, but what about the making of proteins? Well this too is universal in that the mechanisms of transcription and translation are essentially the same in all living organisms. As a result, transferred DNA can be transcribed and translated within the cells of the recipient (transgenic) organism and the proteins it codes for can be manufactured in the same way as they would be within the donor organism. This is all indirect evidence for evolution.

The process of making a protein using the DNA technology of gene transfer and cloning involves a number of stages:

1 **isolation** of the DNA fragments that have the gene for the desired protein
2 **insertion** of the DNA fragment into a vector
3 **transformation**, that is, the transfer of DNA into suitable host cells
4 **identification** of the host cells that have successfully taken up the gene by use of **gene markers**
5 **growth/cloning** of the population of host cells.

Let us consider each stage in detail.

Before a gene can be transplanted, it must be identified and isolated from the rest of the DNA. Given that the required gene may consist

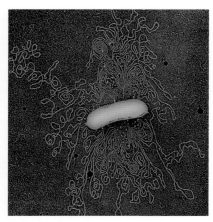

▲ **Figure 1** *An* Escherichia coli *bacterial cell that has been treated so that its DNA is ejected*

of a sequence of a few hundred bases amongst the many millions in human DNA, this is no small feat. There are several methods of producing DNA fragments:

*   conversion of mRNA to cDNA using reverse transcriptase
*   using restriction endonucleases to cut fragments containing the desired gene from DNA
*   creating the gene in a gene machine, usually based on a known protein structure.

Using reverse transcriptase

Retroviruses are a group of viruses, of which the best known is human immunodeficiency virus (HIV) (Topic 5.7). The coded genetic information of retroviruses is in the form of RNA. However, in a host cell they are able to synthesise DNA from their RNA using an enzyme called reverse transcriptase. It is so-named because it catalyses the production of DNA from RNA, which is the reverse of the more usual transcription of RNA from DNA. The process of using reverse transcriptase to isolate a gene is illustrated in Figure 2 and described in Figure 2.

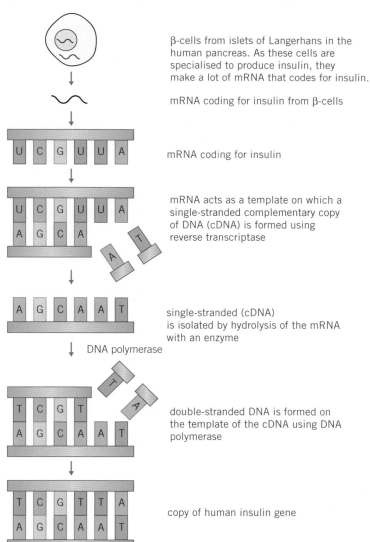

β-cells from islets of Langerhans in the human pancreas. As these cells are specialised to produce insulin, they make a lot of mRNA that codes for insulin.

mRNA coding for insulin from β-cells

mRNA coding for insulin

mRNA acts as a template on which a single-stranded complementary copy of DNA (cDNA) is formed using reverse transcriptase

single-stranded (cDNA) is isolated by hydrolysis of the mRNA with an enzyme

DNA polymerase

double-stranded DNA is formed on the template of the cDNA using DNA polymerase

copy of human insulin gene

> ### Synoptic link
>
> DNA polymerase acts in the same way when forming the second DNA strand during DNA replication, as described in Topic 2.2 , DNA replication.

▲ **Figure 2** *The use of reverse transcriptase to isolate the gene that codes for insulin*

- A cell that readily produces the protein is selected (e.g., the β-cells of the islets of Langerhans from the pancreas are used to produce insulin).
- These cells have large quantities of the relevant mRNA, which is therefore more easily extracted.
- Reverse transcriptase is then used to make DNA from RNA. This DNA is known as **complementary DNA (cDNA)** because it is made up of the nucleotides that are complementary to the mRNA.
- To make the other strand of DNA, the enzyme DNA polymerase is used to build up the complementary nucleotides on the cDNA template. This double strand of DNA is the required gene

Hint

Each restriction endonuclease recognises and cuts DNA at a specific sequence of bases. These sequences occur in the DNA of all species of organisms – but not in the same places!

Using restriction endonucleases

All organisms use defensive measures against pathogens. Bacteria are frequently infected by viruses that inject their DNA into them in order to take over the cell. Some bacteria defend themselves by producing enzymes that cut up the viral DNA. These enzymes are called restriction endonucleases.

There are many types of restriction endonucleases. Each one cuts a DNA double strand at a specific sequence of bases called a recognition sequence. Sometimes, this cut occurs between two opposite base pairs. This leaves two straight edges known as blunt ends. For example, one restriction endonuclease cuts in the middle of the base recognition sequence GTTAAC (Figure 3).

Other restriction endonucleases cut DNA in a staggered fashion. This leaves an uneven cut in which each strand of the DNA has exposed, unpaired bases. An example is a restriction endonuclease that recognises a six-base pair (or six bp) AAGCTT, as shown in Figure 3. In this figure, look at the sequence of unpaired bases that remain. If you read both the four unpaired bases at each end from left to right, the two sequences are opposites of one another, that is, they are a **palindrome**. The recognition sequence is therefore referred to as a

a HpaI restriction endonuclease has a recognition site GTTAAC, which produces a straight cut and therefore blunt ends

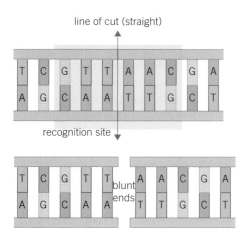

b HindIII restriction endonuclease has the recognition site AAGCTT, which produces a staggered cut and therefore sticky ends

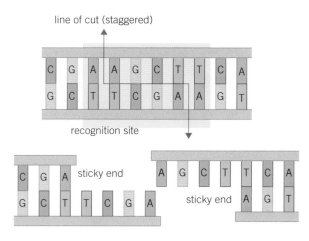

▲ **Figure 3** Action of restriction endonucleases

six bp palindromic sequence. This feature is typical of the way restriction endonucleases cut DNA to leave sticky ends. We shall look at the importance of these sticky ends in Topic 21.2.

The 'gene machine'

It is now possible to manufacture genes in a laboratory in the following manner:

- The desired sequence of nucleotide bases of a gene is determined from the desired protein that we wish to produce. The amino acid sequence of this protein is determined. From this, the mRNA codons are looked up and the complementary DNA triplets are worked out.

- The desired sequence of nucleotide bases for the gene is fed into a computer.

- The sequence is checked for biosafety and biosecurity to ensure it meets international standards as well as various ethical requirements.

- The computer designs a series of small, overlapping single strands of nucleotides, called oligonucleotides, which can be assembled into the desired gene.

- In an automated process, each of the oligonucleotides is assembled by adding one nucleotide at a time in the required sequence.

- The oligonucleotides are then joined together to make a gene. This gene doesn't have introns or other non-coding DNA. The gene is replicated using the polymerase chain reaction (Topic 21.3).

- The polymerase chain reaction also constructs the complementary strand of nucleotides to make the required double stranded gene. It then multiples this gene many times to give numerous copies.

- Using sticky ends (Topic 21.2) the gene can then be inserted into a bacterial plasmid. This acts as a vector for the gene allowing it to be stored, cloned or transferred to other organism in the future.

- The genes are checked using standard sequencing techniques (Topic 20.6) and those with errors are rejected.

The advantages of this process are that any sequence of nucleotides can be produced, in a very short time (as little as 10 days) and with great accuracy. A further advantage is that these artificial genes are also free of introns, and other 'non-coding' DNA, so can be transcribed and translated by prokaryotic cells. The process is shown in Figure 4.

Summary questions

In the following passage replace each number with the most appropriate word or words.

Where the DNA of two different organisms is combined, the product is known as (1) DNA. One method of producing DNA fragments is to make DNA from RNA using an enzyme called (2). This enzyme initially forms a single strand of DNA called (3) DNA. To form the other strand requires an enzyme called (4). Another method of producing DNA fragments is to use enzymes called (5), which cut up DNA. Some of these leave fragments with two straight edges, called (6) ends. Others leave ends with uneven edges, called (7) ends. If the sequence of bases on one of these uneven ends is GAATTC, then the sequence on the other end, if read in the same direction, will be (8).

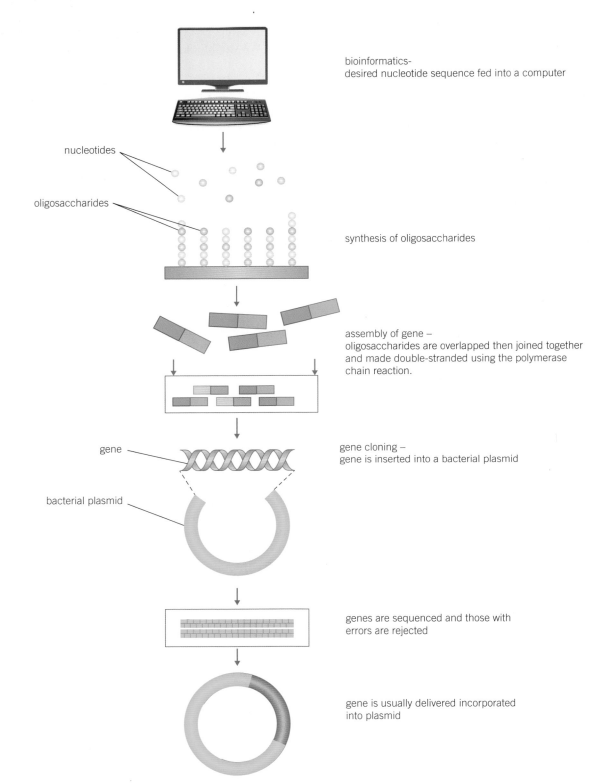

bioinformatics-
desired nucleotide sequence fed into a computer

nucleotides

oligosaccharides

synthesis of oligosaccharides

assembly of gene –
oligosaccharides are overlapped then joined together
and made double-stranded using the polymerase
chain reaction.

gene

gene cloning –
gene is inserted into a bacterial plasmid

bacterial plasmid

genes are sequenced and those with
errors are rejected

gene is usually delivered incorporated
into plasmid

▲ **Figure 4** *Making a gene in a gene machine*

21.2 *In vivo* gene cloning – the use of vectors

Having cut DNA into fragments, it is necessary to find the fragment which has the required gene amongst all the rest. This is done using a DNA probe as described in Topic 21.4. Once the fragment with the gene has been obtained, the next stage is to clone it so that there is a sufficient quantity for medical or commercial use. This can be achieved in two ways:

- *in vivo*, by transferring the fragments to a host cell using a vector
- *in vitro*, using the polymerase chain reaction (see Topic 21.3).

Before we consider how genes can be cloned within living organisms (*in vivo* cloning), let us look at the importance of the sticky ends left when DNA is cut by **restriction endonucleases**.

Learning objectives

- → Explain the importance of sticky ends.
- → Explain how a DNA fragment can be inserted into a vector.
- → Explain how the DNA of the vector is introduced into host cells.
- → Describe the nature of gene markers and explain how they work.

Specification reference: 3.8.4.1

Importance of sticky ends

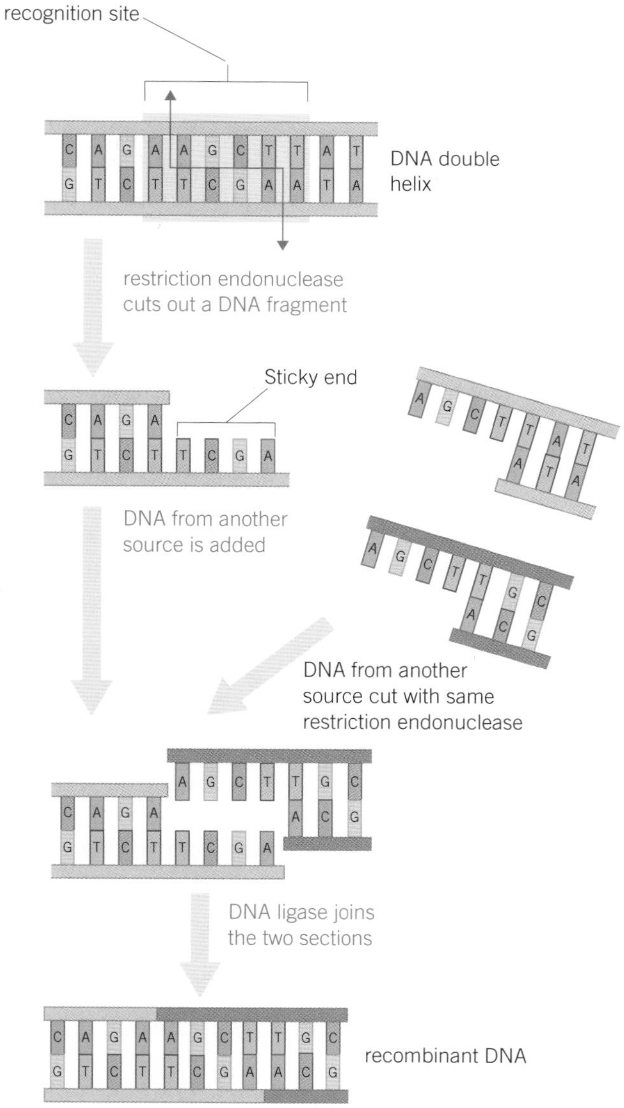

▲ **Figure 1** *The use of sticky ends to combine DNA from different sources*

The sequences of DNA that are cut by restriction endonucleases are called recognition sites. If the recognition site is cut in a staggered fashion, the cut ends of the DNA double strand are left with a single strand which is a few nucleotide bases long. The nucleotides on the single strand at one side of the cut are obviously complementary to those at the other side because they were previously paired together.

If the same restriction endonuclease is used to cut DNA, then all the fragments produced will have ends that are complementary to one another. This means that the single-stranded end of any one fragment can be joined (stuck) to the single-stranded end of any other fragment. In other words, their ends are sticky. Once the complementary bases of two sticky ends have paired up, an enzyme called **DNA ligase** is used to bind the phosphate-sugar framework of the two sections of DNA and so unite them as one.

Sticky ends have considerable importance because, provided the same restriction endonuclease is used, we can combine the DNA of one organism with that of any other organism (see Figure 1).

Preparing the DNA fragment for insertion

The preparation of the DNA fragment involves the addition of extra lengths of DNA. For the transcription of any gene to take place, the enzyme that synthesises mRNA (RNA polymerase) must attach to the DNA near a gene. The binding site for RNA polymerase is a region of DNA, known as a **promoter**. The nucleotide bases of the promoter attach both RNA polymerase and transcription factors (Topic 20.3) and so begin the process of transcription. If we want our DNA fragment to transcribe mRNA in order to make a protein, it is essential that we attach to it the necessary promoter region to start the process.

In the same way as a region of DNA binds RNA polymerase and begins transcription of a gene, another region releases RNA polymerase and ends transcription. This region of DNA is called a **terminator**. Again we need to add a terminator region to the other end of our DNA fragment to stop transcription at the appropriate point.

Insertion of DNA fragment into a vector

Once an appropriate fragment of DNA has been cut from the rest of the DNA and the promoter and terminator regions added, the next task is to join it into a carrying unit, known as a **vector**. This vector is used to transport the DNA into the host cell. There are different types of vector but the most commonly used is the **plasmid**. Plasmids are circular lengths of DNA, found in bacteria, which are separate from the main bacterial DNA. Plasmids almost always contain genes for antibiotic resistance, and restriction endonucleases are used at one of these antibiotic-resistance genes to break the plasmid loop.

The restriction endonuclease used is the same as the one that cut out the DNA fragment. This ensures that the sticky ends of the opened-up plasmid are complementary to the sticky ends of the DNA

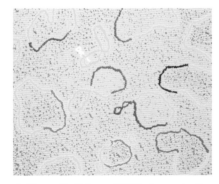

▲ **Figure 2** *Coloured TEM of genetically engineered DNA plasmids from the bacterium* Escherichia coli. *The plasmids (yellow) have had different gene sequences (various colours) inserted into them*

fragment. When the DNA fragments are mixed with the opened-up plasmids, they may become incorporated into them. Where they are incorporated, the join is made permanent using the enzyme DNA ligase. These plasmids now have recombinant DNA. These events are summarised in Figure 4.

Introduction of DNA into host cells

Once the DNA has been incorporated into at least some of the plasmids, they must then be reintroduced into bacterial cells. This process is called **transformation** and involves the plasmids and bacterial cells being mixed together in a medium containing calcium ions. The calcium ions, and changes in temperature, make the bacterial membrane permeable, allowing the plasmids to pass through the cell-surface membrane into the cytoplasm. However, not all the bacterial cells will possess the DNA fragments with the desired gene for the desired protein. Some reasons for this are:

- Only a few bacterial cells (as few as 1%) take up the plasmids when the two are mixed together.
- Some plasmids will have closed up again without incorporating the DNA fragment.
- Sometimes the DNA fragment ends join together to form its own plasmid.

The first task is to identify which bacterial cells have taken up the plasmid. One way to do this is to use the fact that bacteria have evolved mechanisms for resisting the effects of antibiotics, typically by producing an enzyme that breaks down the antibiotic before it can destroy the bacterium. The genes for the production of these enzymes are found in the plasmids.

Some plasmids carry genes for resistance to more than one antibiotic. One example is the R-plasmid, which carries genes for resistance to two antibiotics, ampicillin and tetracycline.

The task of finding out which bacterial cells have taken up the plasmids entails using the gene for antibiotic resistance, which is unaffected by the introduction of the new gene. In Figure 4, this is the gene for resistance to ampicillin. The process works as follows:

- All the bacterial cells are grown on a medium that contains the antibiotic ampicillin.
- Bacterial cells that have taken up the plasmids will have acquired the gene for ampicillin resistance.
- These bacterial cells are able to break down the ampicillin and therefore survive.
- The bacterial cells that have not taken up the plasmids will not be resistant to ampicillin and therefore die.

This is an effective method of showing which of the bacterial cells have taken up the plasmids. However, some cells will have taken up the plasmids and then closed up without incorporating the new gene, and these will also have survived. The next task is to identify the cells without the new gene and eliminate them. This is achieved using marker genes. Gene transfer and cloning are summarised in Figure 3.

Synoptic link

Antibiotic resistance is covered in Topic 9.4, Types of selection.

Marker genes

There are a number of different ways of using marker genes to identify whether a gene has been taken up by bacterial cells. They all involve using a second, separate gene on the plasmid. This second gene is easily identifiable for one reason or another. For example:

* It may be resistant to an antibiotic.
* It may make a fluorescent protein that is easily seen.
* It may produce an enzyme whose action can be identified.

Antibiotic-resistance marker genes

The use of **antibiotic-resistance** genes as markers is a rather old technology and has been superseded by other methods. However, it is an interesting example of how science works, particularly of the way in which scientists use knowledge and understanding to solve new problems, use appropriate methodology and carry out relevant experiments.

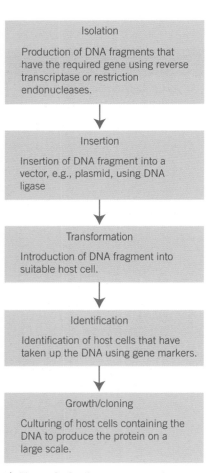

Isolation

Production of DNA fragments that have the required gene using reverse transcriptase or restriction endonucleases.

Insertion

Insertion of DNA fragment into a vector, e.g., plasmid, using DNA ligase

Transformation

Introduction of DNA fragment into suitable host cell.

Identification

Identification of host cells that have taken up the DNA using gene markers.

Growth/cloning

Culturing of host cells containing the DNA to produce the protein on a large scale.

▲ **Figure 3** *Outline summary of gene transfer and cloning*

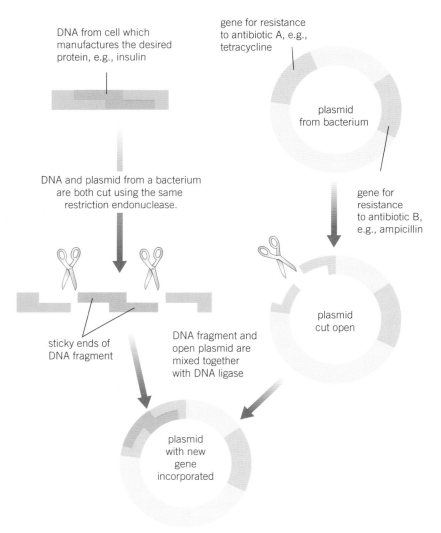

▲ **Figure 4** *Inserting a gene into a plasmid vector*

To identify those cells with plasmids that have taken up the new gene we use a technique called **replica plating**. This process uses the other antibiotic-resistance gene in the plasmid: the gene that was cut in order to incorporate the required gene. In Figure 4 this is the gene for resistance to tetracycline. As this gene has been cut, it will no longer produce the enzyme that breaks down tetracycline. In other words, the bacteria that have taken up the required gene will no longer be resistant to tetracycline. We can therefore identify these bacteria by growing them on a culture that contains tetracycline.

The problem is that treatment with tetracycline will destroy the very cells that contain the required gene. However by using a technique called replica plating it is possible to identify living colonies of bacteria containing the required gene.

Fluorescent markers

A more recent and more rapid method is the transfer of a gene from a jellyfish (Figure 5) into the plasmid. The gene in question produces a green fluorescent protein (GFP). The gene to be cloned is transplanted into the centre of the GFP gene. Any bacterial cell that has taken up the plasmid with the gene that is to be cloned will not be able to produce GFP. Bacterial cells that have not taken up the gene will continue to produce GFP and to fluoresce. Unlike the cells that have not taken up the gene, these cells that have taken it up will not fluoresce. As the bacterial cells with the desired gene are not killed, there is no need for replica plating. Results can be obtained by simply viewing the cells under a microscope and retaining those that do not fluoresce. This makes the process more rapid.

Enzyme markers

Another gene marker is the gene that produces the enzyme lactase. Lactase will turn a particular colourless substrate blue. Again, the required gene is transplanted into the gene that makes lactase. If a plasmid with the required gene is present in a bacterial cell, the colonies grown from it will not produce lactase. Therefore, when these bacterial cells are grown on the colourless substrate they will be unable to change its colour. Where the gene has not been taken up by the bacteria, they will not turn the substrate blue. These bacteria can be discounted.

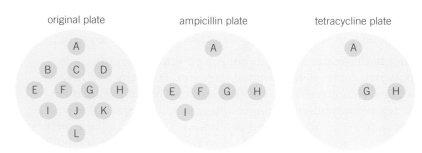

▲ Figure 7

▲ **Figure 5** *The gene in this jellyfish that produces a green fluorescent protein can be transplanted into other organisms and used as a fluorescent marker*

Hint

Interestingly, the gene for the green fluorescent protein has itself been genetically modified by the same techniques it is used to support. As a result, varieties have been engineered that fluoresce more brightly and in a number of different colours.

Summary questions

1   Explain the role of a vector during *in vivo* gene cloning.

2   Explain why gene markers are necessary during *in vivo* gene cloning.

3   Give *one* advantage of using fluorescent gene markers rather than antibiotic gene markers. Explain your answer.

4   Figure 7 shows the results of an experiment using antibiotic-resistance gene markers to find which bacterial cells have taken up a gene X. The circles within each plate represent a colony of growing bacteria. Deduce which colonies on the original plate:

   a   did not take up any plasmids with gene X

   b   contained plasmids possessing gene X.

   Explain your answers.

21.3 *In vitro* gene cloning – the polymerase chain reaction

Learning objectives

→ Describe the polymerase chain reaction.

→ Explain how the polymerase chain reaction is carried out.

→ Summarise the advantages of *in vitro* and *in vivo* cloning.

Specification reference: 3.8.4.1

After looking at *in vivo* cloning in Topic 21.2, let us now consider *in vitro* cloning using the polymerase chain reaction.

Polymerase chain reaction

The polymerase chain reaction (PCR) is a method of copying fragments of DNA. The process is automated, making it both rapid and efficient. The process requires the following:

- **the DNA fragment** to be copied
- **DNA polymerase** – an enzyme capable of joining together tens of thousands of nucleotides in a matter of minutes. One such enzyme, taq polymerase, is obtained from bacteria in hot springs and is therefore tolerant to heat (thermostable) and does not denature during the high temperatures used as part of the process
- **primers** – short sequences of nucleotides that have a set of bases complementary to those at one end of each of the two DNA fragments
- **nucleotides** – which contain each of the four bases found in DNA
- **thermocycler** – a computer-controlled machine that varies temperatures precisely over a period of time (Figure 1).

The polymerase chain reaction is illustrated in Figure 2 and is carried out in three stages:

1 **separation of the DNA strand**. The DNA fragments, primers and DNA polymerase are placed in a vessel in the thermocycler. The temperature is increased to 95 °C, causing the two strands of the DNA fragments to separate due to the breaking of the hydrogen bonds between the two DNA strands.

2 **addition (annealing) of the primers**. The mixture is cooled to 55 °C, causing the primers to join (anneal) to their complementary bases at the end of the DNA fragment. The primers provide the starting sequences for DNA polymerase to begin DNA copying because DNA polymerase can only attach nucleotides to the end of an existing chain. Primers also prevent the two separate strands from simply rejoining.

3 **synthesis of DNA**. The temperature is increased to 72 °C. This is the optimum temperature for the DNA polymerase to add complementary nucleotides along each of the separated DNA strands. It begins at the primer on both strands and adds the nucleotides in sequence until it reaches the end of the chain.

Because both separated strands are copied simultaneously there are now two copies of the original fragment. Once the two DNA strands are completed, the process is repeated by subjecting them to the temperature cycle again, resulting in four strands. The whole temperature cycle takes around two minutes. Over a million copies of the DNA can be made in only 25 temperature cycles and 100 billion copies can be manufactured in just a few hours. The polymerase chain reaction has revolutionised many aspects of science and medicine. Even the tiniest sample of DNA from a single hair or a speck of blood can now be multiplied to allow forensic examination and accurate cross-matching. You will learn more about this in Topic 21.5.

Study tip

Remember DNA polymerase causes nucleotides to join together as a strand, not complementary base pairing.

Hint

The polymerase chain reaction is not the same as semi-conservative replication of DNA in cells.

Study tip

Make sure you can describe the polymerase chain reaction, particularly the importance of the temperature changes.

▲ **Figure 1** *This is a thermocycler, a machine that carries out the polymerase chain reaction (PCR)*

Advantages of *in vitro* and *in vivo* gene cloning

The advantages of *in vitro* gene cloning are:

- **It is extremely rapid.** Within a matter of hours a 100 billion copies of a gene can be made. This is particularly valuable where only a minute amount of DNA is available, for example, at the scene of a crime. This can quickly be increased using the polymerase chain reaction and so there is no loss of valuable time before forensic analysis and matching can take place. A complicating factor is that PCR will also increase massively any other contaminating DNA found at the scene. *In vivo* cloning would take many days or weeks to produce the same quantity of DNA.

- **It does not require living cells.** All that is required is a base sequence of DNA that needs amplification. No complex culturing techniques, requiring time and effort, are needed.

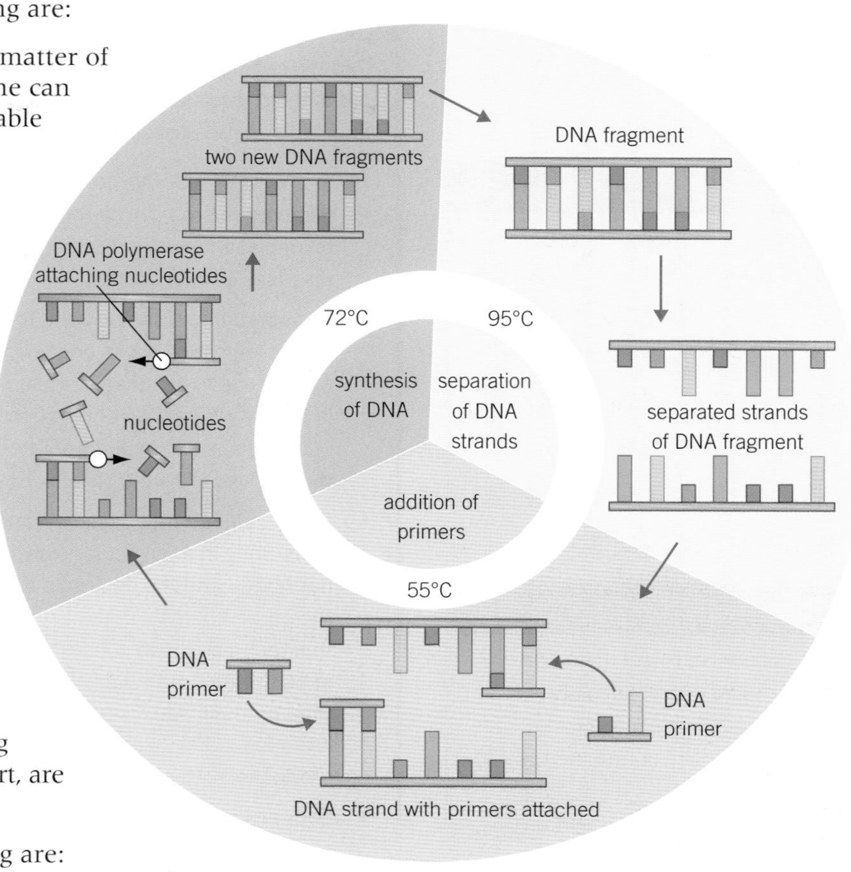

▲ **Figure 2** *The polymerase chain reaction showing a single cycle*

The advantages of *in vivo* gene cloning are:

- **It is particularly useful where we wish to introduce a gene into another organism**. As it involves the use of **vectors**, once we have introduced the gene into a **plasmid**, this plasmid can be used to deliver the gene into another organism, such as a human being (i.e., it can transform other organisms). This is done in a technique called gene therapy.

- **It involves almost no risk of contamination**. This is because a gene that has been cut by the same **restriction endonuclease** can match the sticky ends of the opened-up plasmid. Contaminant DNA will therefore not be taken up by the plasmid. *In vitro* cloning requires a very pure sample because any contaminant DNA will also be multiplied and could lead to a false result.

- **It is very accurate**. The DNA copied has few, if any, errors. At one time, about 20% of the DNA cloned *in vitro* by the PCR was copied inaccurately, but modern techniques have improved the accuracy of the process considerably. However, any errors in copying DNA or any contaminants in the sample will also be copied in subsequent cycles. This problem hardly ever arises with *in vivo* cloning because, although mutations can arise, these are very rare.

- **It cuts out specific genes**. It is therefore a very precise procedure as the culturing of transformed bacteria produces many copies of a specific gene and not just copies of the whole DNA sample.

- **It produces transformed bacteria that can be used to produce large quantities of gene products**. The transformed bacteria can produce proteins for commercial or medical use (e.g., hormones such as insulin).

Summary questions

1. In the polymerase chain reaction (PCR), primers are used. Describe what these are.

2. Explain the role of these primers.

3. Explain why two different primers are required.

4. State what type of bond is broken when DNA strands are separated in the PCR.

5. It is important in the PCR that the fragments of DNA used are not contaminated with any other biological material. Suggest a reason why.

Hint

Ethics is a narrower concept than morals. Ethics are a set of standards that are followed by a particular group of individuals and are designed to regulate their behaviour. They determine what is acceptable in pursuing the aims of the group.

Hint

Social issues relate to human society and its organisation. They concern the mutual relationships of human beings, their interdependence and their cooperation for the benefit of all.

Hint

Evaluating always involves looking at the positives and negatives, that is, the benefits and risks, of a particular issue.

 Evaluation of DNA technology

Genetic engineering undoubtedly brings many benefits to mankind, but it is not without its risks. It is therefore important to evaluate the ethical, moral and social issues associated with its use.

The benefits of recombinant DNA technology

- Microorganisms can be modified to produce a range of substances, for example, antibiotics, hormones and enzymes, that are used to treat diseases and disorders.

- Microorganisms can be used to control pollution, for example, to break up and digest oil slicks or destroy harmful gases released from factories. Care needs to be taken to ensure that such bacteria do not destroy oil in places where it is required, for example, car engines. To do this, a suicide gene can be incorporated that causes the bacteria to destroy themselves once the oil slick has been digested.

- Genetically modified plants can be transformed to produce a specific substance in a particular organ of the plant. These organs can then be harvested and the desired substance extracted. If a drug is involved, the process is called plant pharming. One promising application of this technique is in combating disease. This involves the production of plants that manufacture antibodies to pathogens and the toxins they produce. Alternatively the plants can be modified to manufacture **antigens** which, when injected into humans, induce natural **antibody** production.

- Genetically modified crops can be engineered to have financial and environmental advantages. These include making plants more tolerant to environmental extremes, for example, able to survive drought, cold, heat, salt, or polluted soils, etc. This permits crops to be grown commercially in places where they do not grow at present. Globally, each year, an area of land equal to half the United Kingdom becomes unfit for normal crops because of increases in soil salt concentrations. Growing of genetically modified plants, such as salt-tolerant tomatoes, could bring this land back into productivity. In a world where millions lack a basic nutritious diet, and with a predicted 90 million more mouths to feed by 2025, can we ethically oppose the use of such plant crops?

- Genetically modified crops can help prevent certain diseases. A type of rice, called golden rice, can have a gene for vitamin A production added. Can we justify not developing more vitamin A-enriched crops when 250 million children worldwide are at risk from vitamin A deficiency leading to 500 000 cases of irreversible blindness each year?

- Genetically modified animals are able to produce expensive drugs, antibiotics, hormones and enzymes relatively cheaply.

- Replacing defective genes (gene therapy) might be used to cure certain genetic disorders, such as cystic fibrosis and severe combined immunodeficiency (SCID).

- Genetic fingerprinting can be used in forensic science. Details are given in Topic 21.5.

The risks of recombinant DNA technology

Against the benefits of genetic engineering, must be weighed the risks – both real and potential.

- It is impossible to predict with complete accuracy what the ecological consequences will be of releasing genetically engineered organisms into the environment. The delicate balance that exists in any habitat may be irreversibly damaged by the introduction of organisms with engineered genes. There is often no going back once an organism is released although 'suicide genes' can be inserted or the organism engineered so it can only survive when a supplement is added.

- A recombinant gene may pass from the organism it was placed in, to a completely different one. We know, for example, that viruses can transfer genes from one organism to another. What if a virus were to transfer genes for herbicide resistance and vigorous growth from a crop plant to a weed that competed with the crop plant? What if the same gene were transferred in pollen to other plants? How would we then be able to control this weed?

- Any manipulation of the DNA of a cell will have consequences for the metabolic pathways within that cell. We cannot be sure until after the event what unforeseen by-products of the change might be produced. Could these lead to metabolic malfunctions, cause cancer, or create a new form of disease?

- Genetically modified bacteria often have antibiotic resistance marker genes that have been added. These bacteria might spread antibiotic resistance to harmful bacteria.

- All genes mutate. What then, might be the consequences of our engineered gene mutating? Could it turn the organism into a **pathogen** which we have no means of controlling?

- What will be the long-term consequences of introducing new gene combinations? We cannot be certain of the effects on the future evolution of organisms. Will the artificial selection of 'desired' genes reduce the genetic variety that is so essential to evolution?

- What might be the financial consequences of developing plants and animals to grow in new regions? Developing bananas which grow in Britain could have disastrous consequences for the Caribbean economies that rely heavily on this crop for their income.

- How far can we take the technique of replacing defective genes? It may be acceptable to replace a defective gene to cure cystic fibrosis, but is it equally acceptable to introduce genes for intelligence, more muscular bodies, cosmetic improvements, or different facial features?

- Will knowledge of, and ability to change, human genes lead to eugenics, whereby selection of genes leads to a means of selecting one race rather than another?

- What will be the consequences of the ability to manipulate genes getting into the wrong hands? Will unscrupulous individuals, groups or governments use this power to achieve political goals, control opposition or gain ultimate power?

- Is the financial cost of recombinant DNA technology justified, or would the money be better used fighting hunger and poverty, that are the cause of much human misery. Will sophisticated treatments, with their more high-profile images, be put before the everyday treatment of rheumatoid arthritis or haemorrhoids? Will such treatments only be within the financial reach of the better-off?

- Genetic fingerprinting (Topic 21.5), with its ability to identify an individual's DNA accurately, is a highly reliable forensic tool. How easy would it be for someone to exchange a DNA sample maliciously, leading to wrongful conviction?

- Is it immoral to tamper with genes at all? Should we let nature take its own course in its own time?

- How do we deal with the issues surrounding the **human genome project**? Is it right that an individual or company can patent, and therefore effectively own, a gene?

It is inevitable that we remain inquisitive about the world in which we live, and that we will seek to try to improve the conditions around us. Genetic research is bound to continue, but the challenge will be to develop the safeguards and ethical guidelines that will allow recombinant DNA technology to be used in a safe and effective manner.

> 1 Take any *three* aspects of recombinant DNA technology that are beneficial to humans (as listed above) and present a reasoned argument in each case for the continued use of that technology.
>
> 2 Using the same three aspects, present a reasoned argument that an environmentalist or anti-globalisation activist might make against the continued use of that technology.

Treatment of severe combined immunodeficiency using gene therapy

Severe combined immunodeficiency (SCID) is a rare inherited disorder. People with this condition do not show a cell-mediated immune response (Topic 5.3), nor are they able to produce antibodies (Topic 5.5). The disorder arises when individuals inherit a defect in the gene that codes for the enzyme adenosine deaminase (ADA). This enzyme destroys toxins that would otherwise kill white blood cells. Survival has depended upon patients being given bone marrow transplants and/or injections of ADA. There have been recent attempts to treat the disorder using a technique called gene therapy as follows:

- The normal ADA gene is isolated from healthy human tissue.

- The ADA gene is inserted into a retrovirus.

- The retroviruses are grown with host cells in the laboratory to increase their number and hence the number of copies of the ADA gene.

- The retroviruses are mixed with the patient's T cells into which they inject a copy of the normal ADA gene.

- The T cells are reintroduced into the patient's blood to provide the coded information needed to make ADA.

The effectiveness of this treatment is limited to 6–12 months and so it has to be repeated at intervals. A more long-term treatment involves introducing the normal gene into bone marrow stem cells rather than T cells.

> 1 Outline three ways in which the normal ADA gene might be isolated from human tissue.
>
> 2 Explain what a retrovirus is.
>
> 3 Suggest the likely sequence of events whereby the defective gene could lead to the death of a person with SCID.
>
> 4 Suggest why the effectiveness of the treatment only lasts 6–12 months.
>
> 5 Suggest why introducing the gene into bone marrow stem cells rather than T cells is a more long-term treatment for SCID.

Many human diseases have a genetic origin. These are often the result of a **gene mutation**.

Recombinant DNA technology has enabled us to diagnose and treat many of these genetic disorders. In doing so, it is often necessary to know exactly where a particular DNA sequence (gene) is located. To achieve this we use labelled DNA probes and DNA hybridisation.

DNA probes

A DNA probe is a short, single-stranded length of DNA that has some sort of label attached that makes it easily identifiable. The two most commonly used probes are:

- **radioactively labelled probes**, which are made up of **nucleotides** with the **isotope** ^{32}P. The probe is identified using an X-ray film that is exposed by radioactivity.

- **fluorescently labelled probes**, which emit light (fluoresce) under certain conditions, for instance when the probe has bound to the target DNA sequence.

DNA probes are used to identify particular alleles of genes in the following way:

- A DNA probe is made that has base sequences that are complementary to part of the base sequence of the DNA that makes up the allele of the gene that we want to find.

- The double-stranded DNA that is being tested is treated to separate its two strands.

- The separated DNA strands are mixed with the probe, which binds to the complementary base sequence on one of the strands. This is known as **DNA hybridisation** (see below).

- The site at which the probe binds can be identified by the radioactivity or fluorescence that the probe emits.

Before we can make a specific probe we need to know the base sequence in the particular allele that we are trying to locate. A number of different methods are used to sequence the exact order of bases in a length of DNA.

DNA hybridisation

DNA hybridisation takes place when a section of DNA or RNA is combined with a single-stranded section of DNA which has complementary bases. Before hybridisation can take place, the two strands of the DNA molecule must be separated. This is achieved by heating DNA until its double strand separates into its two complementary single strands (denaturation). When cooled, the complementary bases on each strand recombine (anneal) with each other to reform the original double strand. Given sufficient time,

Learning objectives

→ Describe what DNA probes are and explain how they work.

→ Explain how DNA hybridisation is used to locate specific alleles of genes.

→ Describe the use of labelled DNA probes to screen for heritable conditions or health risks.

→ Consider the use of genetic screening in genetic counselling.

Specification reference: 3.8.4.2

Synoptic link

The material in this topic brings together information from many other topics including DNA replication (Topic 2.2), The genetic code (Topic 8.1), Gene mutations (Topics 9.1 and 20.1), Studying inheritance (Topic 17.1), DNA sequencing (Topic 20.6), and the Polymerase chain reaction (Topic 21.3).

all strands in a mixture of DNA will pair up with their partners. If, however, other complementary sections of DNA are present in the mixture as the DNA cools, these are just as likely to anneal with one of the separated DNA strands as the two strands are with one another.

Locating specific alleles of genes

Using DNA probes and DNA hybridisation, it is possible to locate a specific allele of a gene. For example, we may wish to determine whether someone possesses a mutant allele that causes a particular genetic disorder. The process is as follows:

- We must first determine the sequence of nucleotide bases of the mutant allele we are trying to locate. This can be achieved using DNA sequencing techniques. However, we now have extensive genetic libraries that store the base sequences of most genetic diseases and so we can simply refer to these to obtain the sequence.
- A fragment of DNA is produced that has a sequence of bases that are complementary to the mutant allele we are trying to locate.
- Multiple copies of our DNA probe are formed using the polymerase chain reaction.
- A DNA probe is made by attaching a marker, for example a fluorescent dye, to the DNA fragment.
- DNA from the person suspected of having the mutant allele we want to locate is heated to separate its two strands.
- The separated strands are cooled in a mixture containing many of our DNA probes.
- If the DNA contains the mutant allele, one of our probes is likely to bind to it because the probe has base sequences that are exactly complementary to those on the mutant allele.
- The DNA is washed clean of any unattached probes.
- The remaining hybridised DNA will now be fluorescently labelled with the dye attached to the probe.
- The dye is detected by shining light onto the fragments causing the dye to fluoresce which can be seen using a special microscope.

The process is summarised in Figure 1.

Genetic screening

Many genetic disorders, such as sickle-cell anaemia, are the result of **gene mutations**. Gene mutations may arise if one or more **nucleotide** bases in DNA are changed in any one of a variety of ways. If the mutation results in a **dominant allele**, all individuals will have the genetic disorder. If the allele is **recessive**, it will only be apparent in those individuals that have two recessive alleles, that is, who are **homozygous** recessive. Individuals that are **heterozygous** will not display symptoms of the disease but will carry one copy of the mutant allele. These individuals are known as carriers. They have the capacity to pass the disease to their offspring if the other parent is also heterozygous or homozygous recessive.

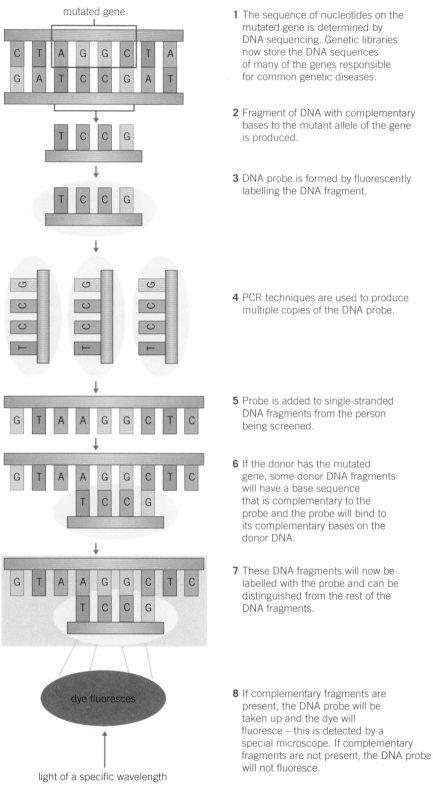

1 The sequence of nucleotides on the mutated gene is determined by DNA sequencing. Genetic libraries now store the DNA sequences of many of the genes responsible for common genetic diseases.

2 Fragment of DNA with complementary bases to the mutant allele of the gene is produced.

3 DNA probe is formed by fluorescently labelling the DNA fragment.

4 PCR techniques are used to produce multiple copies of the DNA probe.

5 Probe is added to single-stranded DNA fragments from the person being screened.

6 If the donor has the mutated gene, some donor DNA fragments will have a base sequence that is complementary to the probe and the probe will bind to its complementary bases on the donor DNA.

7 These DNA fragments will now be labelled with the probe and can be distinguished from the rest of the DNA fragments.

8 If complementary fragments are present, the DNA probe will be taken up and the dye will fluoresce – this is detected by a special microscope. If complementary fragments are not present, the DNA probe will not fluoresce.

▲ **Figure 1** *Summary of the process to locate a specific allele of a gene*

It is important to screen individuals who may carry a mutant allele. Such individuals often have a family history of a disease. Screening can determine the probabilities of a couple having offspring with a genetic disorder. As a result, potential parents who are at risk can obtain advice from a genetic counsellor about the implications of having children, based on their family history and the results of genetic screening.

It is possible to fix hundreds of different DNA probes in an array (pattern) on a glass slide. By adding a sample of DNA to the array, any complementary DNA sequences in the donor DNA will bind to one or more probes. In this way it is possible to test simultaneously for many different genetic disorders by detecting fluorescence that occurs where binding has taken place.

Another area where genetic screening can be valuable is in the detection of oncogenes, which are responsible for cancer. Cancers may develop as a result of mutations that prevent the **tumour suppressor genes** inhibiting cell division. Mutations of both alleles must be present to inactivate the tumour suppressor genes and to initiate the development of a tumour. Some people inherit one mutated tumour suppressor gene. These individuals are at greater risk of developing cancer.

If a mutated gene is detected by genetic screening, individuals who are at greater risk of cancer can then make informed decisions about their lifestyle and future treatment. They can choose to give up smoking, lose weight, eat more healthily and avoid **mutagens** as far as possible. They can also check themselves more regularly for early signs of cancer, which can lead to an early diagnosis and a better chance of successful treatment. They may choose to undergo some form of surgery or other treatment.

Personalised medicine

One of the advantages of genetic screening is personalised medicine. It allows doctors to provide advice and health care based on an individual's genotype. Some people's genes can mean that a particular drug may be either more or less effective in treating a condition. By genetically screening patients, doctors and pharmacists can determine, more exactly, the dose of a drug which will produce the desired outcome. This can save money that would otherwise be wasted on overprescribing the drug. In some cases it avoids medications that could cause harm or avoids raising false hopes.

One example is the prescribing of painkillers. To function effectively many pain medications need a specific enzyme to activate them. About half the population have genes that alter the function of this enzyme. Screening for the presence of these genes allows the dosage to be adjusted to compensate for the ways in which the genes affect an individual's metabolism of the painkiller. This ensures that their use is both safe and effective.

Another example involves vitamin E. It has been shown that among people who have diabetes, vitamin E reduces the risk of cardiovascular disease for those with certain genotypes, but it can increase the risk

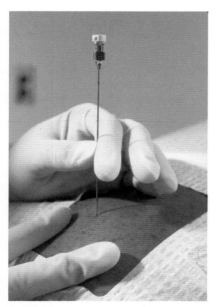

▲ **Figure 2** *Amniotic fluid is being taken from this pregnant woman. The fluid can be used to screen the unborn baby for genetic disorders*

for those with a different genotype. It is clearly advantageous to screen a person who has diabetes before advising on whether or not to take vitamin E supplements.

Genetic screening goes hand in hand with genetic counselling. The expert advice provided by a counsellor helps individuals to understand the results and implications of the screening and so make appropriate decisions.

Genetic counselling

Genetic counselling is like a special form of social work, where advice and information are given that enable people to make personal decisions about themselves or their offspring. One important aspect of genetic counselling is to research the family history of an inherited disease and to advise parents on the likelihood of it arising in their children.

Consider a mother who has a family history of sickle-cell anaemia. If the mother herself is unaffected but carries the gene for sickle-cell anaemia, she must be heterozygous for the condition. Suppose that she wishes to have children with a man who has no family history of sickle-cell anaemia. In this case, it can be assumed that the man does not carry the allele for the disease, and therefore none of the children will develop the disease, although they may be carriers. On the other hand, if the man has a family history of the disease, it is possible that he too carries the allele. In this case, the genetic counsellor can make the couple aware that there is a one in four chance of their children being affected and a two in four chance that their children will be carriers.

A counsellor can also inform the couple of the effects of sickle-cell anaemia and its emotional, psychological, medical, social and economic consequences. On the basis of this advice the couple can then choose whether or not to have children. Counselling can also make them aware of any further medical tests that might give a more accurate prediction of whether their children will have the condition, for example IVF with screening of embryos.

Genetic counselling is closely linked to genetic screening and the screening results provide the genetic counsellor with a basis for informed discussion. For example, in cases of cancer, screening can help to detect:

- oncogene mutations, which can determine the type of cancer that the patient has and hence the most effective drug or radiotherapy to use
- gene changes that predict which patients are more likely to benefit from certain treatments and have the best chance of survival. For example, the drug herceptin is most effective at treating certain types of breast cancer
- a single cancer cell among millions of normal cells, thus identifying patients at risk of relapse from certain forms of leukaemia.

This information can help a counsellor to discuss with the patient the best course of treatment and their prospects of survival.

Synoptic link

To remind yourself about how oncogenes and tumour suppressor genes work, re-read Topic 20.5, Gene expression and cancer.

▲ **Figure 3** *Genetic counsellor talking with potential parents. Genetic counselling is used to advise on the risk of passing on genetic disorders and can be given to couples in certain risk groups who are intending to have children*

Summary questions

1. Explain what a DNA probe is.
2. Outline the process of genetic screening for a disease.
3. Genetic screening shows that a person has one mutant allele of the tumour suppressor gene.
   a. Describe the role of the tumour suppressor gene.
   b. Suggest how the person might use the information revealed by genetic screening.

Synoptic link

Not all DNA sequences carry obvious genetic information. You will have already come across this idea with the processing of DNA described in Topic 8.4, Protein synthesis – transcription and splicing.

Genetic fingerprinting is a diagnostic tool used widely in forensic science, plant and animal breeding, and medical diagnosis. It is based on the fact that the DNA of every individual, except identical twins, is unique.

Genetic fingerprinting

This technique relies on the fact that the genome of most eukaryotic organisms contains many repetitive, non-coding bases of DNA. Indeed, 95% of human DNA is currently not known to code for any characteristic but may yet be found to be functional. DNA bases which are non-coding are known as **variable number tandem repeats (VNTRs)**. For every individual the number and length of VNTRs has a unique pattern. They are different in all individuals except identical twins, and the probability of two individuals having identical sequences of these VNTRs is extremely small. However, the more closely related two individuals are, the more similar the VNTRs will be.

Gel electrophoresis

Gel electrophoresis is used to separate DNA fragments according to their size. The DNA fragments are placed on to an agar gel and a voltage is applied across it (Figure 1). The resistance of the gel means that the larger the fragments, the more slowly they move. Therefore, over a fixed period, the smaller fragments move further than the larger ones. In this way DNA fragments of different lengths are separated. If the DNA fragments are labelled, for example with radioactive DNA probes (Topic 21.4), their final positions in the gel can be determined by placing a sheet of X-ray film over the agar gel for several hours. The radioactivity from each DNA fragment exposes the film and shows where the fragment is situated on the gel. Only DNA fragments up to around 500 bases long can be sequenced in this way. Larger genes and whole genomes must therefore be cut into smaller fragments by **restriction endonucleases**.

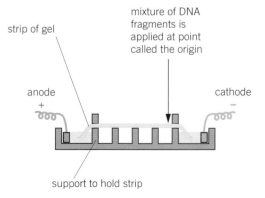

▲ **Figure 1** *Apparatus for carrying out electrophoresis*

▲ **Figure 2** *An electrophoretic gel being loaded with DNA samples*

The making of a genetic fingerprint consists of five main stages: extraction, digestion, separation, hybridisation and development. The complete process of genetic fingerprinting is summarised in Figure 3.

Extraction

Even the tiniest sample of animal tissue, such as a drop of blood or a hair root, is enough to give a genetic fingerprint. Whatever the sample, the first stage is to extract the DNA by separating it from the rest of the cell. As the amount of DNA is usually small, its quantity can be increased by using the polymerase chain reaction (Topic 21.3).

Digestion

The DNA is then cut into fragments, using the same restriction endonucleases (Topic 21.1). The endonucleases are chosen for their ability to cut close to, but not within, the target DNA.

Separation

The fragments of DNA are next separated according to size by gel electrophoresis under the influence of an electrical voltage. The gel is then immersed in alkali in order to separate the double strands into single strands.

Hybridisation

Radioactive (or fluorescent) DNA probes are now used to bind with VNTRs (Topic 21.4). The probes have base sequences which are complementary to the base sequences of the VNTRs, and bind to them under specific conditions, such as temperature and pH. The process is carried out with different probes, which bind to different target DNA sequences

Development

Finally, an X-ray film is put over the nylon membrane. The film is exposed by the radiation from the radioactive probes. (If using fluorescent probes, the positions are located visually.) Because these points correspond to the position of the DNA fragments as separated during electrophoresis, a series of bars is revealed. The pattern of the bands (Figure 5) is unique to every individual except identical twins.

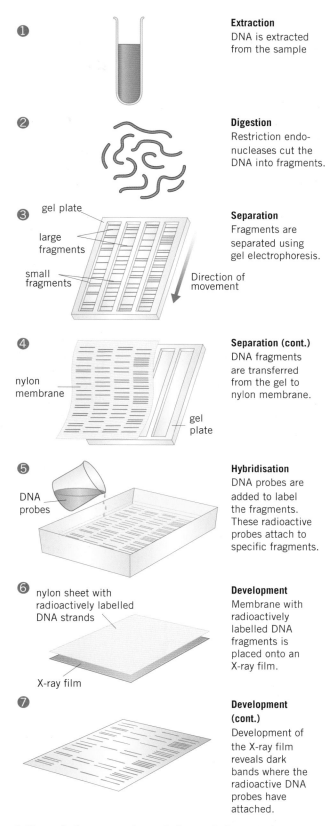

Extraction
DNA is extracted from the sample

Digestion
Restriction endonucleases cut the DNA into fragments.

gel plate
large fragments
small fragments
Direction of movement

Separation
Fragments are separated using gel electrophoresis.

nylon membrane
gel plate

Separation (cont.)
DNA fragments are transferred from the gel to nylon membrane.

DNA probes

Hybridisation
DNA probes are added to label the fragments. These radioactive probes attach to specific fragments.

nylon sheet with radioactively labelled DNA strands
X-ray film

Development
Membrane with radioactively labelled DNA fragments is placed onto an X-ray film.

Development (cont.)
Development of the X-ray film reveals dark bands where the radioactive DNA probes have attached.

▲ **Figure 3** *Summary of genetic fingerprinting technique*

Hint

Remember that, in gel electrophoresis, the smallest DNA fragments travel the furthest.

Hint

In theory, the inheritance of VNTRs does not have any influence on the phenotype of an organism.

Interpreting the results

DNA fingerprints from two samples, for example, from blood found at the scene of a crime and from a suspect, are visually checked. If there appears to be a match, the pattern of bars of each fingerprint is passed through an automated scanning machine, which calculates the length of the DNA fragments from the bands. It does this using data obtained by measuring the distances travelled during electrophoresis by known lengths of DNA. Finally, the odds are calculated of someone else having an identical fingerprint. The closer the match between the two patterns, the greater the probability that the two sets of DNA have come from the same person.

Uses of DNA fingerprinting

DNA fingerprinting has a variety of uses:

Genetic relationships and variability

DNA fingerprinting can be used to help resolve questions of paternity. Individuals inherit half their genetic material from their mother and half from their father. Therefore each band on a DNA fingerprint of an individual should have a corresponding band in one of the parents parents' DNA fingerprint (Figure 4). This can be used to establish whether someone is the genetic father of a child. Genetic fingerprinting is also useful in determining genetic variability within a population. The more closely two individuals are related the closer the resemblance of their genetic fingerprints. A population whose members have very similar genetic fingerprints has little genetic diversity. A population whose members have a greater variety of genetic fingerprints has greater genetic diversity.

Forensic science

DNA is often left at the scene of a crime, for example, blood at the scene of a violent crime, semen at the scene of a rape and hair at the scene of a robbery. Genetic fingerprinting can establish whether a person is likely to have been present at the crime scene, although this does not prove they actually carried out the crime. Even if there is a close match between a suspect's DNA and the DNA found at the crime scene, it does not follow that the suspect carried out the crime. Other possible explanations need to be investigated. For example:

- The DNA may have been left on some other, innocent occasion.
- The DNA may belong to a very close relative.
- The DNA sample may have been contaminated after the crime, either by the suspect's DNA or by chemicals that affected the action of the restriction endonucleases used in preparing the fingerprint.

Finally, the probability that someone else's DNA might match that of the suspect has to be calculated. This calculation is based on the assumption that the DNA which produces the banding patterns is randomly distributed in the community. This may not always be the case, for example, it may not apply where religious or ethnic groups tend to have partners from within their own small community.

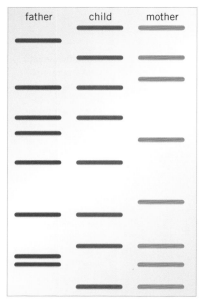

▲ **Figure 4** DNA fingerprints of a child and each parent. Note that each band on the child's fingerprint corresponds to a band on one or other parent's fingerprint . As the child only inherits half the VNTR's of each parent, there are inevitably some bands in each parental fingerprint that do not match to bands in the child's fingerprint

Medical diagnosis

Genetic fingerprints can help in diagnosing diseases such as Huntington's disease. This is a genetic disorder of the nervous system. It results from a three-base sequence (AGC) at one end of a gene on chromosome 4 being repeated over and over again – a sort of genetic stutter. People with fewer than 30 repeats are unlikely to get the disease, while those with more than 38 repeats are almost certain to do so. If they have over 50 repeats, the onset of the disease will occur earlier than average.

A sample of DNA from a person with the allele for Huntington's disease can be cut with restriction endonucleases and a DNA fingerprint prepared. This can then be matched with fingerprints of people with various forms of the disease and those without the disease. In this way, the probability of developing the symptoms, and when, can be determined.

Genetic fingerprints are also used to identify the nature of a microbial infection by comparing the fingerprint of the microbe found in patients with that of known pathogens.

Plant and animal breeding

Genetic fingerprinting can be used to prevent undesirable inbreeding during breeding programmes on farms or in zoos. It can also identify plants or animals that have a particular allele of a desirable gene. Individuals with this allele can be selected for breeding in order to increase the probability of their offspring having the characteristic that it produces. Another application is the determination of paternity in animals and thus establishing the pedigree (family tree) of an individual.

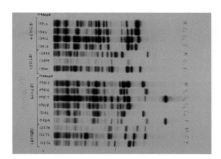

▲ **Figure 5** *The bands in these DNA fingerprints are marked M for mother, C for child and F for father*

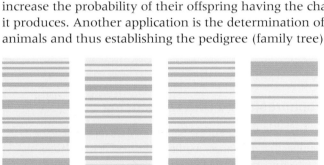

blood at crime scene   suspect A   suspect B   victim

▲ Figure 6

Summary questions

1 Explain why it is often necessary to use the polymerase chain reaction when producing a genetic fingerprint.

2 Figure 6 shows the genetic fingerprints of four DNA samples collected following a crime.

   a Which of the two suspects do you think was present at the scene of the crime? Give a reason for your answer.

   b Suggest why a genetic fingerprint of a DNA sample from the victim was made.

3 Suggest how chemicals that affect the action of restriction endonucleases can alter the genetic fingerprint of a DNA sample.

4 Suggest how the genetic fingerprint of someone with the allele for Huntington's disease might differ from that of someone who does not have the allele.

5 Explain how genetic fingerprinting can be used to ensure that inbreeding is avoided.

Locating DNA fragments

A section of DNA was cut into fragments and these fragments were separated by electrophoresis. Table 1 shows the number of base pairs in each fragment. The position of the fragments after gel electrophoresis is shown in Figure 7.

1 Name an enzyme that could have been used to cut the DNA.
2 Using the letters (A–F) in the boxes in Table 1, indicate which of the fragments (1–6) in Figure 7 are located in each box. Explain your answer.
3 The enzyme used to cut the DNA does so at a particular sequence of nucleotide bases. How many times does this base sequence occur in the original section of DNA?

▼ Table 1

Fragment	Number of base pairs (kilobases)
A	8.02
B	5.43
C	4.78
D	11.31
E	2.46
F	6.12

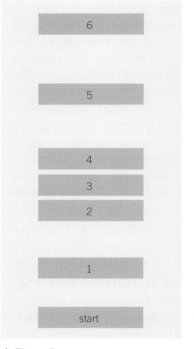

▲ Figure 7

Gel electrophoresis and DNA sequencing

Gel electrophoresis is an integral part of the Sanger method for sequencing DNA. You may remember that the Sanger method uses modified nucleotides, called terminators, which cannot attach to the next base in the sequence when they are being joined together. They therefore end the synthesis of a DNA strand. Four different terminator nucleotides are used, each with one of the four bases adenine, thymine, guanine, or cytosine. As explained in Topic 20.6, depending on exactly where the terminator nucleotide binds to the DNA, its synthesis may be terminated after only a few nucleotides or after a long fragment of DNA has been synthesised. As a result, DNA fragments of varying lengths are produced. Each fragment will end in one of the four bases, adenine, thymine, guanine, or cytosine. These fragments can be identified because the primer attached to the other end of the DNA section is labelled radioactively.

The results of one such experiment in DNA sequencing is shown in Figure 8. This used a DNA fragment that was just eight nucleotides long. The results are read from the bottom up because the shortest fragments move the furthest distance. The smallest fragment (labelled 1) is just one nucleotide long and is therefore nearest the bottom. This fragment has a terminator nucleotide with the base adenine. The second fragment (labelled 2) is two bases long and has a terminator nucleotide with the base guanine, and so on. In this way the whole sequence of bases on the terminator nucleotides

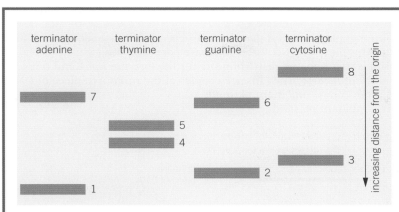

▲ **Figure 8** *Results of a DNA sequencing experiment*

was found to be AGCTTGAC and this is the sequence on one of the strands of the newly formed DNA.

The results of sequencing another fragment of DNA in the same manner are shown in Figure 9.

1 √x̄ using Figure 9 calculate how many adenine bases were present in the fragment of DNA.

2 Determine which nucleotide ends the shortest fragment.

3 Deduce the nucleotide sequence of the longest DNA fragment that has been produced.

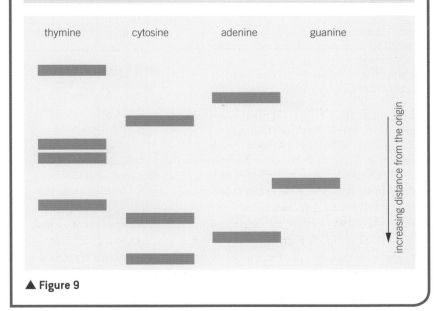

▲ **Figure 9**

1   Scientists used restriction mapping to investigate some aspects of the base sequence of an
     unknown piece of DNA. This piece of DNA was 3 000 base pairs (bp) long.
     The scientists took plasmids that had one restriction site for the enzyme *Kpn*1 and one
     restriction site for the enzyme *Bam*H1. They inserted copies of the unknown piece of DNA
     into the plasmids. This produced recombinant plasmids.
     The diagram shows a recombinant plasmid.

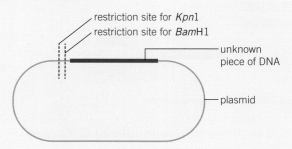

(a) When the scientists digested one of the recombinant plasmids with Kpn1, they
     obtained two fragments. One fragment was measured as 1 000 bp. The other
     fragment was described as "very large".
     (i)   What does this show about the base sequence of the unknown piece
           of DNA?   (*2 marks*)
     (ii) One of the fragments that the scientists obtained was described as
           "very large". What is represented by this very large fragment?   (*1 mark*)
(b) When the scientists digested another of the recombinant plasmids with
     *Bam*H1, they obtained three fragments.
     How many *Bam*H1 restriction sites are there in the unknown piece of DNA? (*1 mark*)
(c) (i)   Scientists can separate fragments of DNA using electrophoresis. Suggest
           how they can use electrophoresis to estimate the number of base pairs
           in the separated fragments.   (*2 marks*)
     (ii) Scientists need to take precautions when they
           carry out restriction mapping. They need to
           make sure that the enzyme they have used
           has completely digested the DNA. One check
           they may carry out is to add the sizes of the
           fragments together. How could scientists use
           this information to show that the DNA has
           not been completely digested? Explain
           your answer.   (*2 marks*)
                                              AQA June 2011

2   Silkworms secrete silk fibres, which are harvested and
     used to manufacture silk fabric.
     Scientists have produced genetically modified (GM)
     silkworms that contain a gene from a spider.
     The GM silkworms secrete fibres made of spider web
     protein (spider silk), which is stronger than normal
     silk fibre protein.
     The method the scientists used is shown in **Figure 5**.
     (a) Suggest why the plasmids were injected into
          the eggs of silkworms, rather than into the
          silkworms.   (*2 marks*)
     (b) Suggest why the scientists used a marker gene
          and why they used the EGFP gene.   (2 marks)
          The scientists ensured the spider gene was
          expressed only in cells within the silk glands.
     (c) What would the scientists have inserted into the
          plasmid along with the spider gene to ensure

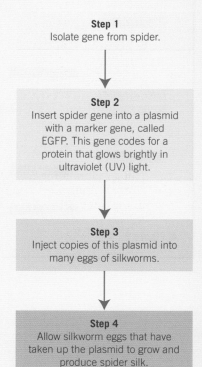

▲ **Figure 5**

that the spider gene was only expressed in the silk glands
of the silkworms? (*1 mark*)

(d) Suggest two reasons why it was important that the spider gene was
expressed only in the silk glands of the silkworms. (*2 marks*)

AQA Specimen 2014

3 Haemophilia is a genetic condition in which blood fails to clot. Factor
IX is a protein used to treat haemophilia. Sheep can be genetically
engineered to produce Factor IX in the milk produced by their
mammary glands. The diagram shows the stages involved in this process.

(a) Name the type of enzyme that is used to cut the gene for Factor IX
from human DNA (Stage **1**). (*1 mark*)

(b) (i) The jellyfish gene attached to the human Factor IX gene
(Stage **2**) codes for a protein that glows green under fluorescent
light. Explain the purpose of attaching this gene. (*2 marks*)

(ii) The promoter DNA from sheep (Stage **3**) causes transcription
of genes coding for proteins found in sheep milk. Suggest the
advantage of using this promoter DNA. (*2 marks*)

(c) Many attempts to produce transgenic animals have failed. Very few
live births result from the many embryos that are implanted.

(i) Suggest **one** reason why very few live births result from the
many embryos that are implanted. (*2 marks*)

(ii) It is important that scientists still report the results from failed
attempts to produce transgenic animals. Explain why. (*2 marks*)

AQA June 2012

5 In gel electrophoresis DNA fragments separate according to their lengths
measured in base pairs. In a gel, standard marker fragments of known
length migrated as follows:

fragment length (bp)	distance migrated (mm)
10000	6.1
8000	9.5
6000	15.5
4000	21.8
2000	36.0
1000	50.0
500	61.2

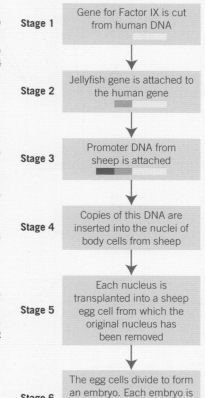

Stage 1 — Gene for Factor IX is cut from human DNA

Stage 2 — Jellyfish gene is attached to the human gene

Stage 3 — Promoter DNA from sheep is attached

Stage 4 — Copies of this DNA are inserted into the nuclei of body cells from sheep

Stage 5 — Each nucleus is transplanted into a sheep egg cell from which the original nucleus has been removed

Stage 6 — The egg cells divide to form an embryo. Each embryo is implanted into the uterus of a different sheep

(a) Plot a calibration graph of $\log_{10}$ fragment length against distance migrated (3 marks)

(b) Use the graph to estimate the length of a fragment which migrates 27.0 mm
in the same gel (2 marks)

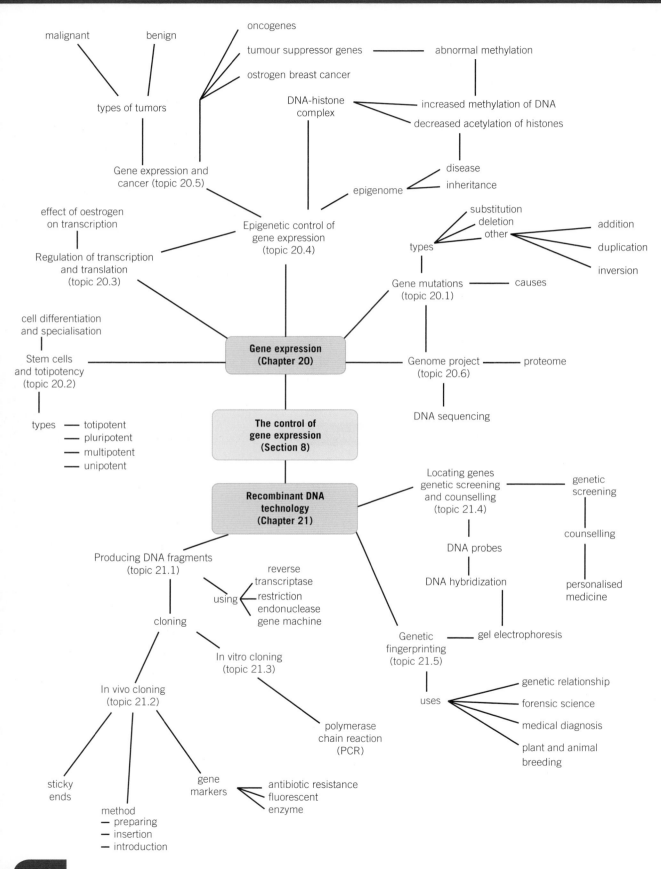

Practical skills

In this section you have met the following practical skills:

- How to carry out an experiment to show the effect of the environment on the phenotype of a plant species.
- How to obtain and evaluate experimental data linking smoking to disease.
- How to carry out *in vivo* and *in vitro* cloning of DNA fragments.
- How to use gene markers to identify whether a gene has been taken up by a bacterial cell.
- How to carry out gel electrophoresis.

Maths skills

In this section you have met the following maths skills:

- Interpret graphical information, for example, about the expression of genes in haemoglobin.
- Interpret and understand tabular information, for example, the results of an experiment to investigate the effects of growth factors on the development of plant tissue cultures.

Extension tasks

Recombinant DNA technology has allowed us to map the genomes of humans and other organisms. This has provided us with vast quantities of information. With information comes power and opportunity; the power to make informed decisions and the opportunity to change what we do. Knowledge of an individual's genetic make-up is perhaps the greatest invasion of an individual's privacy.

Using the internet, find out what legislation governs who is allowed access to an individual's genome and what rules control the use of this information.

Genetic screening allows parents to choose whether to have children that might possess a mutant gene. In time this could lead to the removal of these genes from the human genome. This, along with other human activities that you have covered during your A-level course, reduces genetic diversity.

Does mankind have a responsibility to maintain genetic diversity? Write a short passage in answer to this question justifying your viewpoint.

1 Human Immunodeficiency virus (HIV) particles have a specific protein on their surface. This protein binds to a receptor on the plasma membrane of a human cell and allows HIV to enter. This HIV protein is found on the surface of human cells after they have become infected with HIV.

Scientists made siRNA to inhibit expression of a specific HIV gene inside a human cell. They attached this siRNA to a carrier molecule. The flow chart shows what happens when this carrier molecule reaches a human cell infected with HIV.

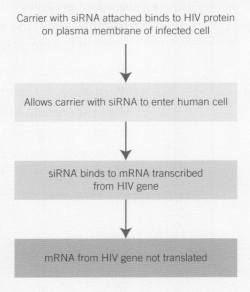

Carrier with siRNA attached binds to HIV protein on plasma membrane of infected cell

Allows carrier with siRNA to enter human cell

siRNA binds to mRNA transcribed from HIV gene

mRNA from HIV gene not translated

(a) When siRNA binds to mRNA, name the complementary base pairs holding the siRNA and mRNA together. One of the bases is named for you:

.... with ......adenine.........

.... with .... (*1 mark*)

(b) This siRNA would **only** affect gene expression in cells infected with HIV. Suggest **two** reasons why. (*4 marks*)

(c) The carrier molecule on its own may be able to prevent the infection of cells by HIV. Explain how. (*2 marks*)

AQA June 2013

2 Scientists wanted to measure how much mRNA was transcribed from allele **A** of a gene in a sample of cells. This gene exists in two forms, **A** and **a**.

The scientist isolated mRNA from the cells. They added an enzyme to mRNA to produce cDNA.

(a) Name the type of enzyme used to produce the cDNA. (*1 mark*)

The scientists used the polymerase chain reaction (PCR) to produce copies of the cDNA. They added a DNA probe for allele **A** to the cDNA copies. This DNA probe had a dye attached to it. This dye glows with a green light **only** when the DNA probe is attached to its target cDNA.

(b) Explain why this DNA probe will only detect allele **A**. (*2 marks*)

(c) The scientists used this method with cells from two people, H and G. One person was homozygous, **AA**, and the other was heterozygous, **Aa**. The scientists used the PCR and the DNA probe specific for allele **A** on the cDNA from both people.

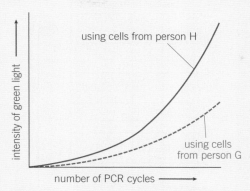

▲ **Figure 1** *shows the scientists' results*

(i) Explain the curve for person H. *(3 marks)*
(ii) Which person, H or G, was heterozygous, **Aa**? Explain your answer. *(2 marks)*

AQA June 2014

3 **(a)** Describe how a gene can be isolated from human DNA. *(2 marks)*
   (b) Describe how an isolated gene can be replicated by the polymerase chain reaction (PCR). *(4 marks)*
   (c) (i) Describe how a harmless virus, genetically engineered to contain a CFTR gene, can be used to insert the gene into a cystic fibrosis sufferer. *(2 marks)*
      (ii) A virus used in gene therapy has RNA as its genetic material and has an enzyme called reverse transcriptase. Inside a human cell, reverse transcriptase uses viral RNA to make viral DNA. Explain why the enzyme is called *reverse transcriptase*. *(1 mark)*

AQA Jan 2004

4 **(a)** Plasmids are often used as vectors in genetic engineering.
      (i)   What is the role of a vector? *(1 mark)*
      (ii) Describe the role of restriction endonucleases in the formation of plasmids that contain donor DNA. *(2 marks)*
      (iii) Describe the role of DNA ligase in the production of plasmids containing donor DNA. *(1 mark)*
   (b) There are many different restriction endonucleases. Each type cuts the DNA of a plasmid at a specific base sequence called a restriction site. Figure 2 shows the position of four restriction sites, **J**, **K**, **L**, and **M**, for four different enzymes on a single plasmid. The distances between these sites is measured in kilobases of DNA.

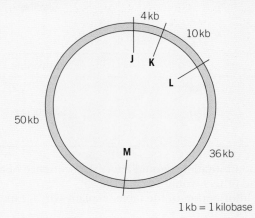

▲ **Figure 2**

The plasmid was cut using only two restriction endonucleases. The resulting fragments were separated by gel electrophoresis. The positions of the fragments are shown in Figure 3.

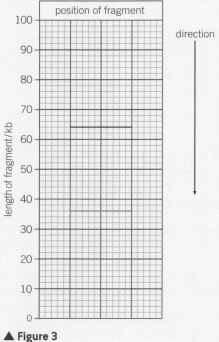

▲ **Figure 3**

(i)   Which of the restriction sites were cut?   (*1 mark*)
(ii) Explain your answer.   (*1 mark*)

AQA June 2006

5   (a) Gene mutations occur naturally.
Give **one** factor that increases the rate of gene mutations.   (*1 mark*)
(b) Table 1 shows the DNA base sequences that code for three amino acids.

▼ **Table 1**

DNA base sequence(s) coding for amino acids	Amino acid
CCA	
CCG	
CCT	Glycine
CCC	
TAC	Methionine
TAA	
TAG	Isoleucine
TAT	

Some substitution mutations would affect the sequence of amino acids in a polypeptide, and others would not.
Using only the information in the table, explain why. (3 marks)

AQA Jan 2007

6   Huntington's disease is a genetic condition that leads to a loss in brain function. The gene involved contains a section of DNA with many repeats of the base sequence CAG. The number of these repeats determines whether or not an allele of this gene will cause Huntington's disease.
   * An allele with 40 or more CAG repeats will cause Huntington's disease.
   * An allele with 36 – 39 CAG repeats may cause Huntington's disease.
   * An allele with fewer than 36 CAG repeats will not cause Huntington's disease.
     The graph shows the age at which a sample of patients with Huntington's disease first developed symptoms and the number of CAG repeats in the allele causing Huntington's disease in each patient.

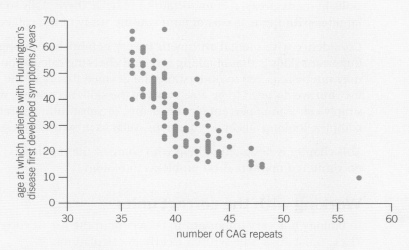

(a) (i) People can be tested to see whether they have an allele for this gene with more than 36 CAG repeats. Some doctors suggest that the results can be used to predict the age at which someone will develop Huntington's disease. Use information in the graph to evaluate this suggestion. (*3 marks*)

(ii) Huntington's disease is always fatal. Despite this, the allele is passed on in human populations. Use information in the graph to suggest why. (*2 marks*)

(b) Scientists took DNA samples from three people, **J**, **K** and **L**. They used the polymerase chain reaction (PCR) to produce many copies of the piece of DNA containing the CAG repeats obtained from each person. They separated the DNA fragments by gel electrophoresis. A radioactively labelled probe was then used to detect the fragments. The diagram shows the appearance of part of the gel after an X-ray was taken. The bands show the DNA fragments that contain the CAG repeats.

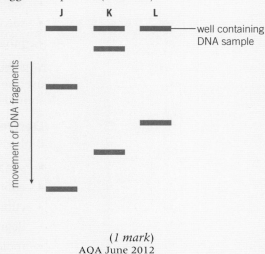

(i) Only one of these people tested positive for Huntington's disease. Which person was this? Explain your answer. (*2 marks*)

(ii) The diagram only shows part of the gel. Suggest how the scientists found the number of CAG repeats in the bands shown on the gel. (*1 mark*)

(iii) Two bands are usually seen for each person tested. Suggest why only one band was seen for Person **L**. (*1 mark*)

AQA June 2012

Section 9 Skills in A level Biology
Chapter 22 Mathematical skills

Biology students are often less comfortable with the application of mathematics compared with students such as physicists, for whom complex maths is a more obvious everyday tool. Nevertheless, it is important to realise that biology does require competent maths skills in many areas. It is important to practise these skills so you are familiar with them as part of your routine study of the subject.

Confidence with mental arithmetic is very helpful, but among the most important skills is that of taking care and checking calculations. We may not be required to understand the detailed theory of the maths we use, but we do need to be able to apply the skills accurately, whether simply calculating percentages or means, or substituting numbers into complex-looking algebraic equations, such as in statistical tests.

This chapter is designed to help with some of the regularly encountered mathematical problems in biology.

Working with the correct units

In biology it is very important to be secure in the use of correct units. These must always be written clearly in calculations.

Base units
The units we use are from the Système Internationale – the SI units. In biology we most commonly use the SI base units:

- metre (m) for length, height, distance
- kilogram (kg) for mass
- second (s) for time
- mole (mol) for the amount of a substance.

You should develop good habits right from the start, being careful to use the correct abbreviation for each unit used. For example, seconds should be abbreviated to s, not 'sec' or 'S'.

Derived units
Biologists also use SI derived units, such as:

- square metres (m^2) for area
- cubic metre (m^3) for volume
- cubic centimetre (cm^3), also written as millilitre (ml), for volume
- degree Celsius (°C) for temperature
- mole per litre (mol dm^{-3}) is usually used for concentration of a substance in solutions (although the official SI derived unit is moles per cubic metre)
- joule (J) for energy
- pascal (Pa) for pressure
- volt (V) for electrical potential.

Maths link $\sqrt{x}$

MS 0.1

Non-SI units

Although examination boards use SI units, you may also encounter non-SI units elsewhere, for example:

- litre (cubic decimetre) (l, L, dm³) for volume;
- Minute (min) for time;
- hour (h) for time;
- svedberg (S) (for sedimentation rate), used for ribosome particle size.

Unit prefixes

To accommodate the huge range of dimensions in our measurements, they may be further modified using appropriate prefixes. For example, one thousandth of a second is a millisecond (ms). This is illustrated in the Table 1.

▼ Table 1

Division	Factor	Prefix	Length		Mass		Volume		Time	
one thousand millionth	10^{-9}	nano	nanometre	nm	nanogram	ng	nanolitre	nl	nanosecond	ns
one millionth	10^{-6}	micro	micrometre	µm	microgram	µg	microlitre	µl	microsecond	µs
one thousandth	10^{-3}	milli	millimetre	mm	milligram	mg	millilitre	ml/cm³	millisecond	ms
one hundredth	10^{-2}	centi	centimetre	cm						
whole unit			metre	m	gram	g	litre	l/L/dm³	second	s
one thousand times	10^{3}	kilo	kilometre	km	kilogram	kg				

Converting between units

You may need to convert between units in order to be able to scale and express numbers in sensible forms. For example, rather than refer to the width of a cell in metres you would use micrometres (µm). This allows your measurements to be understood within the relevant scale of the observation.

Divide by 1000 for each step to convert in this direction ⟹

nano-	micro-	milli-	whole unit	kilo-
e.g. nm	e.g. µm	e.g. mm	e.g. m	e.g. km

⟸ Multiply by 1000 for each step to convert in this direction

▲ Figure 1

Examples:

Convert 1 m to mm: 1 × 1000 = 1000 mm

Convert 1 m to µm: 1 × 1000 = 1000 mm, then 1000 × 1000 = 1 000 000 µm

Convert 1 l to cm³: 1 × 1000 = 1000 cm³

Convert 20 000 µm to mm: 20 000 ÷ 1000 = 20 mm

Converting between square or cube units requires a bit more care.

One m^2 = 1000 × 1000 = 1 000 000 mm^2, so your conversion factor becomes × or ÷ 1 000 000.

One m^3 is 1000 × 1000 × 1000 = 1 000 000 000 mm^3, so your conversion factor now becomes × or ÷ 1 000 000 000.

Examples:

Convert 20 m^2 to km^2: 20 ÷ 1 000 000 = 0.000 02 km^2

Convert 1 m^2 to mm^2: 1 × 1 000 000 = 1 000 000 mm^2

Convert 5 000 000 mm^3 to m^3: 5 000 000 ÷ 1 000 000 000 = 0.005 m^3

Convert 0.000 000 7 m^3 to mm^3: 0.000 000 7 × 1 000 000 000 = 70 mm^3

Decimals and standard form

Maths link $\sqrt{x}$

MS 0.2

When you are using numbers that are very small, such as dimensions of molecules and organelles, it is useful to use **standard form** to express them more easily. Standard form is also commonly called **scientific notation**.

Standard form is essentially expressing numbers in powers of ten. For example, 10 raised to the power 10 means 10 × 10, i.e. 100. This may be written down as 10×10^1 or 1×10^2. To get to 1000 you use 10 × 10 × 10, which would be written as 1×10^3.

An easy way to look at this is to imagine the decimal point moving one place per power of ten. For example, to write down 58 900 000 000 as standard form, you would follow the steps below.

Step 1: write down the smallest number between 1 and 10 that can be derived from the number to be converted. In this case it would be 5.89.

Step 2: write the number of times the decimal place will have to shift to expand this to the original number as powers of ten. On paper this can be done by hopping the decimal over each number like this:

$$5.89000000000$$

▲ **Figure 2**

until the end of the number is reached. In this example, that requires 10 shifts, so the standard form should be written as 5.89×10^{10}.

Going the other way, for example expressing 0.000 007 8 as standard form, write the number in terms of the number of places the decimal place would have to hop forward to make the smallest number between 1 and 10, so to get to 7.8 you would have to hop over six times, so this number is written as 7.8×10^{-6}.

Significant figures

Maths link $\sqrt{x}$

MS 1.1

There are some simple rules to use when working out significant figures.

Rule 1: All non-zero digits are significant.

For example, 78 has two significant figures, 9.543 has four significant figures and 340 has two significant figures.

Rule 2: Intermediate zeros are significant.

For example, 706 has three significant figures and 5.900 76 has six significant figures.

Rule 3: Any leading zeroes are not significant.

For example, 0.005 67 has three significant figures (5,6 and 7; ignore the leading zeroes)

Rule 4: Zeroes at the ends of numbers containing decimal places are significant.

For example, 45.60 has four significant figures and 330.00 has five significant figures.

Significant figures and rounding

Table 2 shows the effect of rounding numbers to decimal places compared with significant figures. Remember that in rounding, when the next number is 5 or more round up, while if it is 4 or less don't round up. For example, 4.35 rounds to 4.4 and 4.34 rounds to 4.3.

Table 2 shows examples of rounding the number 23.336 00 to decimal places and to significant figures.

▼ **Table 2**

Measurements expressed by rounding to decimal places	Number of decimal places	Measurements expressed by rounding to significant figures	Number of significant figures	Measured to the nearest
23.336 00	5	23.336	5	100 thousandth
23.336 0	4	23.34	4	Ten thousandth
23.336	3	23.3	3	Thousandth
23.34	2	23	2	Hundredth
23.3	1	20	1	Tenth
23	0	—	—	Whole number

Significant figures and standard form

In standard form only the significant figures are written as digits, for example 5.600×10^3 has four significant figures. If this were written as a straight number it would be 5600. But according to the rules above, 5600 only has two significant figures – what does this mean?

In a given number, the significant figures are defined as the ones that contribute to its precision. Writing the number as 5600 implies precision only to the nearest whole hundred. The zeroes in the number could mean that it has simply been rounded, e.g. 5600.44 or even 5633. But if this number were actually more precise, for example it had been measured with equipment genuinely sensitive to the nearest hundredth part (2 decimal places) then 5600.00 is actually very precise and the two zeros have significance because they tell us that the measurement is *exactly* 5600 with no tenths or hundredths at all. So using standard form allows this precision to remain clearly as part of the stated number, because all significant figures are written.

Maths link $\sqrt{x}$

MS 1.2 and 1.6

Averages

An average value is actually a measure of central tendency. The most familiar measurement is the arithmetic mean (mean for short), but median or modal values are sometimes more appropriate to the data.

The arithmetic mean

Usually referred to simply as the mean, this is a measure of central tendency that takes into account the number of times each measurement occurs together with the range of the measurements. When repeated measurements are averaged, the mean will approach the true value, which will lie somewhere in the middle of the observed range, more accurately. This is why it is important to repeat experimental measurements, especially in biology where the natural unpredictability of living systems leads to inevitable fluctuations.

The mean is determined by adding together all the observed values and then dividing by the number of measurements made.

For a range of values of x, the mean $\bar{x} = \dfrac{\sum x}{n}$.

$\bar{x}$ is the mean value.

$\sum x$ is the sum of all values of x.

n is the number of values of x.

For example, five mice were weighed, giving masses of 6.2 g, 7.7 g, 6.7 g, 7.1 g and 6.3 g.

The mean mass is $(6.2 + 7.7 + 6.7 + 7.1 + 6.3) \div 5 = 6.8$ g

Be careful with your decimal places when calculating mean values. Your mean should normally have the same level of precision as the original measurements and therefore the same number of decimal places, otherwise you may be implying that the averaged measurements are more precise than they really are. For example, masses in whole grams would not average to a mass with one or more decimal places. Similarly averaging the numbers of whole objects should result in a whole number; if counts of bubbles in a pondweed experiment were averaged to a decimal place it implies you counted a fraction of one bubble, which is impossible!

The median

The median value in a set of data is calculated by placing the values in numerical order then finding the middle value in the range.

For example, the data set 12, 15, 10, 17, 9, 13, 13, 19, 10, 11 rearranges as 9, 10, 10, 11, 12, 13, 13, 15, 17, 19.

The middle of this range is 12.5.

The median value is very useful when data sets have a few values (outliers) at the extremes, which if included in an arithmetic mean could skew the data. It also allows comparison of data sets with similar means but a clear lack of overlap, skewed data and when there are too few measurements to calculate a reliable mean value.

For example, in the data set 1, 3, 3, 11, 12, 12, 12, 13, 14, 15, the median value is 12, a sensible looking mid point, but the mean would be 9.6, skewed to the left by the numbers at the lower extreme.

The mode
The modal value is the most frequent value in a set of data. It is very useful when interpreting data that is qualitative or in situations where the distribution has more than one peak (bimodal).

For example, in the data set 9, 10, 11, 11, 12, 13, 13, 13, 14, 17, 18, 19, the modal value is 13.

In biology, caution should be used because the sets of data are usually small and can introduce confusion. For example, in the data set 9, 10, 11, 11, 12, 13, 13, 14, 17, 18 there are apparently two modal values, 11 and 13, while in the set 11, 12, 13, 14, 17 there is no most frequent number and the mode is effectively every number and therefore of no value at all.

The modal value is not used very often, but it can be usefully applied when data is collected in categories, for example, numbers of moths attracted to lights of different colours.

Percentages
A percentage is simply expressing a fraction as a decimal. It is important to be confident with calculating percentages, which although straightforward are commonly calculated incorrectly.

Maths link

MS 0.3

Percentages as proportions and fractions
For example, two shapes of primrose flowers exist depending on stigma length; 'pin eyed' and 'thrum eyed'. In a survey of two areas of grassland, one area had 323 pin and 467 thrum (total 790 plants), the other had 667 pin and 321 thrum (total 988 plants). The percentage of pin eyed plants in each area is calculated as follows:

Area 1: $fraction = \frac{323}{790}$ which gives $decimal$ 0.41, which multiplied by 100 gives $percentage$ 41%.

The percentage of pin eyed flowers in Area 2 is $\frac{667}{988} \times 100 = 67.5\%$.

Percentages as chance
In genetics the likelihood of different offspring phenotypes should always be expressed as a percentage. For example, in a simple genetic cross between two heterozygous parents carrying the cystic fibrosis allele, one out of every four possible children could potentially be affected by the disorder. The chance of a cystic fibrosis child from these parents is therefore $\frac{1}{4} \times 100 = 25\%$.

Percentage change
This often comes up in osmosis experiments where samples (usually of potato tissue) gain and lose mass in different bathing solutions.

For example, a sample weighed 18.50 g at the start and at the end it weighed 11.72 g.

The actual loss in mass = 18.50 − 11.72 g = 6.78 g

The percentage change = $\frac{\text{mass change}}{\text{starting mass}} \times 100 = \frac{6.78}{18.50} \times 100 = -36.7\%$

Note the use of the minus sign to indicate that this is a loss.

Maths link $\sqrt{x}$

MS 2.2, 2.4 and 2.3

Equations

Substituting into equations

There are several equations (mathematical formulae) that you will need to be able to use in advanced level biology. You do not need to learn the theoretical maths from which they are derived, but you do need to be able to put known numerical values in the right place (this is *substituting into the equation*) and then calculate the result of the equation by performing the different steps in the right order (*this is solving the equation*).

An example that you will encounter during ecology studies is called the Simpson's Index of Diversity, which has the formula $= \dfrac{N(N-1)}{\sum n(n-1)}$.

Each symbol (*term*) in the equation has a specific meaning. In this example:

N means the total number of all individual organisms in a survey.

n means the total number of each different species.

$\sum$ means 'the total of' and requires you to add together all the indicated values.

Brackets indicate sub-calculations that must be done, for example $N - 1$ means the total of all species found -1.

The figures in brackets need to be multiplied by the figures outside them, e.g. $N(N-1)$ means $N \times (N-1)$.

An example of the data to use could be counts of the plant species found in a certain area. To make life easy, use a table like Table 3.

▼ **Table 3**

Plant species	Number of plants of each species found (n)	($n-1$)	$n(n-1)$
A	22	22 – 1 = 21	22 × 21 = 462
B	30	30 – 1 = 29	30 × 29 = 870
C	25	25 – 1 = 24	25 × 24 = 600
D	23	23 – 1 = 22	23 × 22 = 506
Totals of all plants = N	$N = 100$ $N - 1 = 99$		$\sum n(n-1) = 2438$

The brackets in equations always need to be solved first.

Begin by finding $n - 1$ for each plant (see column 3 in Table 3) and $N - 1$ (at the bottom of column 2).

Next work out $n(n-1)$ (column 4 in Table 3).

Now find $\sum n(n-1)$ by totalling the figures in column 4.

Substituting the known values into the equation works like this:

$D = \dfrac{N(N-1)}{\sum n(n-1)}$ becomes $D = \dfrac{100(99)}{2438}$ which calculates to $D = \dfrac{9900}{2438}$, which gives the result D = 4.1.

Rearranging equations

The individual parts or *terms* in equations are all related, but sometimes you might know all the values of the terms except one. The equation can be re-written so that the unknown term can be calculated. This is called rearranging or *changing the subject of* an equation. A very useful example of this arises during the study of microscopy and magnification.

The different terms are magnification, size of image and actual size of the object being observed. The equation that relates them together is:

$$magnification = \frac{size\ of\ image}{size\ of\ real\ object}.$$

You can use the equation to calculate magnification factors quite simply. For example, if you had a photograph of your pet dog, the magnification of the image would be the height of the image of that dog divided by its real height.

Be very careful to use the same units for each measurement! If the dog is 9 cm tall in the photograph and the real dog is 0.4 m tall you would have to convert the units before starting. 0.4 m is 40 cm, so the sum would be 9 ÷ 40 = 0.23. You picture's magnification is ×0.23.

Suppose you only had the photo and the magnification. How would you find out how big the real object was? You may need to do this type of calculation on photomicrographs of cells or parts of cells.

For example, a photograph shows a mitochondrion which is 41 mm long in the picture and is taken at magnification ×34 000. How long is the original mitochondrion?

To find out, rearrange the equation. You might use an equation triangle to help.

On Figure 3 the horizontal line means divide and the vertical line means multiply.

So $magnification = \dfrac{size\ of\ image}{size\ of\ real\ object}$

rearranges as *size of image = magnification × real size*

and $real\ size = \dfrac{size\ of\ image}{magnification}.$

You need to find the real size of the mitochondrion, so your sum will be:

$$real\ size = \frac{41}{34000} = 0.0012\ mm.$$

At this point you need to check that your units are sensible. A mitochondrion is so small that the appropriate unit of measurement is a micrometre (µm). The question may even ask you to use this unit. Earlier in this chapter you saw that 1 µm is 1/1000th of a mm, so to convert you need to multiply by 1000. The real mitochondrion is 0.0012 × 1000 = 1.2 µm long.

Gathering data and making measurements

Estimating results

When measuring and recording data it is useful to be able to make an estimate of the number you should be getting. This will allow you to judge whether the results you actually record seem believable. This

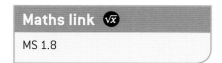

Maths link √x̄

MS 1.8

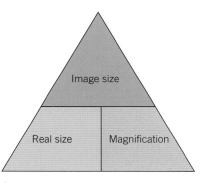

▲ Figure 3

Maths link √x̄

MS 0.4

is especially important when using a calculator, because it is easy to mis-type an entry and get a wrong answer. An estimate is really a sensible guess. It is a good idea to practise this skill, for example when collecting data from practical work in class.

Uncertainties in measurements

When making measurements, even using good quality instruments such as rulers and thermometers, there is a certain level of doubt in the precision of the measurement obtained. This is the *uncertainty of measurement*. The uncertainty can be stated, in which case a margin of error is identified. For example, measurements made using a good mm scale ruler a measurement may be reported as +/- 0.5mm, which is the maximum error likely when using the ruler carefully.

Percentage error is a way of using the maximum error to calculate the possible total error in a given measurement. Some types of instrument have maximum errors written on them, for example a balance may state +/- 0.01g. Other devices such as rulers and thermometers may rely on common sense, e.g. +/- 0.5mm or +/- 0.5 °C when recorded by eye. To find percentage error use the formula:

$$\% \ error = \frac{maximum \ error}{measured \ value \ recorded} \times 100$$

For example, with a ruler the maximum likely error is usually 0.5mm. If an object is measured at 6mm with the ruler, the percentage error = $\frac{0.5}{6} \times 100 = 8.3\%$.

A larger object will have a smaller % error because the +/- 0.5 mm is a lesser part of the total recorded, e.g. an object measured at 87mm has a % error = $\frac{0.5}{87} \times 100 = 0.6\%$.

Working with graphs and charts

Choosing the right type of graph or chart

During your course you will most commonly use line graphs, bar charts and histograms. You need to be able to choose the right one to suit the data and also to be able to draw graphs accurately.

The first part of your decision depends on the type of independent variable that you have measured. When you have used an independent variable that has specific values on a continuous scale, such as temperature, you should use a line graph, e.g. oxygen volume consumed by woodlice in a respirometer at a variety of temperatures. Alternatively your data might be in discrete categories, for example the number of left-handed or right-handed people. For this data a bar chart should be used with a space between each bar. When your categoric data is in groups that can be arranged on a continuous scale, e.g. height categories of plants such as 0 to 1 cm, >1 to 2 cm, >2 to 3 cm and so on, a histogram should be chosen, in which the bars are not separated by gaps.

Plotting the graph or chart

The rules when plotting the graph are:

- Ensure that the graph occupies the majority of the space available (this means more than half the space).
- Mark axes using a ruler and divide them clearly and equidistantly (i.e. 10, 20, 30, 40 not 10, 15, 20, 30, 45.

Maths link

MS 1.11

Maths link

MS 3.1, 3.2, 3.4, 3.5, and 1.3

Study tip

Decisions about data also have a large bearing on the choice of statistics and statistical tests you might need to use.

- Ensure that the dependent variable that you measured is on the *y* axis and the independent variable that you varied is on the *x* axis.

- Ensure that both axes have full titles and units clearly labelled, e.g. pH of solution, not just 'pH'; mean height/m, not just 'height'.

- Plot the points accurately using a sharp pencil and '×' marks so the exact position of the point is obvious and is not obscured when you plot a trend line.

- Draw a neat best-fit line, either a smooth curve or a ruled line. It does not have to pass through all the points. Alternatively use a point to point ruled line, which is often used in biology where observed patterns do not necessarily follow mathematically predictable trends!

- Confine your line to the range of the points. Never extrapolate the line beyond the range within which you measured. Extrapolation is conjecture! A common mistake is to try and force the plotted line to go through the origin.

- Distinguish separate plotted trend lines using a key.

- Add a clear concise title.

- Where data ranges fall a long way from zero, a broken axis will save space. For example, if the first value on the *y* axis is 36 it may be sensible to start the axis from 34 rather than zero. This will avoid leaving large areas of your graph blank.

You will be expected to follow these conventions. If you do, then questions that involve drawing a graph become easy.

Adding range bars and error bars to your plotted points

The position of the point on a graph is always subject to uncertainty. It may be a mean value, which will depend on the values averaged or whether you include or exclude any possible anomalies. A way of indicating the level of certainty in the positioning of your points is to use a range bar or error bar. These are ways of pictorially indicating the possible range of positions of the point and reflect the spread or variability in the original measurements that were averaged. The more spread the measurements, the less certain the position of the mean when plotted.

Table 4 shows some example data from an experiment on gas production by a photosynthesising plant at different temperatures, with which the different styles can be demonstrated.

Maths link $\sqrt{x}$

MS 1.10

Range bars

A range bar is the simplest way of showing the showing the spread in the data. Look for the maximum and minimum values in each set of repeats; they are picked out in bold in the table. After plotting the point on your graph mark the positions of the maximum and minimum values above and below the point using a small bar. Join the two extremes with a neat ruler line running vertically through the plotted point (Figure 4).

▼ Table 4

Temperature (°C)	Time taken to collect 10 cm³ of gas (s)								
	1	2	3	4	5	mean	*s*	*mean +s*	*mean −s*
15	87	95	102	**121**	117	104	14.4	118.4	89.6
20	67	78	**61**	**90**	86	76	12.3	88.3	63.7
25	57	59	**48**	**66**	51	56	7.0	63	49
30	**47**	45	39	42	**21**	39	10.4	49.4	28.6
35	**118**	123	**145**	136	132	131	10.7	141.7	120.3

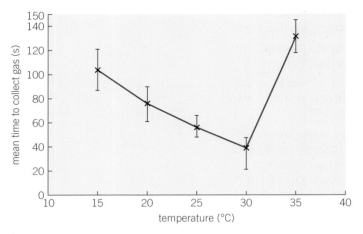

▲ **Figure 4** *Range bars*

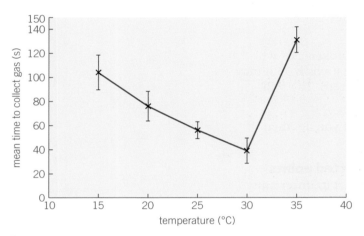

▲ **Figure 5** *Error bars*

Maths link √x̄

MS 3.5 and 3.6

Notice that the range bars are not always symmetrical above and below the plotted points. The tops and bottoms just show the largest or smallest values among the measurements made.

Error bars

To plot error bars you use the standard deviation (a calculated measure of the spread of the data), to indicate your ranges. This is better than using range bars because it reduces the effect of any extreme values in the dataset. In the table the values of standard deviation are shown in the column headed *s*.

Plot the bars by marking the top and lower limits exactly plus and minus one standard deviation above and below the point. These values are also included in the table. The result is shown in Figure 5.

Notice that the error bars are symmetrical above and below the points, which are now indicated with a range of ± one standard deviation. The length of the bars now indicates not the maximum/minimum values but the mathematical spread in the data. The more the data spread out around the mean, the longer the error bar becomes and the less certain you are that the mean is really accurate.

Calculating rates from graphs

When data have been plotted on a line graph relating measured values on the y axis to time on the x axis, it is possible to calculate a rate of change for the y variable. There are two common graph forms that you will encounter, shown in Figure 6.

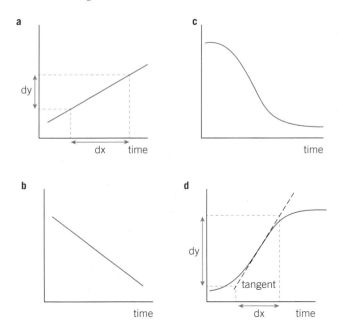

a might represent oxygen production by a photosynthesising plant

b might represent oxygen consumption by a respiring organism

c might represent pH change during a lipid digestion experiment

d might represent growth of a bacterial population in a fermenter

The rate of change is simply the gradient of the graph.

The formula that is used is $\frac{change\ in\ y}{change\ in\ x}$ or $\frac{dy}{dx}$

With a straight line graph follow these steps, which are marked on Figure 6a

- select any two points on the plotted line
- use a ruler to mark construction lines from the two points to the x and y axes
- measure the difference between the two points on the y axis, this is dy
- measure the time difference between the points on the x axis, this is dx
- substitute the values into the equation and remember to quote suitable units for the result, e.g. cm^3 O_2 per minute.

With a curved line the procedure is the same except that you need to start by marking a tangent against the curve, usually at its steepest point to find the maximum rate of change. This takes a bit of practise and is usually done by eye, although it is possible to calculate a position mathematically.

Once the tangent is drawn, select two points on it and proceed using the same steps that were applied to the straight line. Figure 6d has been marked with a tangent as an example.

Scatter diagrams

A scatter diagram is a method of plotting two variables in order to try and identify a correlation between them. The dependent variable is plotted on the y axis and the independent along the x axis. Once the points are plotted a trend line can be added to show a possible relationship. An example might be plotting incidence of lung cancer against number of cigarettes smoked per day. Once such a plot has been made the relationship can be tested using a statistical test, for example the correlation coefficient, r.

Maths link $\sqrt{x}$

MS 1.7

Probability

When data appears to show a pattern, it is possible to determine whether the pattern is simply due to chance or whether it has an underlying cause. For example, 36 throws of die should give near enough six of each possible number. Throwing 7 4's and 5 6's is likely to be a fluke, but if you threw 23 6's then this is definitely against the rules of probability!

Maths link $\sqrt{x}$

MS 1.4

Probability is assessed using a statistical test, for example chi-squared to test how closely observed measurements fit with expectation, such as in genetic cross results, or student t, which compares the means of sets of data to assess whether they differ, e.g. leaf width of sun versus shade grown ivy plants.

Such tests produce a calculated value that may be found in a table of probability. In these tables, it is the *probability that the data observed differ*

Maths link

MS 0.3, 1.5, 1.8, 1.9, 1.10, 2.2, 2.3, 2.4 and 4.1.

Maths link

MS 1.9

by chance alone that is being found. In biology we accept any probability greater than 5% as likely to be just chance or fluke, but probabilities of 5% or below show us that the data do differ significantly and there must be a cause influencing the outcome.

Using statistical tests to calculate probability

Chi-squared (χ^2) test

This test is used to compare the pattern in data collected or observed with the pattern that would be expected by chance. For example, if a die was thrown 36 times, each number should be expected 6 times if only chance determined the outcome. The further the actual numbers thrown deviate from this, the more likely the dice is loaded - a factor other than chance is influencing the outcome.

The test is commonly used to check the results of genetic crosses.

For example, a cross between two heterozygous tall plants has the expected outcome of 3 tall:1 short plant (short is the recessive allele). In the real results, 69 plants were tall and 28 short. We need to test the null hypothesis that there is no significant difference between the observed and expected results.

The formula for the chi-squared (χ^2) test is $\chi^2 = \sum \dfrac{(O-E)^2}{E}$. It is easiest to lay out the calculation in a table.

	Observed (O)	Expected (E)	(O–E)	(O–E)^2	$\dfrac{(O-E)^2}{E}$
Tall pea plants	69	72.75	3.75	14.06	0.19
Dwarf pea plants	28	24.25	3.75	14.06	0.58
				Total :	$\chi^2 = 0.77$

The probability value can now be found by using the table of chi-squared distribution (Table 6).

First, work out the degrees of freedom, which is the number of categories minus 1. Here there are two categories (tall or short), so 1 degree of freedom.

The critical value of χ^2 for one degree of freedom at the $p = 5\%$ level is 3.84. The calculated value of 0.77 is less than 3.84 and gives a p value of between 0.1 and 0.5 (10% and 50%) of the data collected being just chance. In this case we should accept the null hypothesis; there is no evidence of a significant difference between the observed and expected numbers.

Maths link

MS 1.9

Student *t* test

This test is used to judge the significance of any difference between the means of sets of data that are collected from two groups. There must be enough data that a reliable mean can be calculated and it should be normally distributed. The number in each sample does not need to be the same, but will ideally be more than 15 samples in each set.

For example, limpet diameters were measured at two sites to find out whether there was any difference due to aspect. One site faced east and the other west. 28 animals were measured at each site. The null hypothesis was that there would be no difference between the sites.

	Site 1 (east)	Site 2 (west)
n	28	28
Mean limpet diameter / mm	35.64	37.36
Variance (s^2) (= the square of standard deviation) See Table 5.	77.17	74.4

First, calculate the value of t, using this formula:

$$t = \frac{\overline{x}_1 - \overline{x}_2}{\sqrt{\dfrac{s_1^2}{n_1} + \dfrac{s_2^2}{n_2}}}$$

Where s_1^2 and s_2^2 are the variances at each site, n_1 and n_2 are the numbers sampled at each site and $\overline{x}_1$ and $\overline{x}_2$ are the means for each site.

So substituting in the values from the table

$$t = \frac{35.64 - 37.36}{\sqrt{\dfrac{77.17}{28} + \dfrac{74.4}{28}}} = \frac{-1.72}{2.33} \text{ (ignore the } - \text{ sign)} = 0.74$$

Next, calculate the degrees of freedom. For an unpaired test this is $(n_1 + n_2) - 2 = 54$

The t value can be looked up in Table 8.

In this case the probability of the difference between the means being due to chance alone is more than 10%, so chance must have caused the difference. The null hypothesis is accepted and there is no significant difference between the two sites.

Correlation coefficient, r (Pearson's product moment correlation coefficient)

When sampling collects data from two variables it is possible to use a calculation to determine whether the two variables correlate in any way. For example, does rock pool algal diversity increase as pool surface area increases? The variables compared need to be plotted on scatter graphs which will indicate possible relationships that can then be tested.

The formula for correlation coefficient is $r = \dfrac{\Sigma(x - \overline{x}) \times (y - \overline{y})}{\sqrt{\Sigma(x - \overline{x})^2 \times \Sigma(y - \overline{y})^2}}$

In the equation

x = the values of the first variable

$\overline{x}$ = the mean of the values of the first variable

y = the values of the second variable

$\overline{y}$ = the mean of the values of the second variable

Σ = the sum of

For example, data were collected to investigate the possible negative correlation of wrinkling on horse chestnut seeds (conkers) with seed mass. The calculation is most conveniently laid out in a table to make it easy to follow the steps.

Maths link $\sqrt{x}$

MS 1.9

Seed mass (g) $= x$	Number of wrinkles per seed $= y$	$x - \bar{x}$	$y - \bar{y}$	$(x - \bar{x}) \times (y - \bar{y})$	$(x - \bar{x})^2$	$(y - \bar{y})^2$
12	1	5	−14	−70	25	196
10	3	3	−12	−36	9	144
8	8	1	−7	−7	1	49
6	15	−1	0	0	1	0
4	27	−3	12	−36	9	144
2	36	−5	21	−105	25	441
$\Sigma x = 42$	$\Sigma y = 90$			$\Sigma = -254$	$\Sigma = 70$	$\Sigma = 974$
$\bar{x} = 42 \div 6 = 7$	$\bar{y} = 90 \div 6 = 15$					

Now substitute the values from the table into the equation to find r

$$r = \frac{\Sigma(x - \bar{x}) \times (y - \bar{y})}{\sqrt{\Sigma(x - \bar{x})^2 \times \Sigma(y - \bar{y})^2}}$$

$$r = \frac{-254}{\sqrt{70 \times 974}} = \frac{-254}{\sqrt{68180}} = \frac{-254}{261} = -0.97$$

This suggests a negative correlation. The strength of this correlation may be found by looking up the calculated value of r in the table of r values (Table 7).

In this test the number of degrees of freedom = the number of values for the two variables (n) − 2 ($df = n - 2$). In this example $df = 12 - 2 = 10$.

At 10 degrees of freedom a value of 0.97 gives a p value of <0.001 or <0.1% of the observed data being due only to chance. Thus the correlation is 99.9% certain.

▼ **Table 5** *Formulae commonly used in biology*

Circumference of a circle	$\pi \times d$	d = diameter
Surface area of a cube or cuboid	$2(ab) + 2(ac) + 2(bc)$	a, b and c are side lengths
Surface area of a sphere	$4\pi r^2$	r is the radius
Surface area of a cylinder	$2\pi r^2 + (\pi d \times h)$	h is the length or height of a cylinder
Volume of a cube or cuboid	$a \times b \times c$	a, b and c are side lengths
Volume of a sphere	$\frac{4}{3}\pi r^3$	r is the radius
Volume of a cylinder	$\pi r^2 h$	r is the radius
magnification	$magnification = \frac{image\ size}{real\ size}$	Rearrange to find the other quantities
pH	$pH = -log_{10}[H^+]$	$[H^+]$ is the concentration of the hydrogen ion in moles per litre
Pulmonary ventilation rate	$PVR = tidal\ volume \times breathing\ rate$	
Cardiac output	$CO = stroke\ volume \times heart\ rate$	

Species diversity index	$D = \dfrac{N(N-1)}{\sum n(n-1)}$	N is the grand total of all species sampled n is the number of each individual species sampled
Lincoln index	$N = \dfrac{S_1 \times S_2}{R}$	S_1 = total captured and marked, S_2 = total number recaptured, R = number of marked animals recaptured
Efficiency of energy transfer	$\dfrac{energy\ transferred}{energy\ intake} \times 100\%$	
Net Primary Production	$NPP = GPP - R$	GPP is gross primary production and R is energy loss in respiration.
Net production by consumers, N	$N = I - (F + R)$	I = potential energy in ingested food, F = energy lost in faeces and urine, R = energy lost in respiration.
Standard deviation, s Used to assess spread or dispersion in a set of data	$s = \sqrt{\dfrac{\sum x^2 - \dfrac{(\sum x)^2}{n}}{n-1}}$	x refers to the values of the measurements taken n = the number of measurements $\sum$ means "the sum of"
An alternative sum for standard deviation	$s = \sqrt{\dfrac{\sum(x - \bar{x})^2}{n-1}}$	$\bar{x}$ is the mean value of the values of x
Chi-squared (χ^2) test, used to compare agreement between sample and expectation	$x^2 = \sum \dfrac{(O - E)^2}{E}$	O are the values you actually measure E are the values you expected to see
Student t test, used to compare the means of two sets of data	$t = \dfrac{\bar{x}_1 - \bar{x}_2}{\sqrt{\dfrac{s_1^2}{n_1} + \dfrac{s_2^2}{n_2}}}$	s^2 is the variance of a set of data Subscript denotes the data being compared, e.g. n_1 is the number of values of the first set s_2^2 is the variance of the second set $\bar{x}$ is the mean value of the values of x
Variance s^2, which is also a measure of dispersion in a set of data	$s^2 = \dfrac{\sum(x - \bar{x})^2}{n-1}$	
Correlation coefficient, r	$r = \dfrac{\sum(x - \bar{x}) \times (y - \bar{y})}{\sqrt{\sum(x - \bar{x})^2 \times \sum(y - \bar{y})^2}}$	use this formula to test correlation between two variables. Values will lie between 1 (perfect positive correlation) and −1 (perfect negative correlation).
Hardy-Weinberg formula	$p^2 + 2pq + q^2 = 1$	P = frequency of dominant allele, q = frequency of recessive allele

Statistical test tables

Table 6 *Table of values of chi-squared*

df	p values								df
	0.99	0.95	0.90	0.50	0.10	0.05	0.01	0.001	
1	0.0016	0.0039	0.016	0.46	2.71	3.84	6.63	10.83	1
2	0.02	0.10	0.21	1.39	4.60	5.99	9.21	13.82	2
3	0.12	0.35	0.58	2.37	6.25	7.81	11.34	16.27	3
4	0.30	0.71	1.06	3.36	7.78	9.49	13.28	18.46	4
5	0.55	1.14	1.61	4.35	9.24	11.07	15.09	20.52	5
6	0.87	1.64	2.20	5.35	10.64	12.59	16.81	22.46	6
7	1.24	2.17	2.83	6.35	12.02	14.07	18.48	24.32	7
8	1.65	2.73	3.49	7.34	13.36	15.51	20.09	26.12	8
9	2.09	3.32	4.17	8.34	14.68	16.92	21.67	27.88	9
10	2.56	3.94	4.86	9.34	15.99	18.31	23.21	29.59	10
11	3.05	4.58	5.58	10.34	17.28	19.68	24.72	31.26	11
12	3.57	5.23	6.30	11.34	18.55	21.03	26.22	32.91	12
13	4.11	5.89	7.04	12.34	19.81	22.36	27.69	34.53	13
14	4.66	6.57	7.79	13.34	21.06	23.68	29.14	36.12	14
15	5.23	7.26	8.55	14.34	22.31	25.00	30.58	37.70	15

Table 7 *Values of r, correlation coefficient*

df	p values			
	0.1	0.05	0.01	0.001
1	0.9877	0.9969	0.9999	1.0000
2	0.9000	0.9500	0.9900	0.9990
3	0.8054	0.8783	0.9587	0.9912
4	0.7293	0.8114	0.9172	0.9741
5	0.6694	0.7545	0.8745	0.9507
6	0.6215	0.7067	0.8343	0.9249
7	0.5822	0.6664	0.7977	0.8982
8	0.5494	0.6319	0.7646	0.8721
9	0.5214	0.6021	0.7348	0.8471
10	0.4973	0.5760	0.7079	0.8233
11	0.4762	0.5529	0.6835	0.8010
12	0.4575	0.5324	0.6614	0.7800
13	0.4409	0.5139	0.6411	0.7603
14	0.4259	0.4973	0.6226	0.7420
15	0.4124	0.4821	0.6055	0.7246
16	0.4000	0.4863	0.5897	0.7084
17	0.3887	0.4555	0.5751	0.6932
18	0.3783	0.4438	0.5614	0.6787
19	0.3687	0.4329	0.5487	0.6652
20	0.3598	0.4227	0.5368	0.6524

Table 8 *Values of t*

Degree of freedom (df)	p values			
	0.10	0.05	0.01	0.001
1	6.31	12.71	63.66	636.60
2	2.92	4.30	9.92	31.60
3	2.35	3.18	5.84	12.92
4	2.13	2.78	4.60	8.61
5	2.02	2.57	4.03	6.87
6	1.94	2.45	3.71	5.96
7	1.89	2.36	3.50	5.41
8	1.86	2.31	3.36	5.04
9	1.83	2.26	3.25	4.78
10	1.81	2.23	3.17	4.59
12	1.78	2.18	3.05	4.32
14	1.76	2.15	2.98	4.14
16	1.75	2.12	2.92	4.02
18	1.73	2.10	2.88	3.92
20	1.72	2.09	2.85	3.85
α	1.64	1.96	2.58	3.29

Chapter 23 Practical skills

Practical skills are at the heart of Biology, a good foundation in practical skills will help you take your skills to a higher level. Biology is a dynamic subject in which our understanding constantly changes, largely as a result of developments in practical research. In the A level specification there is a separate practical endorsement which requires you to carry out 12 practicals across the two years of the A level course. The practical endorsement is not graded as part of your A level qualification. It is assessed by your teachers as a pass or a fail, a pass will be reported separately on your A level certificate. You will also be required to apply your understanding and knowledge of practicals to the written exams. Practical-based questions account for 15% of the total assessment – the majority of these questions are in paper 3.

By undertaking the set practical activities in this course, it will not only develop your manipulative skills with specific apparatus and techniques but will also help you to gain a deeper understanding into the processes of scientific investigations. Skills such as researching, planning, implementing by making and processing measurements safely, analysing, and evaluating results will be reinforced and enhanced.

It is advantageous for you to answer practical questions when you have completed the practical – any questions on practical skills will have been written with the expectation that you will have carried out the practical activities. Having undertaken the practical, this helps with the teaching and learning of concepts in the specification. A richer practical experience will be gained if you do more practicals than the following twelve set practical activities in Table 1. Table 1 shows the 12 practicals which will be assessed in exams. For the practical endorsement the 12 practicals can consist of the required practicals or teacher devised practicals. For each activity, Table 1 references the relevant topic(s) in this book, the first six you will have already covered in your AS year of study.

▼ **Table 1** *A level required practical activities*

	Practical	Topic
1	Investigation into the effect of a named variable on the rate of an enzyme-controlled reaction	1.8 Factors affecting enzyme action
2	Preparation of prepared squashes of cells from plant root tips; set-up and use of an optical microscope to identify the stages of mitosis in these stained squashes and calculation of a mitotic index	3.1 Methods of studying cells 3.7 Mitosis
3	Production of a dilution series of a solute to produce a calibration curve with which to identify the water potential of plant tissue	4.3 Osmosis
4	Investigation into the effect of a named variable on the permeability of cell-surface membranes	4.1 Structure of the cell-surface membrane
5	Dissection of animal or plant gas exchange systems, a mass transport system or of an organ within such a system	6.2 Gas exchange in insects 6.3 Gas exchange in fish 6.4 Gas exchange in the leaf of a plant 6.6 Mammalian lungs
6	Use of aseptic technique to investigate the effect of antimicrobial substances on microbial growth	9 Genetic diversity and adaptation
7	Use of chromatography to investigate the pigments isolated from leaves of different plants, e.g., leaves from shade-tolerant and shade intolerant plants or leaves of different colours	3.5.1 Photosynthesis
8	Investigation into the effect of a named factor on the rate of dehydrogenase activity in extracts of chloroplasts	3.5.1 Photosynthesis

▼ **Table 1** *continued*

	Practical	Topic
9	Investigation into the effect of a named variable on the rate of respiration of cultures of single-celled organisms	3.5.2 Respiration
10	Investigation into the effect of an environmental variable on the movement of an animal using either a choice chamber or a maze	3.7.4 Populations in ecosystems
11	Production of a dilution series of a glucose solution and use of colorimetric techniques to produce a calibration curve with which to identify the concentration of glucose in an unknown 'urine' sample	3.6.4.2 Control of blood glucose
12	Investigation into the effect of a named environmental factor on the distribution of a given species	3.7.2 Populations

Practical questions

The following questions are designed to give you some practice at this practical style of question. If you haven't completed the practical yet, just think of similar practicals you have done or when you have used the apparatus and this will help you.

Practical 1 – The effect of pH on catalases

A celery extract was liquidised and prepared by the technician as a source of the enzyme catalase. A burette had been filled up to the 50 cm³ mark with hydrogen peroxide. 10 cm³ of celery extract was added and the height of the upper level of the frothing liquid was recorded. The class was asked to repeat the procedure adding the following to the H_2O_2:

- Add 2 drops HCl / 2 drops distilled water
- Add 4 drops HCl
- Add 2 drops NaOH / 2 drops distilled water
- Add 4 drops NaOH

The pH of each solution was tested before starting the experiment.

1 (a) Sketch a graph of your expected results. Remember to label your axes.
   (b) List all variables that need to be controlled and how you would control them.

Describe how you could change the method to make it:

2 (a) more reliable
   (b) more valid

Practical 2 – The mitotic index

A student performed a root tip squash on tissue from a garlic root tip using acetic orcein stain. She counted the number of cells she could see in one of the stages of mitosis. In total there were 150 cells and 12 cells were undergoing mitosis.

1 (a) Calculate the mitotic index for these cells. Show your working.   (*2 marks*)
   (b) Another student didn't follow the exact procedure and as a result did not see any cells undergoing mitosis.
        Suggest two reasons why she did not see any cells in any stages of mitosis.   (*2 marks*)

In a further investigation to see the effect of cells environment on cell division, cells were taken from varying distances from the root tip and the number of cells undergoing mitosis was counted. To make the results quantitative, the student calculated the mitotic index for each sample and plotted it on graph 1.

Graph 1

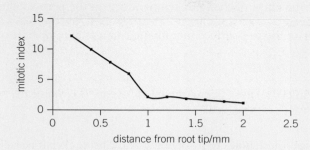

distance from root tip/mm

2 (a) Using graph 1, describe the results.   (2 marks)
   (b) Suggest one reason why this relationship exists?   (1 mark)

Practical 3 – Water potential in plant tissue

A practical was carried out to estimate the water potential in plant tissue. Six solid cylinders of potato were prepared each with identical dimensions. The students were given 100 cm³ of a stock 0.5 mol dm⁻³ solution sucrose. They were instructed to make up a series of six different dilutions. The mass (g) of the potato cylinder was measured before being submerged in the solution for 1 hour and then measured again after 1 hour.

Concentration of sucrose ($mol\,dm^{-3}$)	Mass before submerging in solution (g)	Mass after submerging in solution (g)	Percentage change in mass of potato tissue (%)
0.0	4.5	5.0	
0.1	3.9	4.3	
0.2	4.3	4.5	
0.3	4.1	4.2	
0.4	4.4	3.7	
0.5	4.4	3.6	

1   Construct a table to show how to make up 20 cm³ of each of the six dilutions required.

2   (a) Calculate the % change in mass for all results in the results table. Show your working.
    (b) Why is it important to calculate a % change of mass in this experiment?

3   (a) Plot a graph of sucrose concentration against % change in mass.
    (b) Use this graph to find the concentration of sucrose.
    (c) Describe how you can estimate the water potential with this value.

Practical 4 - Effect of temperature on beetroot cell-surface membranes

Core samples of beetroot were washed and put into tubes containing distilled water. Each tube was left in a different temperature for 20 minutes. The distilled water in the tube became coloured and was transferred to a colorimeter and a reading was taken.

Graph 2

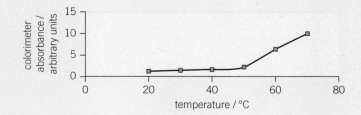

temperature / °C

1 **(a)** Explain why the distilled water in the tubes becomes coloured?
   (b) Using the graph, describe the effect of changing temperature on the permeability of the cell-surface membrane.
   (c) Explain how the structure of the cell-surface membrane changes at temperatures above 50°C?

Practical 5 – Dissection

A live locust was being examined in class. Using a magnifying glass, tiny holes on each side of the segments were visible.

1 **(a)** Name these small holes?
   (b) Further observation showed these holes to be opening and closing. What benefit does this give the locust?
   (c) Some other insects have hairs around these holes.
   What environmental condition does this help them survive in and how does it aid their survival?

The teacher dissected a locust and exposed the inside of the body cavity. The teacher located the tracheal tubes and mounted a small sample of tissue onto a slide. The tracheal tubes were seen to be highly branched.

2 **(a)** Give one advantage of the tubes being highly branched.
   (b) What substances cause the ring thickening of the tracheae?
   (c) The tracheae branch into tubes of a much smaller diameter with little thickening. Suggest a reason for this structure.

Practical 6 – Aseptic technique

The technician set up a petri dish with nutrient agar jelly using equipment that had been in an autoclave. He inoculated it with non-pathogenic bacteria and left it to incubate for 48 hours.

1 **(a)** Why did he use equipment that had been in an autoclave?
   (b) Describe the steps he took to transfer the bacteria from the bottle to the petri dish. Only make reference to steps concerning keeping conditions aseptic.
   (c) Explain why he disinfected the work surface and washed his hands after the experiment was finished?

Practical 7

A student used paper chromatography to separate the pigments from a sample of petals. The procedure followed by the student is shown in the diagram below.

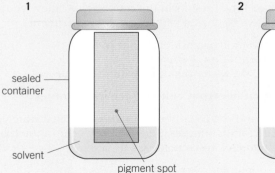
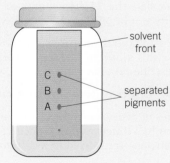

(a) State one precaution which the student should have taken in this investigation.
(*1 mark*)

(b) Outline the chemical principle illustrated by this technique.
(*2 marks*)

The table shows the distance moved by the solvent and pigments.

Substance	Distance moved (mm)
Solvent front	93
Pigment A	18
Pigment B	35
Pigment C	36

(c) (i) Define the term Rf value. *(2 marks)*

(ii) Calculate the Rf values for pigments B and C. Show your working. *(2 marks)*

(iii) Suggest one way by which greater separation of pigments B and C could have been achieved. *(1 mark)*

Practical 8

In photosynthesis the light-dependent reactions produce a reducing agent. This normally reduces NADP, but in this experiment the electrons are accepted by the blue dye DCPIP. Reduced DCPIP is colourless. Spinach leaves were blended and placed in a cold isolation medium then filtered. The filtrate was then centrifuged until there was a small pellet of chloroplasts. The pellet was then suspended in solution. The pellet now distributed in the solution was divided between 5 test tubes and set up as below then DCPIP was added. Tubes 1, 2, 4 and 5 were placed near a light source, tube 3 was placed in darkness. The time taken for the DCPIP to decolourise in each tube was measured.

Tube	Leaf extract (cm^3)	Supernatant (cm^3)	Isolation medium (cm^3)	Distilled water (cm^3)	DCPIP solution (cm^3)
1	0.5	–	–	–	5
2	–	–	0.5	–	5
3	0.5	–	–	–	5
4	0.5	–	–	5	–
5	–	0.5	–	–	5

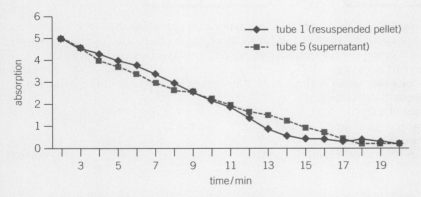

Tube 3 (incubated in the dark) gave a reading of 4.9 absorption units after 10 minutes. Tube 2 (DCPIP with no leaf extract) was 6.4 absorption units

1 Describe and explain the changes observed in the five tubes. Compare the results and make some concluding comments about what they show. *(5 marks)*

2 The rate of photosynthesis in intact leaves can be limited by several factors including light, temperature and carbon dioxide. Which of these factors will have little effect on the reducing capacity of the leaf extract? *(2 marks)*

3 Describe how you might extend this practical to investigate the effect of light intensity on the light-dependent reactions of photosynthesis. *(2 marks)*

Practical 9

An investigation was carried out to determine the effect of temperature on the rate of cellular respiration in yeast. Five experimental groups, each containing five fermentation tubes, were set up. The fermentation tubes all contained the same quantities of water, glucose, and yeast. Each group of five tubes was placed in a water bath at a different temperature. After 30 minutes, the volume of gas produced (D) in each fermentation tube was measured in millilitres. The mean for each group was calculated. A sample setup and the data collected are shown below.

Mean volume of gas produced (D) after 30 minutes at various temperatures

Group	Temperature (°C)	D (cm³)
1	5	0
2	20	5
3	40	12
4	60	6
5	80	3

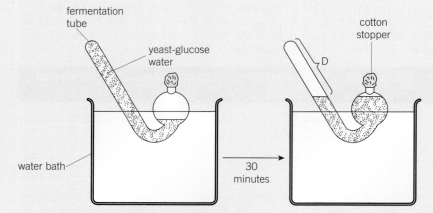

(a) Using the information in the data table, construct a line graph to show the relationship between temperature and the volume of gas produced. *(3 marks)*

(b) Deduce from the graph the temperature at which the maximum rate of cellular respiration in yeast occurred. *(1 mark)*

(c) Compared to the other tubes at the end of 30 minutes, state which of the following the tubes in group 3 contained:

(1) smallest volume of CO_2

(2) smallest quantity of glucose

(3) smallest quantity of ethanol

(4) same quantities of glucose, ethanol, and CO_2 *(2 marks)*

Practical 10

The diagram shows a choice chamber that can be used to investigate woodlice behaviour.

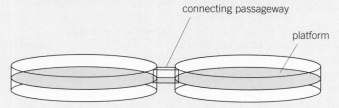

(a) Name two environmental factors that might attract woodlice to live underneath dead leaves. *(2 marks)*

(b) Design an experiment to show how you would use the choice chamber to investigate one of the factors you have named. *(3 marks)*
(c) Predict the results of your experiment *(1 mark)*
(d) Suggest one advantage to the woodlice of behaving in the way you predicted. *(1 mark)*

Practical 11

A student wanted to identify the concentration of glucose in an unknown sample. They used the following method

- A range of known glucose solutions were used starting with 20g dm^{-3}
- Each solution was heated with Benedict's solution. Once there was no more colour change the liquid was then cooled and filtered.
- The absorbance of the liquid was measured with a colorimeter.
- The student's results are shown below

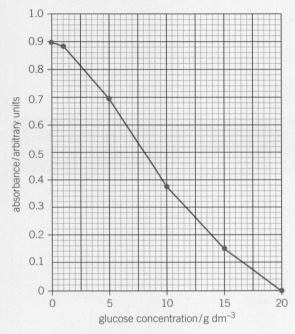

(a) State two precautions that the student should have taken during the procedure to ensure that the results give a valid comparison between the different glucose solutions *(2 marks)*
(b) The student tested the unknown sample and the absorbance reading obtained was 0.60 arbitrary units. Use the graph to determine the concentration of the sample. *(1 mark)*
(c) The procedure used does **not** test for non-reducing sugars such as sucrose. How could the student alter the procedure to determine the concentration of non-reducing sugar in the sample? *(2 marks)*

Practical 12

A group of students decided to investigate the relationship between light intensity and the distribution of herbs in a deciduous woodland. Light intensity was measured at randomly chosen sampling points within the woodland. At each location a sampling quadrat of diameter 1.0 m² was set up and the number of three herb species within the quadrat was recorded. The outline of the woodland and the sampling pattern used are shown in the diagram below.

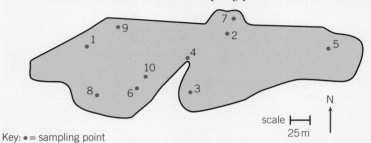

Key: ● = sampling point

(a) Outline a suitable procedure for randomly selecting sampling points in this investigation. *(2 marks)*

(b) State one advantage and one disadvantage of random sampling. *(2 marks)*

(c) A section of one student's records are shown in the table below.

Sample point	Number of herbs			Mean light intensity (% of maximum incident)
	Species A	Species B	Species C	
1	27	18	4	70
2	25	22	0	85
3	18	18	3	65
4	19	14	7	60
5	29	24	6	90
6	10	12	20	30
7	38	26	4	80
8	0	2	20	15
9	39	29	4	80
10	0	0	16	5

(i) Plot a graph showing the relationship between mean light intensity and the number of herb species A and C. *(5 marks)*

(ii) State the relationship between average light intensity and the number of herb species C. *(1 mark)*

(iii) Suggest one possible explanation for the distribution of herb species C. *(1 mark)*

(d) Suggest two adaptations which may be seen in the leaves of plants growing in low light intensities. *(2 marks)*

1 (a) Gas exchange in fish takes place in gills. Explain how two features of gills allow efficient gas exchange. *(2 marks)*

(b) A zoologist investigated the relationship between body mass and rate of oxygen uptake in four species of mammal. The results are shown in the graph.

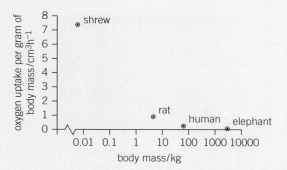

(i) The scale for plotting body mass is a logarithmic scale. Explain why a logarithmic scale was used to plot body mass. *(1 mark)*

(ii) Describe the relationship between body mass and oxygen uptake. *(1 mark)*

(iii) The zoologist measured oxygen uptake per gram of body mass. Explain why he measured oxygen uptake per gram of body mass. *(2 marks)*

(iv) Heat from respiration helps mammals to maintain a constant body temperature. Use this information to explain the relationship between body mass and oxygen uptake shown in the graph. *(3 marks)*

AQA, Jan 2010

2 (a) Describe how DNA is replicated. *(6 marks)*

(b) The graph shows information about the movement of chromatids in a cell that has just started metaphase of mitosis.

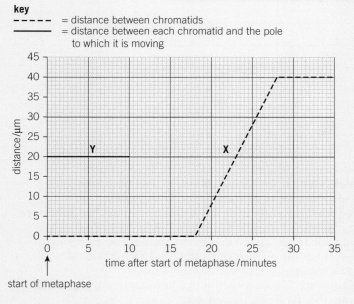

(i) What was the duration of metaphase in this cell? *(1 mark)*

(ii) Use line **X** to calculate the duration of anaphase in this cell. *(1 mark)*

(iii) Complete line **Y** on the graph. *(2 marks)*

(c) A doctor investigated the number of cells in different stages of the cell cycle in two tissue samples, **C** and **D**. One tissue sample was taken from a cancerous tumour. The other was taken from non-cancerous tissue. The table shows his results.

Stage of the cell cycle	Percentage of cells in each stage of the cell cycle	
	Tissue sample C	Tissue sample D
Interphase	82	45
Prophase	4	16
Metaphase	5	18
Anaphase	5	12
Telophase	4	9

(i)   In tissue sample **C**, one cell cycle took 24 hours. Use the data in the table to calculate the time in which these cells were in interphase during one cell cycle. Show your working. *(2 marks)*

(ii)   Explain how the doctor could have recognised which cells were in interphase when looking at the tissue samples. *(1 mark)*

(iii) Which tissue sample, **C** or **D**, was taken from a cancerous tumour? Use information in the table to explain your answer. *(2 marks)*

AQA Jan 2013

3   Taxol is a drug used to treat cancer. Research scientists investigated the effect of injecting taxol on the growth of tumours in mice. Some of the results are shown in **Figure 3**.

Number of days of treatment	Mean volume of tumour / mm^3	
	Control group	Group injected with taxol in saline
1	1	1
10	7	2
20	21	11
30	43	20
40	114	48
50	372	87

▲ **Figure 3**

(a)   Suggest how the scientists should have treated the control group. *(2 marks)*

(b)   Suggest and explain **two** factors which should be considered when deciding the number of mice to be used in this investigation. *(2 marks)*

(c)   The scientists measured the volume of the tumours. Explain the advantage of using volume rather than length to measure the growth of tumours. *(1 mark)*

(d)   The scientists concluded that taxol was effective in reducing the growth rate of the tumours over the 50 days of treatment. Use suitable calculations to support this conclusion. *(2 marks)*

(e)   In cells, taxol disrupts spindle activity. Use this information to explain the results in the group that has been treated with taxol. *(3 marks)*

(f)   The research scientists then investigated the effect of a drug called OGF on the growth of tumours in mice. OGF and taxol were injected into different mice as separate treatments or as a combined treatment. **Figure 4** and **Figure 5** show the results from this second investigation.

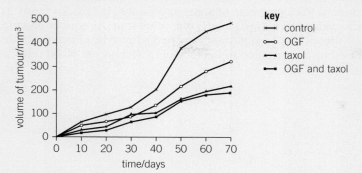

▲ Figure 4

Treatment	Mean volume of tumour following 70 days treatment / mm³ (± standard deviation)
OGF	322 (± 28.3)
Taxol	207 (± 22.5)
OGF and taxol	190 (± 25.7)
Control	488 (± 32.4)

▲ Figure 5

(i) What information does standard deviation give about the volume of the tumours in this investigation? *(1 mark)*

(ii) Use **Figure 4** and **Figure 5** to evaluate the effectiveness of the two drugs when they are used separately and as a combined treatment. *(4 marks)*

AQA Jan 2010

4   Read the following passage.

Gluten is a protein found in wheat. When gluten is digested in the small intestine, the products include peptides. Peptides are short chains of amino acids. These peptides cannot be absorbed by facilitated diffusion and leave the gut in faeces. Some people have coeliac disease. The epithelial cells of people with coeliac disease do not absorb the products of digestion very well. In these people, some of the peptides from gluten can pass between the epithelial cells lining the small intestine and enter the intestine wall. Here, the peptides cause an immune response that leads to the destruction of microvilli on the epithelial cells.   5

Scientists have identified a drug which might help people with coeliac disease. It reduces the movement of peptides between epithelial cells. They have carried out trials of the drug with patients with coeliac disease.   10

Use the information in the passage and your own knowledge to answer the following questions.

(a) Name the type of chemical reaction which produces amino acids from proteins. *(1 mark)*

(b) The peptides released when gluten is digested cannot be absorbed by facilitated diffusion (lines 2 – 3). Suggest why. *(3 marks)*

(c) The epithelial cells of people with coeliac disease do not absorb the products of digestion very well (lines 4 – 5). Explain why. *(3 marks)*

(d) Explain why the peptides cause an immune response (lines 7 – 8). *(1 mark)*

(e) Scientists have carried out trials of a drug to treat coeliac disease (lines 10 – 11). Suggest **two** factors that should be considered before the drug can be used on patients with the disease. *(2 marks)*

AQA June 2012

5 **(a)** Haemoglobin contains iron. One type of anaemia is caused by a lack of iron. This type of anaemia can be treated by taking tablets containing iron. A number of patients were given a daily dose of 120 mg of iron. **Figure 8** shows the effect of this treatment on the increase in the concentration of haemoglobin in their red blood cells.

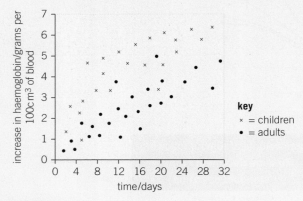

▲ **Figure 8**

(i) Give one difference in the response of adults and children to this treatment. *(1 mark)*

(ii) You could use the graph to predict the effect of this treatment on the increase in haemoglobin content of an adult after 40 days. Explain how. *(2 marks)*

(iii) Haemoglobin has a quaternary structure. Explain what is meant by a quaternary structure. *(1 mark)*

(b) (i) Pernicious anaemia is another type of anaemia. One method of identifying pernicious anaemia is to measure the diameter of the red blood cells in a sample of blood that has been diluted with an isotonic salt solution. Explain why an isotonic salt solution is used to dilute the blood sample. *(3 marks)*

(ii) A technician compared the red blood cells in two blood samples of equal volume.
One sample was from a patient with pernicious anaemia, the other was from a patient who did not have pernicious anaemia. **Figure 9** shows some of the results she obtained.

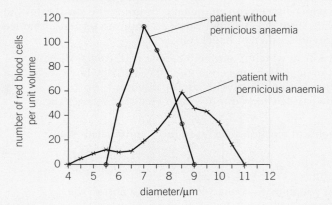

▲ **Figure 9**

Describe **two** differences between the blood samples. *(2 marks)*

(c) Scientists' analysis of blood proteins has indicated a lack of genetic diversity in populations of some organisms. Describe the processes that lead to a reduction in the genetic diversity of populations of organisms. *(6 marks)*

AQA June 2010

6 Students investigated the effect of different concentrations of sodium chloride solution on discs cut from an apple. They weighed each disc and then put one disc into each of a range of sodium chloride solutions of different concentrations. They left the discs in the solutions for 24 hours and then weighed them again. Their results are shown in the table.

Concentration of sodium chloride solution / mol dm^{-3}	Mass of disc at start / g	Mass of disc at end / g	Ratio of mass at start to mass at end
0.00	16.1	17.2	0.94
0.15	19.1	20.2	0.95
0.30	24.3	23.2	1.05
0.45	20.2	18.7	1.08
0.60	23.7	21.9	
0.75	14.9	13.7	1.09

(a) (i) Calculate the ratio of the mass at the start to the mass at the end for the disc placed in the 0.60 mol dm^{-3} sodium chloride solution. *(1 mark)*

(ii) The students gave their results as a ratio. What is the advantage of giving the results as a ratio? *(2 marks)*

(iii) The students were advised that they could improve the reliability of their results by taking additional readings at the same concentrations of sodium chloride. Explain how. *(2 marks)*

(b) (i) The students used a graph of their results to find the sodium chloride solution with the same water potential as the apple tissue. Describe how they did this. *(2 marks)*

(ii) The students were advised that they could improve their graph by taking additional readings. Explain how. *(2 marks)*

AQA Jan 2010

7 The diagram shows the structure of the cell-surface membrane of a cell.

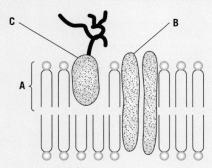

(a) Name **A** and **B**. *(2 marks)*

(b) (i) **C** is a protein with a carbohydrate attached to it. This carbohydrate is formed by joining monosaccharides together. Name the type of reaction that joins monosaccharides together. *(1 mark)*

(ii) Some cells lining the bronchi of the lungs secrete large amounts of mucus. Mucus contains protein.
Name **one** organelle that you would expect to find in large numbers in a mucus-secreting cell and describe its role in the production of mucus. *(2 marks)*

AQA June 2013

8 Students investigated the effect of removing leaves from a plant shoot on the rate of water uptake. Each student set up a potometer with a shoot that had eight leaves. All the shoots came from the same plant. The potometer they used is shown in the diagram.

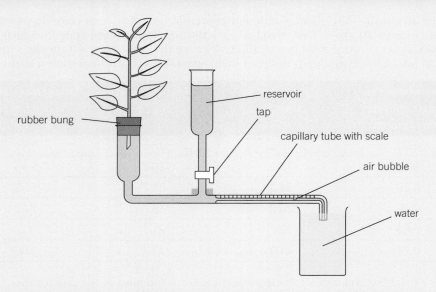

(a) Describe how the students would have returned the air bubble to the start of the capillary tube in this investigation. *(1 mark)*

(b) Give **two** precautions the students should have taken when setting up the potometer to obtain reliable measurements of water uptake by the plant shoot. *(2 marks)*

(c) A potometer measures the rate of water uptake rather than the rate of transpiration. Give two reasons why the potometer does not truly measure the rate of transpiration. *(2 marks)*

(d) The students' results are shown in the table.

Number of leaves removed from the plant shoot	Mean rate of water uptake / cm³ per minute
0	0.10
2	0.08
4	0.04
6	0.02
8	0.01

Explain the relationship between the number of leaves removed from the plant shoot and the mean rate of water uptake. *(3 marks)*

AQA Jan 2013

9 Read the following passage.

Some foods contain substances called flavenoids. Flavenoids lower blood cholesterol concentration and reduce the risk of developing coronary heart disease.

Some types of dark chocolate have a high concentration of flavenoids. One group of scientists investigated the effect of eating dark chocolate on the risk of developing coronary heart disease.   5

The scientists randomly divided healthy volunteers into two groups. Every day one group was given dark chocolate containing flavenoids to eat. The other group acted as a control.

The scientists measured the diameter of the lumen of the main artery in the arms of the volunteers every week. At the end of a month, the diameter of the lumen of the main artery in the arm of the volunteers who had eaten dark chocolate containing flavenoids had increased.   10

Use information from the passage and your own knowledge to answer the questions.

(a) High blood cholesterol concentration is a risk factor associated with coronary heart disease.

Give **two** other risk factors associated with coronary heart disease. *(2 marks)*

(b) (i) The scientists used healthy volunteers in this investigation (line 7). Why was it important that the volunteers were healthy? *(1 mark)*

(ii) The scientists randomly divided the volunteers into two groups (line 7). Explain why they divided them randomly. *(1 mark)*

(c) (i) Describe how the control group should have been treated. *(2 marks)*

(ii) Why was it important to have a control group in this investigation? *(1 mark)*

(d) Suggest why an increase in the diameter of the lumen of the main artery in the arm (lines 11–12) is associated with a reduced risk of coronary heart disease. *(3 marks)*

AQA June 2010

10 (a) What is intraspecific variation? *(1 mark)*

(b) Schizophrenia is a mental illness. Doctors investigated the relative effects of genetic and environmental factors on the development of schizophrenia. They used sets of identical twins and non-identical twins in their investigation. At least one twin in each set had developed schizophrenia.

- Identical twins are genetically identical.
- Non-identical twins are not genetically identical.
- The members of each twin pair were raised together.

The table shows the percentage of cases where both twins had developed schizophrenia.

Type of twin	Percentage of cases where both twins had developed schizophrenia
Identical	50
Non-identical	15

(i) Explain why both types of twin were used in this investigation. *(2 marks)*

(ii) What do these data suggest about the relative effects of genetic and environmental factors on the development of schizophrenia? *(1 mark)*

(iii) Suggest two factors that the scientists should have taken into account when selecting the twins to be used in this study. *(2 marks)*

AQA Jan 2013

11 (a) Students measured the rate of transpiration of a plant growing in a pot under different environmental conditions. Their results are shown in the table.

Conditions		Transpiration rate/gh^{-1}
A Still air	15 °C	1.2
B Still air	15 °C	1.7
C Still air	25 °C	2.3

During transpiration, water diffuses from cells to the air surrounding a leaf.

(i) Suggest an explanation for the difference in transpiration rate between conditions **A** and **B**. *(2 marks)*

(ii) Suggest an explanation for the difference in transpiration rate between conditions **A** and **C**. *(2 marks)*

(b) Scientists investigated the rate of water movement through the xylem of a twig from a tree over 24 hours. The graph shows their results. It also shows the light intensity for the same period of time.

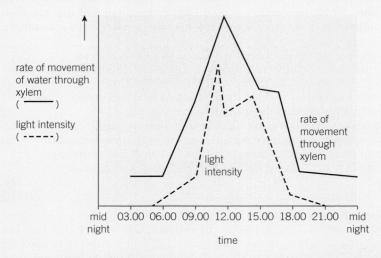

(i) Describe the relationship between the rate of water movement through the xylem and the light intensity. *(1 mark)*

(ii) Explain the change in the rate of water movement through the xylem between 06.00 and 12.00 hours. *(2 marks)*

(iii) The scientists also measured the diameter of the trunk of the tree on which the twig had been growing. The diameter was less at 12.00 than it was at 03.00 hours. Explain why the diameter was less at 12.00 hours. *(2 marks)*

AQA Jan 2011

12 Imatinib is a drug used to treat a type of cancer that affects white blood cells. Scientists investigated the rate of uptake of imatinib by white blood cells. They measured the rate of uptake at 4 °C and at 37 °C.

Their results are shown in the table.

Concentration of imatinib outside cells / µmol dm^{-3}	Mean rate of uptake of imatinib into cells / µg per million cells per hour	
	4 °C	**37 °C**
0.5	4.0	10.5
1.0	10.7	32.5
5.0	40.4	420.5
10.0	51.9	794.6
50.0	249.9	3156.1
100.0	606.9	3173.0

(a) The scientists measured the rate of uptake of imatinib in µg per million cells per hour. Explain the advantage of using this unit of rate in this investigation. *(2 marks)*

(b) Calculate the percentage increase in the mean rate of uptake of imatinib when the temperature is increased from 4 °C to 37 °C at a concentration of imatinib outside the cells of 1.0 µmol dm^{-3}.
Give your answer to one decimal place. *(2 marks)*

(c) Imatinib is taken up by blood cells by active transport.

(i) Explain how the data for the two different temperatures support this statement. *(2 marks)*

(ii) Explain how the data for concentrations of imatinib outside the blood cells at 50 and 100 µmol dm^{-3} at 37 °C support the statement that imatinib is taken up by active transport. *(2 marks)*

AQA June 2013

1 Metastatic melanoma (MM) is a type of skin cancer. It is caused by a faulty receptor protein in cell-surface membranes. There have been no very effective treatments for this cancer.

Dacarbazine is a drug that has been used to treat MM because it appears to increase survival time for some people with MM.

Doctors investigated the use of a new drug, called ipilimumab, to treat MM. They compared the median survival time (ST) for two groups of patients treated for MM:

- a control group of patients who had been treated with dacarbazine
- a group of patients who had been treated with dacarbazine and ipilimumab.

The ST is how long a patient lives after diagnosis.

The doctors also recorded the percentage of patients showing a significant reduction in tumours with each treatment.

The total number of patients in the investigation was 502.

Table 2 shows the doctors' results.

▼ Table 2

Treatment	Median survival time (ST) / months	Percentage of patients showing significant reduction in tumours
Dacarbazine	9.1	10.3
Dacarbazine and ipilimumab	11.2	15.2

1 (a) The doctors compared median survival times for patients in each group. How would you find the median survival time for a group of patients? *(2 marks)*

(b) In many trials of new drugs, a control group of patients is given a placebo that does not contain any drug. The control group in this investigation had been treated with dacarbazine. Suggest why they had not been given a placebo. *(1 mark)*

(c) A journalist who read this investigation concluded that ipilimumab improved the treatment of MM. Do the data in **Table 2** support this conclusion? Give reasons for your answer. *(4 marks)*

(d) MM is caused by a faulty receptor protein in cell-surface membranes. Cells in MM tumours can be destroyed by the immune system. Suggest why they can be destroyed by the immune system. *(3 marks)*

AQA Specimen 2014

2 (a) Insect pests of crop plants can be controlled by chemical pesticides or biological agents. Give two advantages of using biological agents. *(2 marks)*

Two-spotted mites are pests of strawberry plants. Ecologists investigated the use of predatory mites to control two-spotted mites. They released predatory mites on strawberry plants infested with two-spotted mites. They then recorded the percentage of strawberry leaves occupied by two-spotted mites and by predatory mites over a 16-week period. The results are shown on the graph.

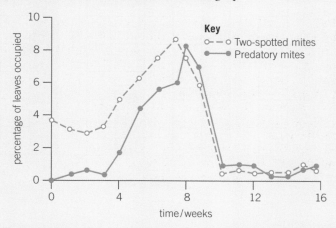

(b) Describe how the percentage of leaves occupied by predatory mites changed during the period of this investigation. *(2 marks)*

(c) The ecologists concluded that in this investigation control of the two-spotted mite by a biological agent was effective. Explain how the results support this conclusion. *(2 marks)*

(d) Farmers who grow strawberry plants and read about this investigation might decide **not** to use these predatory mites. Suggest **two** reasons why.

(e) The ecologists repeated the investigation but sprayed chemical pesticide on the strawberry plants after 10 weeks. After 16 weeks no predatory mites were found but the population of two-spotted mites had risen significantly. Suggest an explanation for the rise in the two-spotted mite population. *(2 marks)*

AQA June 2012

3 Mountains are harsh environments. The higher up the mountain, the lower the temperature becomes. The diagram shows a forest growing on the side of a mountain. The upper boundary of the forest is called the tree line. Trees do not grow above the tree line.

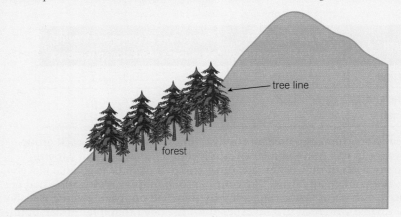

(a) **(i)** The position of the tree line is determined by abiotic factors. What is meant by an abiotic factor? *(1 mark)*

   (ii) Other than temperature, suggest **one** abiotic factor that is likely to affect the position of the tree line on the mountain. *(1 mark)*

(b) Scientists measured the concentration of carbon dioxide in the air in one part of the forest. They took measurements at different times of day and at two different heights above the ground. Their results are shown in the bar chart.

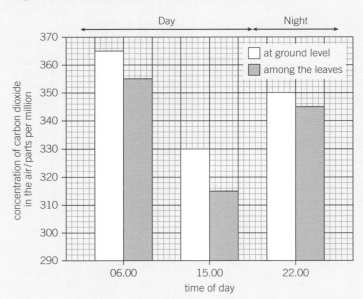

Use your knowledge of photosynthesis and respiration to explain the data in the bar chart. *(4 marks)*

(c) The population of trees in the forest evolved adaptations to the mountain environment. Use your knowledge of selection to explain how. *(3 marks)*

AQA Jan 2012

4 Plant physiologists attempted to produce papaya plants using tissue culture. They investigated the effects of different concentrations of two plant growth factors on small pieces of the stem tip from a papaya plant. Their results are shown in the table.

Concentration of auxin / $\mu mol\ dm^{-3}$	Concentration of cytokinin / $\mu mol\ dm^{-3}$		
	5	25	50
0	No effect	No effect	Leaves produced
1	No effect	Leaves produced	Leaves produced
5	Leaves produced	No effect	Leaves and some plantlets produced
10	Callus produced	Leaves and some plantlets produced	Plantlets produced
15	Callus produced	Callus and some leaves produced	Callus and some leaves produced

Callus is a mass of undifferentiated plant cells. Plantlets are small plants.

(a) Explain the evidence from the table that cells from the stem tip are totipotent. *(2 marks)*

(b) Calculate the ratio of cytokinin : auxin that you would recommend to grow papaya plants by this method. *(2 marks)*

(c) (i) Papaya plants reproduce sexually by means of seeds. Papaya plants grown from seeds are very variable in their yield. Explain why. *(2 marks)*

(ii) Explain the advantage of growing papaya plants from tissue culture rather than from seeds. *(1 mark)*

AQA June 2011

5 A Sri Lankan scientist investigated the effect of human disturbance on the organisms living on a rocky seashore. He chose three areas for the study. These areas had different amounts of human disturbance. The scientist measured human disturbance by walking from one end of the beach to the other. He recorded the number of people he encountered. Figure 1 shows his results.

	Site R	Site G	Site U
Mean number of people encountered per hour (± standard deviation)	2.2 (± 2.1)	17.6 (± 9.6)	34.6 (± 11.6)

▲ Figure 1

(a) (i) What conclusions can you draw about the number of people visiting Site **R** compared with the number of people visiting the other two sites? Give evidence from **Figure 1** to support your answer. *(2 marks)*

(ii) The scientist believed that there was a significant difference between the numbers of people visiting site R and the other sites. Select and carry out a suitable statistical test to compare the sites. *(3 marks)*

(iii) Comment on the scientist's conclusion about visitor numbers. *(2 marks)*

(b) The scientist used quadrats to find the number of species at each of the three sites. He carried out a preliminary investigation and recorded the total number of species in an increasing number of quadrats. **Figure 2** shows the results.

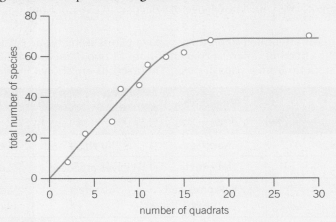

▲ **Figure 2**

   (i)   Use **Figure 2** to explain why 10 would not be an appropriate number of quadrats to use. *(1 mark)*

   (ii) Use **Figure 2** to explain why 25 would not be an appropriate number of quadrats to use. *(1 mark)*

(c) The scientist measured the dry biomass of seaweeds at each of sites **R**, **G** and **U**. He collected all the organisms of a particular species in a quadrat and incubated them in an oven at a temperature of 80°C.

The scientist incubated the seaweeds at 80°C. Suggest why incubating them at a higher temperature would **not** produce valid results. *(1 mark)*

As well as measuring the dry biomass of the seaweeds, the scientist measured the dry mass of the animals present. He also measured the abundance of each species. **Figure 3** shows the data he collected.

	Site R	Site G	Site U
Mean number of people per hour	2.2	17.6	34.6
Mean number of species of seaweed per quadrat	4.2	2.1	1.3
Ratio of dry biomass of animals to dry biomass of seaweeds	0.15	0.06	0.03
Ratio of dry biomass of animals to abundance of animals	0.20	0.10	0.09
Ratio of dry biomass of seaweeds to abundance of seaweeds	0.79	1.57	3.24

▲ **Figure 3**

(d) The ratio of the dry biomass of animals to the dry biomass of seaweeds is always a lot less than one. Explain why. *(2 marks)*

(e) (i)   Conservation officers were working on the beaches used in this investigation. They noticed that there were fewer larger seaweeds on beaches used by a large number of people than on beaches visited by only a few people. Explain how the data in **Figure 3** support this. *(2 marks)*

   (ii) What conclusions can you draw from the data in **Figure 3** about the effect of human disturbance on the animals living on the seashore? Explain your answer. *(4 marks)*

AQA June 2010 (apart from 5a (ii) and (iii))

6   Shrews are small mammals. Three species of shrew live in mainland Britain. The table
    shows the body mass of ten shrews in three different shrew species.

	Body mass (g)							
Common shrew	10.2	11.6	7.9	12.7	13.5	6.5	8.7	9.7
Pygmy shrew	5.9	5.6	4.9	4.1	4.9	5.9	4.6	4.8
Water shrew	14.2	13.7	12	13.9	12.5	12.3	13.1	12.8

a   (i)   Calculate mean body mass values for the three species.   (2 marks)
    (ii) Calculate the standard deviation (see page 579) for each of the
          three means.   (3 marks)
    (iii) Comment on the variation in the three populations   (2 marks)
A team of biologists investigated a method of estimating the abundance of shrews. They
used plastic tubes, called hair tubes. Some of the hairs from a shrew that enters one of
these tubes stick to glue in the tube. These hairs can be used to identify the species of
shrew. The diagram shows a set of these hair tubes.

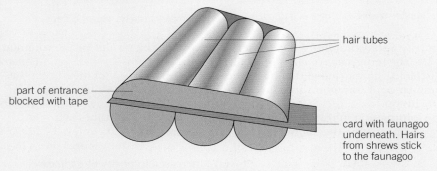

part of entrance
blocked with tape

hair tubes

card with faunagoo
underneath. Hairs
from shrews stick
to the faunagoo

(b) (i)   Faunagoo is a glue that remains sticky after wetting and drying. Explain
           the advantage of using Faunagoo in these hair tubes.   (1 mark)
     (ii) The diagram shows that the biologists partly blocked the entrances to the
           tubes with tape. Suggest why they partly blocked the entrances. (1 mark)
(c) The biologists needed to find a way of distinguishing between the hairs of the three
     species of shrew. They collected hairs from shrews of each species. For each species,
     they selected hairs at random and made different measurements.

     Explain why the biologists selected the hairs at random.   (1 mark)
(d) Repeatable measurements are measurements of the same feature that are very
     similar. In this investigation, each measurement was made by two observers. This
     helped the team to check the repeatability of these measurements.
     (i)   Explain why it was important to check the repeatability of the measurements.
                                                                           (2 marks)
     (ii) You could use a scatter diagram to check the repeatability of measurements
           made by two observers. Describe how.   (2 marks)
(e) The biologists used hair tubes to find the abundance of shrews along the edges of
     some fields. They also used traps that caught shrews without harming them. They
     selected areas where all three species of shrew were present.
     •   They put sets of hair tubes at 5 m intervals along the edges of the fields. They
         inspected the tubes one week later and recorded the number of sets of tubes that
         contained shrew hairs. They called this the hair tube index.
     •   At each site where they used hair tubes, they set traps immediately after using the
         hair tubes. They recorded the number of different shrews caught in these traps.
     (i)   The research team found the hair tube index. Explain why they could
           not use the hair tubes to find the total number of shrews present. (1 mark)
     (ii) The research team set the traps immediately after using the hair tubes.
           Explain why setting the traps immediately after using the hair tubes

would make comparisons between the two methods more reliable. *(2 marks)*
The graphs are types of scatter diagram called bubble plots. They show hair tube index plotted against the number of shrews caught in traps. The area of the bubble is proportional to the number of records plotted.

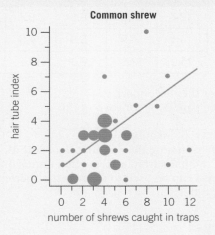

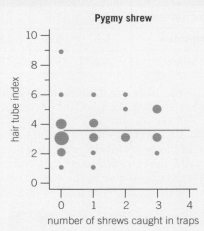

(f) Explain why a statistical test was necessary in analysing the results for the common shrew. Use the terms chance and probability in your answer. *(2 marks)*

(g) **(i)** The biologists concluded that hair tubes were a reliable way of measuring the abundance of common shrews. Give evidence from the graph to support this conclusion. *(1 mark)*

(ii) Use information in this question to evaluate the use of hair tubes as a way of measuring the abundance of pygmy shrews. *(2 marks)*

AQA Jan 2010 (apart from 6 a (i)–(iii))

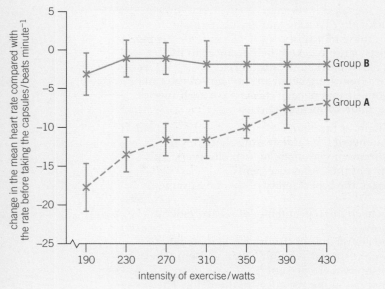

7 **(a)** Increased intensity of exercise leads to an increased heart rate. Explain how. *(3 marks)*

(b) Scientists investigated the effect of taking omega-3 fatty acids in fish oil on heart rate during exercise. They recruited two large groups of volunteers, **A** and **B**. For each group, they measured the mean heart rates at different intensities of exercise. The volunteers were then given capsules to take for 8 weeks.

- Group **A** was given capsules containing omega-3 fatty acids in fish oil.
- Group **B** was given capsules containing olive oil.

After 8 weeks, they repeated the measurements of mean heart rates at different intensities of exercise. The graph shows their results. The bars represent the standard deviations.

(i) Group B was given capsules containing olive oil. Explain why. *(1 mark)*

(ii) The scientists concluded that omega-3 fatty acids lower the heart rate during exercise. Explain how the information in the graph supports this conclusion. *(3 marks)*

AQA June 2012

Question A

Sickle-cell anaemia is a disease caused by a gene mutation in the gene which codes for haemoglobin.

In the DNA molecule that produces one of the amino acid chains in haemoglobin, the normal DNA triplet on the template strand is changed from CTC to CAC. As a result, the mRNA produced has a different code.

1   **(a)**   Identify the type of gene mutation that causes sickle cell anaemia.
     (b)   Deduce the
        (i)   normal mRNA codon produced from the DNA.
        (ii)   mRNA codon produced as a result of the mutation.

The changed mRNA codes for the amino acid valine rather than for glutamic acid. This produces a molecule of haemoglobin (called haemoglobin S) that has a 'sticky patch'. At low oxygen concentrations haemoglobin S molecules tend to adhere to one another by their sticky patches causing them to form long fibres within the red blood cells. These fibres distort the red blood cells, making them inflexible and sickle (crescent) shaped. These sickle cells are unable to carry oxygen and may block small capillaries.

2   Suggest
     (a)   how a change in a single amino acid might lead to the change in protein structure described.
     (b)   why sufferers of sickle-cell anaemia easily become tired.

Sickle-cell anaemia disables and kills individuals and so the gene causing it has been eliminated from most populations. However, the gene is relatively common among black populations of African origin. This is because the malarial parasite, *Plasmodium*, is unable to exist in sickled red blood cells.

3   Suggest a process that might have eliminated the mutant gene from most populations.

Sickle cell anaemia is the result of a gene that has two codominant alleles, Hb^A (normal) and Hb^S (sickled).

4   What is meant by the term 'codominant'?

The three possible genotype combinations of these two codominant alleles and their corresponding phenotypes are as follows:

- homozygous for haemoglobin S (Hb^{S}Hb^S). Individuals suffer from sickle-cell anaemia and are considerably disadvantaged if they do not receive medical attention. They rarely live long enough to pass their genes on to the next generation.
- homozygous for haemoglobin A (Hb^{A}Hb^A). Individuals lead normal healthy lives, but are susceptible to malaria in areas of the world where the disease is endemic.
- heterozygous for haemoglobin (Hb^{A}Hb^S). Individuals are said to have sickle-cell trait, but are not badly affected. Sufferers may become tired more easily but, in general, the condition is symptomless. They do, however, have resistance to malaria.

5   **(a)**   In parts of the world where malaria is prevalent, the heterozygous state (Hb^{A}Hb^S) is selected for at the expense of both homozygous states. Consider the information above and suggest why this is the case.
     (b)   Name the type of selection taking place. Explain your answer.

6   **(a)**   If two heterozygous individuals produce offspring, calculate the chance of any one of them having sickle-cell anaemia.
     (b)   Genetic screening for sickle-cell anaemia can be carried out. Explain why the advice given by a genetic counsellor to individuals with the same genotypes might differ depending on where they live.

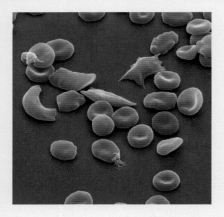

7   In a population of 175 individuals the frequency of the Hb^A allele is 0.6 and the frequency of the Hb^S allele is 0.4.
   (a) Calculate the frequencies of the Hb^A and Hb^S alleles that would be expected in the next generation if the individuals mated randomly.
   (b) Using the Hardy–Weinberg equation, calculate the number of individuals with each phenotype. Show your working.

Question B

Figure 1 shows the energy flow through a freshwater ecosystem.

1   **(a)** **(i)** State which organisms in this food chain are the primary consumers.
   (ii) Calculate the energy lost in respiration and waste products by the freshwater snails. Show your working.
   (iii) Calculate the net primary production of the leeches. Show your working.
   (iv) Calculate the percentage efficiency of energy transfer from leeches to sticklebacks. Show your working and give your answer to three significant figures.

The sun's energy for food chains like the one shown in Figure 1 is converted to chemical energy by the process of photosynthesis. Photosynthesis has two distinct stages, the light independent stage and the light dependent stage. Figure 2 is a simplified sequence of the light dependent stage.

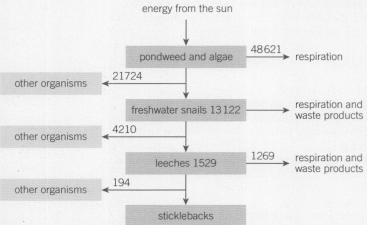

The units shown are in kilojoules per metre squared per year ($kJ\,m^{-2}\,year^{-1}$)

▲ **Figure 1**

2   **(a)** Name the substances P, Q, R and S.
   (b) State the number of carbon atoms in ribulose bisphosphate.

The products of photosynthesis are transported from the leaves to the regions of the plant using or storing them through the tissue shown in Figure 3.

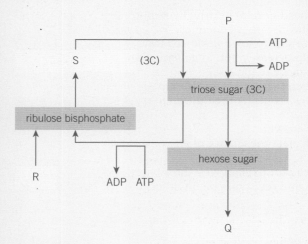

▲ **Figure 2**

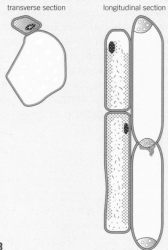

transverse section   longitudinal section

▶ **Figure 3**

3 **(a)** **(i)** Name the tissue shown in Figure 3.
(ii) Figure 3 shows two types of cells that are typical of this tissue. Name the two types of cell.
(b) **(i)** Name two organic substances that are transported in this tissue.
(ii) Outline four pieces of evidence that support the view that organic substances are transported in the phloem.
(c) Explain the advantages of studying cells such as these with an electron microscope rather than a light microscope.

Question C

Figure 1 shows a nephron from a mammalian kidney.

1 **(a)** Name the parts labelled A – H
(b) **(i)** Name a substance found in the blood plasma in the afferent arteriole but **not** present in the structure labelled H.
(ii) Explain why this substance is absent from structure H.
(c) **(i)** Name a substance found in the structure labelled B but absent from the fluid in the structure labelled E.
(ii) Explain why this substance is absent from structure E.
2 Name the two blood vessels through which blood passes on its journey from the heart to the afferent arteriole.
Figure 2 represents a cell from the wall of structure H.
3 **(a)** Calculate the magnification of the cell in Figure 2. Show your working.
(b) **(i)** Name the structures labelled X.
(ii) Describe the function of the structures labelled X.
(c) **(i)** Name the structures labelled Y.
(ii) Explain why these cells have a large number of structure Y.
Goodpasture's disease is a very rare condition in which antibodies attack an antigen that is found in cells of the glomerulus of the kidney. One symptom of the disease is blood in the urine.
4 **(a)** Suggest an explanation for the sufferers of the disease having blood in the urine.
(b) **(i)** Name the type of cell that produces antibodies in the body.
(ii) Antibodies do not directly destroy cells with the antigen they are complementary to, but rather prepare these cells for destruction. Explain the role of antibodies in preparing cells with the antigen for destruction.

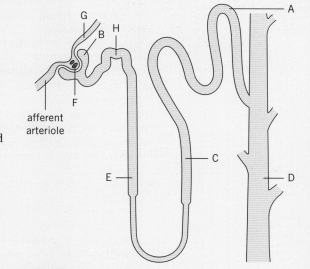

▲ **Figure 1**

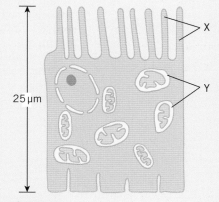

$25\,\mu m$

▲ **Figure 2**

Question D

Cystic fibrosis is caused by a mutant recessive allele in which three DNA bases, adenine-adenine-adenine, are missing from the cystic fibrosis trans-membrane-conductance regulator (CFTR) gene. The deletion results in a single amino acid being left out of the protein produced by this gene. As a result the protein is unable to perform its role of transporting chloride ions across epithelial membranes. CFTR is a chloride-ion channel protein that transports chloride ions out of epithelial cells, and water naturally follows keeping epithelial membranes moist.

1 Explain how it is possible for parents without cystic fibrosis to have a child who suffers from the disease.
2 Name the process by which water will follow chloride ions across epithelial membranes.
In a patient with cystic fibrosis, the epithelial membranes are dry and the mucus they produce remains viscous and sticky. The symptoms this causes include:
• mucus congestion in the lungs leading to a much higher risk of infection, breathing difficulties and less efficient gaseous exchange
• accumulation of thick mucus in the pancreatic ducts, preventing pancreatic enzymes, such as lipases, from reaching the duodenum.

3   **(a)** Explain why mucus congestion in the lungs can lead to a higher risk of infection.

    (b) Describe precisely the action of lipases.

Where there is a history of the disease in both families, parents may choose to be genetically screened to see whether they carry the allele. Genetic screening involves trying to determine if the mutant allele of the CFTR gene is present.

4   Outline how the presence of a mutant allele of the CFTR might be detected using DNA probes and DNA hybridisation.

Research is taking place to treat cystic fibrosis using a technique called gene therapy. This involves replacing or supplementing the defective gene with a healthy one. One method is to introduce cloned normal genes into the epithelial cells of the lungs but the treatment needs to be repeated as often as every few days. The long-term aim is to target the stem cells that give rise to lung epithelial tissue.

5   **(a)** State the meaning of the word 'cloned'.

    (b) Explain why the treatment of lung epithelial tissue has to be repeated frequently.

    (c) Suggest an advantage of delivering the healthy gene to stem cells rather than mature lung epithelial tissue.

Viruses make useful vectors for the transfer of the normal CFTR gene into the epithelial cells. They are grown in epithelial cells in the laboratory along with plasmids that have had the normal CFTR gene inserted. The CFTR gene becomes incorporated into the DNA of the viruses which are isolated and purified before being introduced into the nostrils of the patients.

6   **(a)** Define each of the following terms:

      (i) virus

      (ii) vector

      (iii) plasmid.

    (b) From your knowledge of viruses, suggest a reason why they are used to introduce the healthy CFTR gene into lung epithelial cells.

    (c) Suggest a possible disadvantage of using viruses in this way.

An alternative method of delivering plasmids containing the healthy CFTR gene into lung epithelial cells is to wrap them in lipid molecules to form a liposome. The liposomes are sprayed into the nostrils of the patient as an aerosol and are drawn down into the lungs during inhalation.

7   **(a)** Describe the process of inhalation in a human in terms of the structures involved and the associated pressure and volume changes.

    (b) Suggest a reason why liposomes are able to deliver the CFTR gene into lung epithelial cells.

Answers to questions

11.1

1 carbon dioxide and water

2 glucose and oxygen

3 a grana / thylakoids

   b stroma

4 a reduced NADP, ATP and oxygen

   b sugars and other organic molecules

11.2

1 on the thylakoid membranes (of the grana in the chloroplast)

2 Water molecules are split to form electrons, protons and oxygen, as a result of light exciting electrons / raising the energy levels of electrons in chlorophyll molecules.

3 a reduction   b reduction   c oxidation

Chloroplasts and the light-dependent reaction

1 A = (double) membrane of chloroplast / chloroplast envelope

   C = granum   D = stroma

2 C (granum)

3 starch

4 Any 2 from: the light-dependent reaction does not produce sufficient ATP for the plants' needs / photosynthesis does not take place in the dark / cells without chlorophyll cannot produce ATP in this way and ATP cannot be transported around the plant.

5 Length X–Y on Figure 4 = 24 mm (= 24 000 µm)

   Actual length X–Y = 2 µm

   Magnification = $\dfrac{24\,000}{2}$ = 12 000 times

11.3

1 It accepts / combines with a molecule of CO_2 (to produce 2 molecules of glycerate-3-phosphate).

2 It is used to reduce (donate hydrogen) glycerate-3-phosphate to triose phosphate.

3 ATP

4 Stroma of the chloroplasts.

5 The Calvin cycle requires ATP and reduced NADP in order to operate. Both are the products of the light-dependent reaction, which needs light. No light means no ATP or reduced NADP are produced and so the Calvin cycle cannot continue once any ATP or reduced NADP already produced have been used up.

Factors affecting photosynthesis

1 Volume of oxygen produced / CO_2 absorbed.

2 Light intensity – because an increase in light intensity produces an increase in photosynthesis over this region of the graph.

3 Raising the CO_2 level to 0.1% – because this increases the rate of photosynthesis more than increasing the temperature to 35 °C.

4 Because light is limiting photosynthesis and so an increase in temperature will not increase the rate of photosynthesis.

5 More CO_2 is available to combine with RuBP to form more GP, then more triose phosphate and ultimately more glucose.

Measuring photosynthesis

1 Because any air escaping from or entering the apparatus will respectively decrease or increase the volume of gas measured, which will give an unreliable result.

2 So that any changes in the rate of photosynthesis can be said to be the result of changes in light intensity and not changes in temperature.

3 To ensure there is sufficient CO_2 and so it does not limit the rate of photosynthesis.

4 To prevent other light falling on the plant as this may fluctuate and will affect the light intensity and hence the rate of photosynthesis, leading to an unreliable result.

5 To prevent photosynthesis and to allow any oxygen produced before the experiment begins, to disperse.

6 Because the volume of oxygen produced will be less than that produced by photosynthesis as some of the oxygen will be used up in cellular respiration / dissolved oxygen (and other gases) may be released from, or absorbed by the water.

Using a lollipop to work out the light-independent reaction

1 To allow the substances into which it becomes incorporated to be identified / to allow the sequence of substances produced to be identified.

2 The radioactive carbon is initially found in glycerate 3-phosphate (5 seconds) and is next found in triose phosphate (10 seconds).

3 The high temperature and / or the methanol denature the enzymes that catalyse reactions.

4 The quantity of GP begins to decrease almost immediately. The rate of decrease becomes less until, after about 4.5 minutes, the quantity of GP becomes constant, but at around a quarter of its original level. The quantity of RuBP rises almost immediately. The rate of increase is steady

at first, but then slows, peaking at 3.5 minutes. The quantity of RuBP then falls until it becomes constant at around 4.5 minutes, but at around double its original level. The quantities of GP and RuBP are the same after 2.5 minutes.

5   RuBP combines with CO_2 to form GP during the light-independent reaction / Calvin cycle of photosynthesis. GP is ultimately used to regenerate RuBP. When the CO_2 level is decreased, there is less to combine with RuBP and so less GP is formed but it is still being used up and so its level falls. There is still some CO_2 and so some GP is made, but much less than originally. With less CO_2 to combine with, the RuBP accumulates because it cannot be converted to GP. Its quantity rises to a new higher level due to the lower level of CO_2.

12.1

1	cytoplasm	6	triose phosphate
2	glucose	7	hydrogen
3	phosphate	8	NAD
4	ATP	9	pyruvate
5	phosphorylated glucose	10	ATP

12.2

1   3   2   acetylcoenzyme A

3   matrix of mitochondria

4   True – a, b, c, d, g, h, l, n
    False – e, f, i, j, k, m, o, p, q, r

Coenzymes in respiration

1   To show that the yeast suspension was responsible for any changes that occurred and the glucose did not change methylene blue nor did methylene blue change by itself.

2   a   Yeast uses glucose as a respiratory substrate producing hydrogen atoms that are taken up by methylene blue causing it to become reduced and changing from blue to colourless.

    b   As in (2a), except that the yeast uses stored carbohydrate as a respiratory substrate that has to be converted to glucose and so the production of hydrogen atoms is slower / reduced.

3   Contents of tube might have remained blue because the enzymes involved in respiration are denatured at 60 °C and so respiration, and hence the reduction of methylene blue, ceases / the enzymes involved in hydrogen transport have been denatured and so the indicator is not reduced by hydrogen.

4   Air contains oxygen, which would re-oxidise methylene blue, turning it blue.

5   This is a single experiment. The same results would need to be obtained on many occasions to increase reliability.

12.3

1   The movement of electrons along the chain is due to oxidation. The energy from the electrons combines inorganic phosphate and ADP to form ATP = phosphorylation.

2   It provides a large surface area of membrane incorporating the coenzymes (NAD / FAD) and electron carriers that transfer the electrons along the chain.

3   Oxygen is the final acceptor of the electrons and hydrogen ions (protons) in the electron transfer chain. Without it the electrons would accumulate along the chain and respiration would cease.

4   water molecule

Sequencing the chain

1   Sequence – C, A, D, B

    Explanation – Electron carriers become reduced by electrons from glycolysis and the Krebs cycle. Enzymes catalyse the transfer of these electrons to the next carrier. If an enzyme is inhibited all molecules prior to that enzyme will not be able to pass on their electrons and so will be reduced and those after it will be oxidised. The first molecule in the chain will be reduced with all inhibitors, the second with 2 out of 3 inhibitors, the third with 1 out of 3 and the last in the chain with none (i.e. it is always oxidised).

12.4

1 a D	b A, C, D	c A, D	d A, B
e A, D	f B, C, D	g A	

Investigating where certain respiratory pathways take place in cells

1   Homogenate is spun at slow speed. Heavier particles (e.g. nuclei) form a sediment. Supernatant is removed, transferred to another tube and spun at a greater speed. Next heaviest particle is removed. Process is repeated.

2   Nuclei and ribosomes – because neither CO_2 nor lactate (products of respiration) are formed in any of the samples.

3   a   mitochondria

    b   Krebs cycle produces CO_2 and results show that CO_2 is produced when mitochondria only are incubated with pyruvate.

4   (remaining) cytoplasm (Note: The complete homogenate is not a 'portion' of the homogenate.)

5 Cyanide prevents electrons passing down the transport chain. Reduced NAD therefore accumulates and blocks Krebs cycle where CO_2 is produced. Glycolysis can still occur because the reduced NAD it produces is used to make lactate. Glucose can therefore be converted to lactate, but not into CO_2, in the presence of cyanide.

6 The conversion of glucose to CO_2 involves glycolysis (occurs in cytoplasm) and Krebs cycle (occurs in mitochondria). Only the complete homogenate contains both cytoplasm and mitochondria.

7 ethanol and CO_2

8 xylem vessel – no mitochondria as mature xylem vessels are dead and the cell contents have been lost.

liver cell–many mitochondria as it is metabolically very active and requires much ATP.

epithelial cell of intestine – many mitochondria needed to provide the ATP required for the active transport of glucose, amino acids etc.

myofibril – many mitochondria to provide ATP for contraction of fibre.

9 The absence of mitochondria leaves extra space for haemoglobin and so increases the oxygen carrying capacity of red blood cells. As mitochondria carry out oxidative phosphorylation, they could use up some oxygen leaving less to be carried to the tissues by the red blood cell.

13.1

1 dragonfly nymphs

2 unicellular and filamentous algae

3 sticklebacks

4 the direction of energy flow

5 saprobionts/decomposers

13.2

1 Any 3 from: some of the organism is not eaten; some parts are not digested and so are lost as faeces; some energy is lost as excretory materials; some energy is lost as heat

2 The proportion of energy transferred at each trophic level is small (less than 20%). After four trophic levels there is insufficient energy to support a large enough breeding population.

3 $40000 \div 25 = 1600 \, kJ \, m^{-2} \, year^{-1}$

Calculating the efficiency of energy transfers

1 a $\dfrac{1250 \times 100}{6300} = 19.84\%$

b $\dfrac{50 \times 100}{42\,000} = 0.12 \, (0.119)\%$

Adding up the totals

1 saprobionts/decomposers

2 insect-eating birds

3 $\dfrac{42\,500 \times 100}{1.7 \times 106} = 2.5\%$

4 Any 3 from: most (90%+) solar energy is reflected by clouds, dust or absorbed by the atmosphere / not all light wavelengths are used in photosynthesis / much of the light does not fall on the chloroplast/chlorophyll molecule / factors may limit the rate of photosynthesis or photosynthesis is inefficient / respiration by producers means energy is lost (as heat).

5 $4120 - (1010 + 810) = 2300 \, kJ \, m^{-2} \, year^{-1}$

Productivity and farming practices

1 A longer dark period means more time is spent resting, less energy is expended, and more energy is converted into body mass.

2 The pesticide might kill beneficial organisms (e.g. ones that prey on organisms that are harmful to the farmed organism).

If the pesticide kills most of the pests then the population of organisms (predators) feeding on it will fall. With no predators controlling it, the pest population will increase again, possibly to a level higher than before. The crop will be even more affected by the pest, leading to lower productivity.

3 **As the number of weeds increases, the productivity of wheat decreases. The reduction in productivity is initially large, between 0 and 40 weeds m^{-2}, but lessens as the number of weeds increases, from 40 to 50 m^{-2}, the curve then flattens out.**

4 Soya bean because it has an increase in productivity of 50% (1000 to 1500 kg ha^{-1}) while wheat only increases by 33% (4500 to 6000 kg ha^{-1}).

5 No. Cost of herbicide per hectare = £100. Reducing weeds from 40–20 m^{-1} increases wheat productivity from 4500 to 5000 kg ha^{-1} – an increase of 500 kg or half a tonne. Wheat is sold at £150 per tonne, so increased income is £75 per hectare which is £25 per hectare less than the cost of treating with herbicide.

A mighty problem

1 Description – In both experiments the spider mite populations rise slowly during the first 15 days and then very rapidly up to around 50 days.

In experiment 1 the spider mite population remains high up to 150 days but fluctuates (between 400 and 900). (Note: The scale is the square root of the numbers and so the figures on the y-axis need to be squared to give actual numbers.)

In experiment 2 the spider mite population falls over the period 50–150 days until it reaches the starting level.

Explanation – In experiment 1 the population of the spider mite increases until some factor (e.g. food supply) limits its size. It remains fairly constant as an equilibrium is reached with the limiting factor, fluctuating slightly as the factor fluctuates.

In experiment 2 the population of the spider mite increases up to 50 days, by which time the population of the predatory mite has increased considerably. The predatory mites feed on the spider mites, causing their population to drop to a very low level by 150 days.

2   Predatory mites are effective in controlling the population of spider mites as their presence reduces the spider mite population from around 400–900 when the predatory mite is absent to around 4 when the predatory mite is present.

3   The two populations will probably remain small as they remain in balance. They will fluctuate because, as the spider mite population falls, there will be less food for the predatory mite and so, a short time later, its population is also likely to fall. The fall in the predatory mite's population means there will be less predation on the spider mite, whose population is likely to increase, followed in turn by an increase in the predatory mite's population.

13.3

1	nitrogen fixation	8	nitrifying
2	plants	9	nitrate
3	nitrate ions	10	denitrifying
4	root hairs	11 A	absorption
5	proteins / amino acids / nucleic acids	B	feeding and digestion
6	saprobionts/ decomposers	C	excretion and decomposition
7	ammonia / ammonium ions	D	erosion
		E	excretion

13.4

1   Crops are grown repeatedly and intensively on the same area of land. Mineral ions are taken up by the crops, which are transported and consumed away from the land. The mineral ions they contain are not returned to the same area of land and so the levels in the soil are reduced, which can limit the rate of photosynthesis. Fertilisers need to be applied to replace them if photosynthesis / productivity is to be maintained.

2   100 kg ha^{-1} – although 150 kg ha^{-1} gives a slightly better yield, this is marginal and the cost of using 50% more fertiliser makes it uneconomical.

3   Some other factor is limiting photosynthesis, e.g. light, CO_2, and only the addition of this factor will increase photosynthesis and hence productivity.

4   Natural fertilisers are organic and come from living organisms in the form of dead remains, urine or faeces (manure).

Artificial fertilisers are inorganic and are mined from rocks and deposits.

Different forms of nitrogen fertilisers

1   manure, bone meal and urea

2   To act as a control to show that any changes in productivity were the result of the nitrogen-containing fertiliser being added.

3   Nitrogen is needed for proteins / amino acids / chlorophyll and DNA and therefore for plant growth. Nitrogen shortage may limit the production of proteins and DNA and hence growth. Its addition increases productivity.

4   Some forms of fertilisers contain more actual nitrogen than others and so different masses are added to ensure that the total nitrogen added was always the same (140 kg ha^{-1}).

5   The data do not support the view. While ammonium nitrate brings about the greatest increase in productivity, ammonium sulphate produces a smaller increase than both urea and bone meal. Therefore the investigation suggests that only some ammonium salts are better.

6   The farmer should spread the manure a few months before the main growing season for the crop.

13.5

1   Eutrophication is the process by which salts build up in bodies of water.

2   The concentration of algae near the surface becomes so dense that no light penetrates to deeper levels. No light means no photosynthesis and hence no carbohydrate for respiration and so plants at lower levels die.

3   Dead plants are used as food by saprobionts. With an increased supply of this food, the population of saprobionts increases exponentially. Being aerobic they use up the oxygen in the water leading to the death of the fish, which cannot respire without it.

Troubled waters

1   It has taken 10 days for the fertiliser that has dissolved in the rainwater to leach through the soil and into the lake.

2   In normal circumstances, a low level of nitrate (or other ions) is the limiting factor to algal growth.

The fertilizer leaching into the lake contains nitrate (and other ions) and removes this limit on growth. The algal population grows rapidly, increasing in density.

3   Description – As the density of the algae increases so the clarity of the water decreases, i.e. there is a negative correlation. For the first 20 days the algal density (30 cells cm^{-3}) and water clarity (Secchi = 9 m) remain constant.

From day 20 to day 100 the algal density increases from 30 to 120 cells cm^{-3} while the water clarity decreases from 9 to 1 m (Secchi depth). However, there is an anomaly between day 40 and day 50 when the water clarity suddenly falls from 7 to 4 m.

Explanation – As the density of algae increases, more light is absorbed / reflected by them and so less light penetrates / water clarity is reduced. Between day 40 and day 50 some factor (e.g. water turbulence stirring up sediment) other than algal density is reducing the water clarity.

4   Days 0–10: oxygen level is constant (at 10 ppm) because there is a balance between oxygen produced in photosynthesis of plants and algae, and oxygen used up in respiration of all organisms.

Days 10–25: oxygen level rises (up to around 13 ppm) due to increased photosynthesis by the larger population of algae.

Days 25–100: oxygen level decreases (more rapidly at first and then less so down to around 3 ppm) due to higher density of algae blocking out the light to lower depths and reducing the rate of photosynthesis of plants / algae at these depths. In time, light is blocked out altogether at lower depths → no photosynthesis → plants / algae die → saprobionts decompose them → their population increases → they use up much oxygen in respiration → oxygen levels fall.

14.1

1   (Negative chemo-) taxis – wastes are often removed from an organism because they are harmful. Moving away prevents the waste harming the organism and so increases its chance of survival.

2   (Positive chemo-) taxis – increases the chances of sperm cells fertilising the egg cells of other mosses and so helps to produce more moss plants / future generations. Cross-fertilisation increases genetic variability, making species better able to adapt to future environmental changes.

3   (Negative gravi-) tropism – takes the seedlings above the ground and into the light, where they can photosynthesise.

More photosynthesis means more carbohydrate and so a better chance of survival.

14.2

1   More IAA moves towards the shaded side of shoots than the light side when the light is unidirectional. In response to this uneven distribution of IAA, the cells on the shaded side elongate faster than those on the light side and the shoot bends towards the light. This ensures that the shoot and the leaves attached to it have a greater chance of being well illuminated. As light is essential for photosynthesis, the process by which organic material for respiration is manufactured, the plant has a greater chance of survival.

2   Response ensures that roots grow down into the soil, anchoring the plant firmly and bringing them closer to water (needed for photosynthesis).

3   The fact that IAA is readily absorbed, easily synthesized and is lethal to plants in low concentrations makes it useful as a herbicide. The fact that it more readily kills broad-leaved plants than narrow-leaved ones is an advantage because many agricultural crops are narrow-leaved while the weeds that compete with them are broad-leaved. As a result, application of IAA at appropriate concentrations will kill only the weeds with little, or no, harm to the crop. As IAA is not easily broken down means it will persist in the soil and continue to act as a selective weedkiller for some time. This may prevent a broad-leaved crop being grown on the land for some time after application of IAA. There is also a danger that IAA might accumulate along food chains with possible harm to animals in those chains.

Discovering the role of IAA in tropisms

1   Experiment 1

2   As mica conducts electricity it will not prevent electrical messages passing from the shoot tip but it will prevent chemical messages passing. As there is no response, the message must be chemical and must pass down the shaded side.

3   Displacement of the tip means that the chemical initially only moves down the side of the shoot that is in contact with the tip. This side grows more rapidly, causing bending away from that side.

4   It prevents chemicals / IAA, but not light, passing from one side to the other.

5   Results support the hypothesis that IAA is transported from the lighter side to the shaded side of the shoot.

Experiment 8 shows that the total IAA produced and collected is the same whether the shoot is in the light or the dark. This discounts the theory that light destroys IAA or inhibits its production.

Experiment 9 shows that the amount of IAA produced at either side of the tip is the same. The glass plate prevents any sideways transfer.

Experiment 10 shows that the IAA is transferred from the light to the shaded side of the shoot soon after it is produced because more than twice as much IAA is found on the shaded side of the shoot than on the light side.

14.3

1 brain / spinal cord
2 brain / spinal cord
3 motor
4 sensory
5 involuntary
6 stimulus
7 (temperature) receptor
8 sensory
9 intermediate
10 motor
11 effectors

14.4

1 Stretch-mediated sodium channel – a special type of sodium channel that changes its permeability to sodium when it changes shape / is stretched.

2 pressure on Pacinian corpuscle → corpuscle changes shape → stretches membrane of neurone → widens stretch mediated sodium ion channels → allows sodium ions into neurone → changes potential of (depolarises) membrane → produces generator potential

3 Only rod cells are stimulated by low-intensity (dim) light. Rod cells cannot distinguish between different wavelengths / colours of light, therefore the object is perceived only in a mixture of black and white, i.e. grey.

4 Light reaching Earth from a star is of low intensity. Looking directly at a star, light is focused on to the fovea, where there are only cone cells. Cone cells respond only to high light intensity so they are not stimulated by the low light intensity from the star and it cannot be seen. Looking to one side of the star means that light from the star is focused towards the outer regions of the retina, where there are mostly rod cells. These are stimulated by low light intensity and therefore the star is seen.

14.5

1 Autonomic nervous system – controls the involuntary activities of internal muscles and glands.

2 Sympathetic nervous system stimulates effectors and so speeds up an activity; prepares for stressful situations, e.g. the fight or flight response.

Parasympathetic nervous system inhibits effectors and slows down an activity; controls activities under resting conditions, conserving energy and replenishing the body's reserves.

3 Blood pressure remains high because the parasympathetic system is unable to transmit nerve impulses to the SA node, which decreases heart rate and so lowers blood pressure.

4 a Heart rate remains as it was before taking exercise – after exercise, blood pressure increases and CO_2 concentration of blood rises (causing blood pH to be lowered). The changes are detected by pressure and chemical receptors in the wall of the carotid arteries. As the nerve from here to the medulla oblongata is cut, no nerve impulse can be sent to the centres that control heart rate.

   b Blood CO_2 concentration increases as a result of increased respiration during exercise.

15.1

1 (nerve) impulses / action potentials
2 nucleus
3 rough endoplasmic reticulum
4 dendrites
5 Schwann cells
6 insulation
7 myelin
8 motor
9 sensory
10 intermediate
11 Hormone response is slow, widespread and long-lasting. Nervous response is rapid, localised and short-lived.

Aging in neurones

1 dendrites become longer with age.
2 dendrites are fewer, are much shorter and are less branched.
3 After 10 years (age 60) there will be
$2000 - (\frac{5}{100} \times 2000) = 1900$ neurones remaining.
After a further 10 years (age 70) there will be 1900 $- (\frac{5}{100} \times 1900) = 1805$ neurones remaining.

15.2

1 Active transport of sodium ions out of the axon by sodium–potassium pumps is faster than active transport of potassium ions into the axon. Potassium ions diffuse out of the axon but few, if any, sodium ions diffuse into the axon because the sodium 'gates' are closed. Overall, there are more positive ions outside than inside and therefore the outside is positive relative to the inside.

2 A = closed B = open C = closed D = closed
   E = closed F = open

Measuring action potentials

1 sodium and potassium ions

2 At resting potential (0.5 ms) there is a positive charge on the outside of the membrane and a negative charge inside, due to the high concentration of sodium ions outside the membrane. The energy of the stimulus causes the sodium voltage-gated channels in the axon membrane to open and therefore sodium ions diffuse in through the channels, along their electrochemical gradient. Being positively charged, they begin a reversal in the potential difference across the membrane. As sodium ions enter, so more sodium ion channels open, causing an even greater influx of sodium ions and an even greater reversal of potential difference: from −70 mv up to +40 mv at 2.0 ms.

3 Two action potentials take place in 10 ms.

Each action potential takes 10 ÷ 2 = 5 ms / action potentials are 5ms apart.

There are 1000 (10^3) ms in 1 second.

Therefore there are 1000 ÷ 5 = 200 action potentials in 1 second (2×10^2 ms^{-1})

15.3

1 **a** node of Ranvier

 b Because the remainder of the axon is covered by a myelin sheath that prevents ions being exchanged / prevents a potential difference being set up.

 c It moves along in a series of jumps from one node of Ranvier to the next.

 d saltatory (conduction)

 e It is faster than in an unmyelinated axon.

2 It remains the same / does not change.

15.4

1 During the refractory period the sodium voltage-gated channels are closed so no sodium ions can move inwards and no action potential is possible. This means there must be an interval between one impulse and the next.

2 All-or-nothing principle – There is a particular level of stimulus that triggers an action potential. At any level above this threshold, a stimulus will trigger an action potential that is the same regardless of the size of the stimulus (the 'all' part). Below the threshold, no action potential is triggered (the 'nothing' part).

3 Mammals have myelinated neurons and so have saltatory conduction. Mammals are endothermic and their constant, usually higher, body temperature increases the rate of diffusion of ions across the axon membrane and hence the speed of conduction of the action potential.

Different axons different speeds

1 The greater the diameter of an axon the faster the speed of conductance. Comparing the data for the two myelinated axons shows that the 20 μm diameter axon conducts at 120 ms^{-1} while the 10 μm diameter axon conducts at only 50 ms^{-1}. Likewise, the data for the two unmyelinated axons show that the 500 μm diameter axon conducts at 25 ms^{-1} while the 1μm diameter axon conducts at 2 ms^{-1}.

2 In myelinated axons, the myelin acts as an electrical insulator. Action potentials can only form where there is no myelin (at nodes of Ranvier). The action potential therefore jumps from node to node (= saltatory conduction) which makes its conductance faster.

3 Schwann cells

4 The presence of myelin has the greater effect because a myelinated human sensory axon conducts an action potential at twice the speed of the squid giant axon, despite being only 1/50th of its diameter. (Note: Similar comparisons can be made between other types of axon, e.g. squid and human motor axons.)

5 Temperature affects the speed of conductance of action potentials. The higher the temperature, the faster the conductance. The conductance of action potentials in the squid will therefore change as the environmental temperature changes. It will react more slowly at lower temperatures.

6 Area of circle = πr^2. Radius of axon = $\frac{500}{2}$ = 250 μm. Area of axon = $\pi \times 250^2 = 196\,349$ μm^2.

Expressed as mm^2 to five significant figures 0.19635 mm^2.

15.5

1 It possesses many mitochondria and large amounts of endoplasmic reticulum.

2 It has receptor molecules for neurotransmitters e.g., acetylcholine, on its membrane.

3 Neurotransmitter is released from vesicles in the presynaptic neurone into the synaptic cleft when an action potential reaches the synaptic knob. The neurotransmitter diffuses across the synapse to receptor molecules on the postsynaptic neurone to which it binds, thereby setting up a new action potential.

4 Only one end can produce neurotransmitter and so this end alone can create a new action potential in the neurone on the opposite side of the synapse. At the other end there is no neurotransmitter that can be released to pass across the synapse and so no new action potential can be set up.

5 a The relatively quiet background noise of traffic produces a low-level frequency of action potentials in the sensory neurones from the ear. The amount of neurotransmitter released into the synapse is insufficient to exceed the threshold in the postsynaptic neurone and to trigger an action potential and so the noise is 'filtered out' / ignored. Louder noises create a higher frequency and the amount of neurotransmitter released is sufficient to trigger an action potential in the postsynaptic neurone and so there is a response. This is an example of temporal summation. (Note: An explanation in terms of spatial summation is also valid: many sound receptors with a range of thresholds → more receptors respond to the louder noise → more neurotransmitter → response.)

b Reacting to low-level stimuli (background traffic noise) that present little danger can overload the (central) nervous system and so organisms may fail to respond to more important stimuli. High-level stimuli (sound of horn) need a response because they are more likely to represent a danger.

6 As the inside of the membrane is more negative than at resting potential, more sodium ions must enter in order to reach the potential difference of an action potential, i.e. it is more difficult for depolarisation to occur. Stimulation is less likely to reach the threshold level needed for a new action potential.

7 a Increase in speed $64 - 40 = 24$ ms^{-1}

Percentage increase $\frac{24}{40} \times 100 = 37.5\%$

b Refex arcs allow rapid responses to potentially harmful situations. Information passes across synapses relatively slowly compared to the speed it passes along an axon. The fewer synapses there are, the shorter the overall time taken to respond to a stimulus – an advantage where a rapid response is required.

15.6

1 a sodium ions **c** ATP

b acetylcholine **d** calcium ions

2 To recycle the choline and ethanoic acid; to prevent acetylcholine from continuously generating a new action potential in the postsynaptic neurone.

Effects of drugs on synapses

1 They will reduce pain.

2 They act like endorphins by binding to the receptors and therefore preventing action potentials being created in the neurones of the pain pathways.

3 Prozac might prevent the elimination of serotonin from the synaptic cleft (Note: Any biologically

accurate answer that results in more serotonin in the synaptic cleft is acceptable.)

4 By increasing the concentration of serotonin in the synaptic cleft, its activity is increased, reducing depression, which is caused by reduced serotonin activity.

5 It will reduce muscle contractions (cause muscles to relax).

6 Valium increases the inhibitory effects of GABA so therefore there are fewer action potentials on the nerve pathways that cause muscles to contract.

7 The molecular structure of Vigabatrin is similar to GABA so it may be a competitive inhibitor (compete) for the active site of the enzyme that breaks down GABA. As less GABA is broken down by the enzyme, more of it is available to inhibit neurone activity. Or Vigabatrin might bind to GABA receptors on the neurone membrane and mimic its action, thereby inhibiting neuronal activity.

15.7

1 Muscles require much energy for contraction. Most of this energy is released during the Krebs cycle and electron transport chain in respiration. Both these take place in mitochondria.

2 A = Z-line B = H-zone C = I-band (isotropic band) D = A-band (anisotropic band).

3 The actin and myosin filaments lie side by side in a myofibril and overlap at the edges where they meet. If cut where they overlap, both filaments can be seen. If cut where they do not overlap, we see one or other filament only.

4 Slow-twitch fibres contract more slowly and provide less powerful contractions over a longer period.

Fast-twitch fibres contract more rapidly and produce powerful contractions but only for a short duration.

5 Slow-twitch fibres have myoglobin to store oxygen, much glycogen to provide a source of metabolic energy, a rich supply of blood vessels to deliver glucose and oxygen, and numerous mitochondria to produce ATP.

Fast-twitch fibres have thicker and more numerous myosin filaments, a high concentration of enzymes involved in anaerobic respiration and a store of phosphocreatine to rapidly generate ATP from ADP in anaerobic conditions.

15.8

1 Myosin is made of two proteins. The fibrous protein is long and thin in shape, which enables it to combine with others to form a long thick filament along which the actin filament can move. The globular protein forms two bulbous structures (the head) at the end of a filament (the tail). This shape allows it to exactly fit recesses in the actin molecule, to which it can become attached. Its shape also means it can be moved at an angle. This allows it to change its angle when attached to actin and so move it along, causing the muscle to contract.

2 Phosphocreatine stores the phosphate that is used to generate ATP from ADP in anaerobic conditions. A sprinter's muscles often work so strenuously that the oxygen supply cannot meet the demand. The supply of ATP from mitochondria during aerobic respiration therefore ceases. Sprinters with the most phosphocreatine have an advantage because ATP can be supplied to their muscles for longer, and so they perform better.

3 A single ATP molecule is enough to move an actin filament a distance of 40 nm.

Total distance moved by actin filament = 0.8 μm (= 800 nm).

Number of ATP molecules required = 800 ÷ 40 = 20.

4 One role of ATP in muscle contraction is to attach to the myosin heads, thereby causing them to detach from the actin filament and making the muscle relax. As no ATP is produced after death, there is none to attach to the myosin, which therefore remains attached to actin, leaving the muscle in a contracted state, i.e. rigor mortis.

16.1

1 Homeostasis is the maintenance of a constant internal environment in organisms.

2 Maintaining a constant temperature is important because enzymes function within a narrow range of temperatures.

Fluctuations from the optimum temperature mean enzymes function less efficiently. If the variation is extreme, the enzyme may be denatured and cease to function altogether. A constant temperature means that reactions occur at a predictable and constant rate.

3 Maintaining a constant blood glucose concentration is important in ensuring a constant water potential. Changes to the water potential of the blood and tissue fluids may cause cells to shrink and expand (even to bursting point), due to water leaving or entering by osmosis. In both instances the cells cannot operate normally. A constant blood glucose concentration also ensures a reliable source of glucose for respiration by cells.

Thermoregulation in ectotherms and endotherms

1 It allows accurate comparisons to be made even though the animals have different body masses. An increase in body size or body mass means there is increased heat generation.

2 a Both increase proportionally up to 25 °C. Above 25 °C, heat generation increases more rapidly (gradient / slope of line increases), whereas evaporative heat loss increases at the same rate (gradient / slope of line remains the same).

b In a mammal, the relationship is the inverse / opposite, i.e. as evaporative heat loss increases, heat generation decreases.

3 Above 25 °C, the metabolic heat generation in reptiles becomes much more rapid. They therefore generate heat faster than they can lose it. As a result, their body temperature increases and enzymes may be denatured, leading to death. As reptiles have no physiological means of cooling, they must seek shade in order to reduce their body temperature.

4 Sweating or panting increases.

16.2

1 If the information is not fed back once an effector has corrected any deviation and returned the system to the set point, the receptor will continue to stimulate the effector and an over-correction will lead to a deviation in the opposite direction from the original one.

2 It gives a greater degree of homeostatic control.

Positive feedback

1 Positive feedback means the contractions get stronger and more frequent over time leading to the birth of the baby. Negative feedback would mean the contractions became weaker and less frequent until they stopped altogether and the baby would not be born.

Negative feedback in temperature control

1 The blood temperature would be become progressively colder (because positive feedback increases the current process i.e. blood cooling).

2 Cutting the nerves would mean that the thermoreceptors would not be able to communicate the rise in blood temperature to the heat loss centre. The centre would not be able to initiate actions that could lower blood temperature. Blood temperature would rise. If the increase were great enough, enzymes vital to keep the organism alive e.g. respiratory enzymes would be denatured, cease to function and death would result.

3 The heat loss centre might be connected to effectors, other than the skin, which could lower the blood temperature e.g. regions of the brain that control behaviour, and so the individual might be able to reduce blood temperature by resting, sheltering from the sun etc.

4 vena cava, pulmonary artery, pulmonary vein, aorta.

16.3

1	respiration	8	gluconeogenesis
2	brain	9	glycogen
3	osmotic / water potential	10	respiration
4	carbohydrate	11	islets of Langerhans
5	glycogen	12	insulin
6	muscles	13	glucagon
7	amino acids	14	adrenaline

16.4

1 Type I is caused by an inability to produce insulin.

Type II is caused by receptors on body cells losing their responsiveness to insulin.

2 Type I is controlled by the injection of insulin.

Type II is controlled by regulating the intake of carbohydrate in the diet and matching this to the amount of exercise taken.

3 Diabetes is a condition in which insulin is not produced by the pancreas. This leads to fluctuations in the blood glucose level. If the level is below normal, there may be insufficient glucose for the release of energy by cells during respiration. Muscle and brain cells in particular may therefore be less active, leading to tiredness.

4 Match your carbohydrate intake to the amount of exercise that you take. Avoid becoming overweight by not consuming excessive quantities of carbohydrate and by taking regular exercise.

Effects of diabetes on substance concentrations in the blood

1 adrenaline

2 The rise in insulin level is both greater and more rapid in group Y than in group X.

3 Glucose is removed from blood by cells using it during respiration.

4 Glucose concentration rises at first because the glucose that is drunk is absorbed into the blood (glucose line on graph rises). This rise in blood glucose causes insulin to be secreted from cells (B cells) in the pancreas (insulin line rises steeply). Insulin causes increased uptake of glucose into liver and muscle cells, activates enzymes that convert glucose into glycogen and fat, and

increases cellular respiration. The effect of all these actions is to reduce glucose concentration (glucose line falls from 2.5 hours onwards). As the glucose concentration rises after 1 hour, so the glucagon concentration falls. The reduction in glucagon concentration decreases glucose production from other sources (glycogen, amino acids, and glycerol) and so also helps to reduce blood glucose concentration. As the blood glucose concentration falls (after 2.5 hours) so the glucagon concentration increases to help maintain the blood glucose at its optimum concentration.

5 Group X has diabetes and therefore the glucose intake does not stimulate insulin production (insulin concentration shown on the graph is low). The glucose concentration in the blood therefore continues to rise (glucose line rises steeply) as there is no insulin to reduce its concentration. Blood glucose concentration remains high, falling only slightly as it is respired by cells.

6 As it is respired by cells, the glucose concentration will decrease steadily until it falls below the optimum concentration.

16.5

1	renal (Bowman's) capsule	6	epithelial cells
2	glomerulus	7	microvilli
3	afferent	8	loop of Henle
4	podocytes	9	distal
5	proximal	10	collecting duct

Control of blood water potential

1 As sweating involves a loss of water from the blood, its water potential will decrease (be lower or more negative).

2 a osmotic cells (in the hypothalamus)

b kidney

3 Being a hormone, it is transported in the blood plasma.

4 Absorption (taking in or consumption or drinking) of water because water has been lost during sweating. As the water potential of the blood returns to normal, the lost water must have been replaced. However, the kidney only excretes less water, it does not replace it. Therefore process X must be the way in which water is replaced.

5 negative feedback

The glomerulus – a unique capillary network

1 The efferent arteriole later divides up into a second capillary bed that surrounds the loop of Henle. These then combine to form the venule.

2 By looking at the structure of its wall. Arterioles have thicker walls with more muscle tissue than venules.

16.6

1 proximal convoluted tubule

2 glomerulus; renal capsule; proximal convoluted tubule; loop of Henle; distal convoluted tubule; collecting duct

3 microvilli to provide a large surface area to reabsorb substances from the filtrate; infoldings at their bases to give a large surface area to transfer reabsorbed substances into blood capillaries; a high density of mitochondria to provide ATP for the active transport.

4 Animals in dry environments would have longer loops of Henle to give a longer counter current multiplier and so more absorption of water by the collecting duct.

16.7

1 hypothalamuss

2 a less–because the water drunk causes a rise in water potential of the blood

  b more–because intense exercise leads to sweating and the loss of water leading to a fall in water potential of the blood

3 ADH binds to receptors on the cell-surface membrane of the cells lining the collecting duct and activate phosphorylase within the cell. The activation of phosphorylase causes vesicles containing pieces of plasma membrane that have numerous water channels/aquaporins to fuse with the cell-surface membrane. This increases the number of water channels and makes the cell-surface membrane much more permeable to water.

4. 0.05 g dm^{-1}

The significance of glucose in the urine

1 Insulin – lowers blood sugar level

  Glucagon – raises blood sugar level

  Adrenaline – raises blood sugar level.

2 The graph measures the concentration of substances and not their actual quantities. Because water is progressively removed from filtrate as it passes along the nephron, there is the same amount of urea but in a smaller volume of water. Therefore the **concentration** of urea increases.

3 Glucose because it is totally reabsorbed in proximal convoluted tubule.

4 Sodium ions because they enter the descending limb of loop of Henle initially and so increase in concentration. They are then actively transported out of the ascending limb and so decrease in concentration. The removal of water from the collecting duct causes their concentration to slowly rise again.

5 The reabsorption of water from the collecting ducts depends on there being a large water potential gradient between the fluid in the collecting duct and that in the blood capillaries. The presence of glucose in the fluid in the collecting duct reduces this gradient and leads to more water being lost in urine leading to dehydration.

17.1

1 genotype   **6** locus

2 mutation   **7** homozygous

3 phenotype   **8** heterozygous

4 nucleotides/bases **9** recessive

5 polypeptides   **10** codominant

17.2

1 Let allele for Huntington's disease = H

  Let allele for normal condition = h

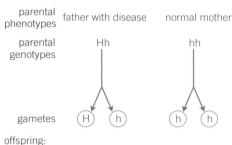

offspring:

		father's gametes
mother's gametes	(H)	(h)
(h)	Hh	hh
(h)	Hh	hh

half (50%) of offspring will have Huntington's disease (Hh).

half (50%) of offspring will be normal (hh).

2 **a** Let allele for black coat = B

Let allele for red coat = b

parental phenotypes   female with black coat   male with red coat

parental genotypes   BB   bb

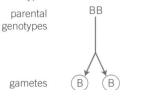

gametes   Ⓑ Ⓑ   ⓑ ⓑ

offspring:

	male gametes	
female gametes	Ⓑ	Ⓑ
Ⓑ	Bb	Bb
Ⓑ	Bb	Bb

All (100%) offspring will have black coats (Bb).

b Let allele for black coat = B

Let allele for red coat = b

parental phenotypes   female with black coat   male with black coat

parental genotypes   Bb   Bb

gametes   Ⓑ ⓑ   Ⓑ ⓑ

offspring:

	male gametes	
female gametes	Ⓑ	ⓑ
Ⓑ	BB	Bb
ⓑ	Bb	bb

3 offspring (75%) with black coat (BB, Bb and Bb).

1 offspring (25%) with red coat (bb).

probability of offspring having red coat = 1 in 4 (25%/0.25)

17.3

1 **a** homozygous dominant (**GG**)

b We cannot be absolutely certain because if the unknown genotype were heterozygous (**Gg**) the gametes produced would contain alleles of two types: either dominant (**G**) or recessive (**g**). It is a matter of chance which of these gametes fuses with those from our recessive parent – all these gametes have a recessive allele (**g**). It is just possible that, in every case, it is the gametes with the dominant allele that fuse and so all the offspring show the dominant

character. Provided the sample of offspring is large enough, however, we can be reasonably sure that the unknown genotype is homozygous dominant.

2 **a** heterozygous (**Gg**)

b We can be certain because 7 of the offspring display the recessive character (in our case yellow pods). These plants are homozygous recessive and must have obtained one recessive allele from each parent. Our unknown parental genotype must therefore have a recessive allele and be heterozygous (in our case **Gg**). It is theoretically possible that the plants with yellow pods were due to a mutation but this is most unlikely. The unexpectedly low number of plants with yellow pods is the result of random fusion of the gametes.

c 50%   **d** 7.29%

17.4

1 Red eyes and normal wings are dominant because these characteristics are expressed in the F_1 generation while pink eyes and vestigial wings are not expressed in the F_1 generation and so these are recessive. Also red eyes and normal wings appear around 3 times more often in the F_2 generation than pink eyes and vestigial wings.

2 **R** for red eyes and **r** for pink eyes, **N** for normal wings, **n** for vestigial wings.

3 Parental cross

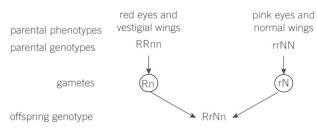

parental phenotypes: red eyes and vestigial wings | pink eyes and normal wings
parental genotypes: RRnn | rrNN

gametes: (Rn) | (rN)

offspring genotype: RrNn

F₁ intercross

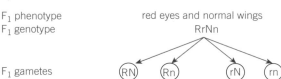

F₁ phenotype: red eyes and normal wings
F₁ genotype: RrNn

F₁ gametes: (RN) (Rn) (rN) (rn)

F₂ offspring

		♂ gametes			
		(RN)	(Rn)	(rN)	(rn)
♀ gametes	(RN)	RRNN	RRNn	RrNN	RrNn
	(Rn)	RRNn	RRnn	RrNn	Rrnn
	(rN)	RrNN	RrNn	rrNN	rrNn
	(rn)	RrNn	Rrnn	rrNn	rrnn

9 red eyes and normal wings (R-N-)
3 red eyes and vestigial wings (R-nn)
3 pink eyes and normal wings (rrN-)
1 pink eyes and vestigial wings (rrnn)

Better late than never

1 The F₁ ratio is 3:1 is a typical ratio for a monohybrid cross and supports Mendel's law of segregation. The F₂ ratio of 9:7 looks initially unusual but has 16 genotypes and so could be a modified 9:3:3:1 ratio that is the typical ratio for a dihybrid cross and supports Mendel's law of independent assortment.

2 The two different varieties when self-fertilised produce three green offspring for each white one. The green offspring must have allele **A** and allele **B** i.e. must be **A-B-** (where – is either dominant or recessive). The white offspring must lack either allele **A** or allele **B** or both i.e. must have the genotype **aabb**, **aaB-** or **A-bb**. To obtain white offspring the green parent must provide a recessive allele so that a double recessive can occur in the offspring. There are three possible genotypes that fulfil these criteria: **AABb**, **AaBB** and **AaBb**. The latter however, when self-fertilised produces a 9:7 ratio of green to white (see later) and not a 3:1 ratio. This leaves the remaining two. As there are two different varieties that produce this 3:1 ratio

when self-fertilised, these must be the genotypes of the parents. Answer = **AABb** and **AaBB**.

proof for variety 1 (AABb)

parental genotypes: AABb | AABb

gametes: (AB) (Ab) | (AB) (Ab)

F₁ offspring

♀ gametes		♂ gametes	
		(AB)	(Ab)
	(AB)	AABB	AABb
	(Ab)	AABb	AAbb

3 offspring have alleles A and B and are therefore green.
1 offspring (AAbb) lacks allele B and is therefore white.

proof for variety 2 (AaBB)

parental genotype: AaBB | AaBB

gametes: (AB) (aB) | (AB) (aB)

F₁ offspring

♀ gametes		♂ gametes	
		(AB)	(aB)
	(AB)	AABB	AaBB
	(aB)	AaBB	aaBB

3 offspring have alleles A and B and are therefore green.
1 offspring (aaBB) lacks allele A and is therefore white.

3

parental genotype: variety 1 — AABb | variety 2 — AaBB

gamete: (AB) (Ab) | (AB) (aB)

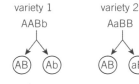

F₁ offspring		variety 1 gametes	
variety 2 gametes		(AB)	(Ab)
	(AB)	AABB	AABb
	(aB)	AaBB	AaBb

All offspring have both allele A and allele B and so are green

4 Examination of the results of cross 2 shows that two of the four genotypes (**AABb** and **AaBB**) were self-fertilised in cross 1 and produced a 3 green to 1 white ratio and so can be discounted. Self-fertilisation of one of the two remaining genotypes, **AABB**, clearly will only produce green offspring. This leaves just the genotype **AaBb**. The results of self-fertilisation of plants with this genotype are as follows:

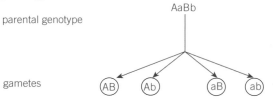

parental genotype AaBb

gametes AB Ab aB ab

F_2 offspring

♀ Gametes	♂ Gametes			
	AB	Ab	aB	ab
AB	AABB	AABb	AaBB	AaBb
Ab	AABb	AAbb	AaBb	Aabb
aB	AaBB	AaBb	aaBB	aaBb
ab	AaBb	Aabb	aaBb	aabb

Nine boxes have a genotype with both allele A and allele B and so can produce chlorophyll and the plants are green.

The seven other boxes lack either allele A, allele B or both and so cannot produce chlorophyll and the plants are white.

5 Each gene controls the production of a polypeptide. Dominant alleles of genes A and B are required to form their respective polypeptides. These polypeptides could form part of a single enzyme or two different enzymes in the biochemical pathway that forms chlorophyll. Only plants carrying both dominant alleles will be able to synthesise chlorophyll.

17.5

1 The man is not the father.

Reasons – child has blood group AB and therefore has alleles I^AI^B. The mother is blood group A and therefore either I^AI^O or I^AI^A. In either case she could have provided the I^A alleles to the child but not the I^B allele. The I^B allele must have come from the real father. The supposed father is blood group O and therefore has alleles I^OI^O. He cannot provide an I^B allele and so cannot be the father.

2

parental phenotypes: mildly frizzled cockerel   frizzled hen

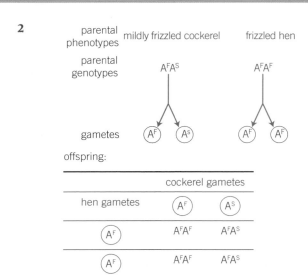

parental genotypes: A^FA^S A^FA^F

gametes: A^F A^S A^F A^F

offspring:

	cockerel gametes	
hen gametes	A^F	A^S
A^F	A^FA^F	A^FA^S
A^F	A^FA^F	A^FA^S

half (50%) frizzled fowl (A^FA^F)
half (50%) mildly frizzled fowl (A^FA^S)

Coat of many colours

1

parental phenotypes: bull with white coat   cow with roan coat

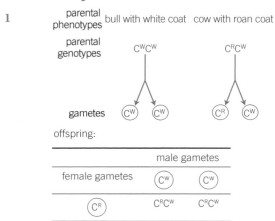

parental genotypes: C^WC^W C^RC^W

gametes: C^W C^W C^R C^W

offspring:

	male gametes	
female gametes	C^W	C^W
C^R	C^RC^W	C^RC^W
C^W	C^WC^W	C^WW^W

half (50%) roan coat ($C^R C^W$)
half (50%) white coat ($C^W C^W$)

2 **a** 100%   **b** 50%   **c** 50%   **d** 50%

3 Kittens develop inside their mother and so are kept warm / at a uniform temperature. As the kitten's coat is light-coloured, tyrosinase must have been denatured / inactivated at this warm temperature. After birth, a kitten is exposed to cooler environmental temperatures and its extremities (ears, face, feet and tail) will be the coolest as they are furthest from the main body, where heat is generated and have a large surface area to volume ratio. Cooler temperature means tyrosine is activated / not denatured. Tyrosinase therefore catalyses the production of dark pigment in these areas.

17.6

1 E = **XX**   F = **XY**

2 A = not colour blind/normal vision B = not colour blind/normal vision D = colour blind

3 G = $\mathbf{X^R X^r}$ H = $\mathbf{X^R Y}$ I = $\mathbf{X^R X^R}$ J = $\mathbf{X^r Y}$

4 0% – because sons inherit their X chromosome from their mother and she has only alleles for normal vision ($\mathbf{X^R}$).

5 By mutation (of the **R** allele).

A right royal disease

1 Because the ancestors from whom they are descended (Edward VII and Victoria) did not have, or carry, alleles for haemophilia.

2 a The disease of haemophilia only occurs in males and not females.

   b Parents without the disease are shown to have children with the disease. Alexandra and Tsar Nicholas II do not have the disease but their son Tsarevitch Alexis does. (Note: There are many other examples.)

3 a $\mathbf{X^H X^H}$ b $\mathbf{X^h Y}$ c $\mathbf{X^H X^h}$

4 Anastasia could have either genotype $\mathbf{X^H X^h}$ or $\mathbf{X^H X^H}$, depending on whether she inherited an $\mathbf{X^H}$ or an $\mathbf{X^h}$ from her mother Alexandra. Waldemar's genotype must be $\mathbf{X^h Y}$. Therefore:

   a Sons would inherit a **Y** from Waldemar and either an $\mathbf{X^H}$ or $\mathbf{X^h}$ from Anastasia (mother). Therefore the possible genotypes are $\mathbf{X^H Y}$ or $\mathbf{X^h Y}$.

   b Daughters must inherit $\mathbf{X^h}$ from Waldemar (father) and either $\mathbf{X^H}$ or $\mathbf{X^h}$ from Anastasia (mother). Therefore the possible genotypes are $\mathbf{X^h X^h}$ or $\mathbf{X^H X^h}$.

17.7

1 In sex-linkage the linked genes are on the same sex chromosome (usually the X chromosome) whereas in autosomal linkage they are on any chromosome other than the sex chromosomes.

2

parental phenotype	short, white fur	long, grey fur
parental genotype	hh gg	Hh Gg
Gametes produced by meiosis	(hg)	(HG) (hg)

	Male gametes	
Female gametes	(HG)	(hg)
(hg)	HhGg	hhgg

1 rabbit with long, grey fur (HhGg)

1 rabbit with short, white fur (hhgg)

Tales of the unexpected

1 a **Y** = allele for yellow flowers

      y = allele for white flowers

      R = allele for red fruit

      r = allele for yellow

2 **YyRr**

3 3 yellow flowers and red fruit

   1 white flowers and yellow fruit.

4 9 yellow flowers and red fruit

   3 yellow flowers and yellow fruit

   3 white flowers and red fruit

   1 white flowers and yellow fruit

5 a yellow flowers and yellow fruit

      white flowers and red fruit

   b crossing over between chromatids giving rise to recombinants

17.8

1 a mouse 1 = albino(white); mouse 2 = agouti

   b **AABb**; **AaBb**; **AAbb**; **Aabb**; **AaBb**; **aaBb**; **Aabb**; **aabb**

   c 4 albino : 3 agouti : 1 black

2 a **AaBb**

   b F$_1$ phenotype   Purple seeds

      F$_1$ genotype   AaBb

      F$_1$ gametes   (AB) (Ab) (aB) (ab)

      F$_2$ offspring

		male gametes			
		(AB)	(Ab)	(aB)	(ab)
Female	(AB)	AABB	AABb	AaBB	AaBb
gametes	(Ab)	AABb	AAbb	AaBb	Aabb
	(aB)	AaBB	AaBb	aaBB	aaBb
	(ab)	AaBb	Aabb	aaBb	aabb

   c i **AABB**; **AABb**; **AaBB**; **AaBb**; **AABB**; **AaBb**; **AaBB**; **AaBb**; **AaBb**.

      ii They all possess at least one dominant allele **A** and one dominant allele B.

   d The production of anthocyanin uses a biochemical pathway that requires two functional enzymes each coded for by the dominant allele of both genes A and B. If either gene is represented by two recessive alleles the enzyme it codes for is non-functional and the pathway cannot be completed. This is an example of epistasis because it affects the other gene in that, even if it is functional and produces its enzyme, its effects cannot be expressed because no pigment can be manufactured.

17.9

1 There is no significant difference between the observed and the expected results.

2 Three degrees of freedom.

3 Chi-squared value = 9.11.

4 The value of 9.11 lies between 7.82 and 9.84, which is equivalent to a probability of 0.05 (5%) and 0.02 (2%) that the deviation is due to chance alone.

5 This deviation is significant and we must reject the null hypothesis.

18.1

1 Allelic frequency is the number of times an allele occurs within the gene pool.

2 The proportion of dominant and recessive alleles of any gene in the population remains the same from one generation to the next.

3 No mutations arise.

The population is isolated / no flow of alleles into, or out of, the population.

No natural selection occurs / all alleles are equally advantageous.

The population is large.

Mating within the population is random.

4 p + q = 1.0 and p = 0.942

Therefore q = 1.0 − 0.942 = 0.058

Frequency of the heterozygous genotype = 2pq

= 2 × 0.942 × 0.058

= 0.109

As a percentage = 0.109 × 100 = 10.9%.

Not as black and white as it seems

1 It is not sex-linked because the number of males and females of each wing colour are approximately equal – the small difference is due to statistical error.

2 0.254 (25.4%)

Number of moths with dark-coloured wings (having two recessive alleles) = 562. Total sample = 1653 + 562 = 2215. Proportion with two recessive alleles = 562 ÷ 2215 = 0.254.

3 a 0.254 of the sample population has two recessive alleles

Therefore q^2 = 0.254 and $q = \sqrt{0.254}$ = 0.504.

b q = 0.504 and p + q = 1.0.

Therefore p = 1.0 − 0.504 = 0.496.

c Frequency of heterozygotes = 2pq = 2 × 0.496 × 0.504 = 0.5

Therefore % of heterozygotes = 50%

4 Capture a sample of moths and mark them in some way. Release them back into the population. Sometime later randomly recapture a given number of moths. Record the number of marked and unmarked moths in this second sample. Calculate the size of the population as follows:

$$\frac{\text{total number of moths in first sample} \times \text{total number of moths in second sample}}{\text{number of marked moths recaptured}}$$

5 234 people. As tongue rolling is dominant, only those who are homozygous recessive for the allele will be unable to roll their tongue = 26 people. The proportion of the total sample population that are non-rollers is therefore $\frac{26}{416}$ = 0.0625. In the Hardy Weinberg equation ($p^2 + 2pq + q^2 = 1.0$) q = the frequency of the recessive allele. As the non-rollers have two recessive alleles (q^2) then their proportion of the population is q^2 = 0.0625. Therefore $q = \sqrt{0.0625}$ = 0.25. We know that p + q = 1.0, therefore p = 1.0 − 0.25 = 0.75. Homozygous dominant tongue rollers have two dominant alleles (p^2) their proportion in the population is 0.75 × 0.75 = 0.5625. In a sample of 416 people this is equal to 0.5625 × 416 = 234.

18.2

1 mutation, meiosis and random fusion of gametes

2 mutation only

3 a environmental; b genetic; c genetic:
   d environmental; e environmental;

18.3

1 The total number of all the alleles of all the genes of all the individuals within a particular population at a given time.

2 predation/competition for (food/water/space)/ disease/natural disasters

3 In malarial regions, the disadvantages of having the disease will be offset by the advantages of having resistance to malaria and so there will be little if any selection against the gene and its frequency will be relatively high. In non-malarial regions there is no advantage in having resistance to malaria and so individuals with sickle cell anaemia will be at a disadvantage; they will be selected against and the frequency of the gene will be low.

How genetic variation leads to natural selection – copper tolerance in grasses

1 Normal distribution curve

2 Polygenic

3 Where the soil is contaminated, the non-tolerant species are poisoned by the copper and die so there

is less competition and so the tolerant species' population becomes larger. Where there is no contamination, all varieties can survive, there is greater competition and so the populations are smaller.

4 Cross pollinate the plant many times with other varieties of *Agrostis capillaris*. If fertile offspring result they are the same species and therefore a variety. If not, it is likely that they are separate species. The more often they fail to produce fertile offspring, the more likely they are to be separate species.

5 a Total sample = 450 of which 72 are copper tolerant and therefore homozygous recessive. $\frac{72}{450}$ = 16% (or 0.16) are copper tolerant (genotype = tt). If p = frequency of T and q = frequency of t then q^2 = 0.16. q is therefore $\sqrt{0.16}$ = 0.4. We know p + q = 1, therefore p = 0.6. Using Hardy-Weinberg equation frequencies are:

TT = p^2 = 0.6^2 = 0.36 = 36%

Tt = 2pq = 2 × 0.6 × 0.4 = 0.48 = 48%

tt = q^2 = 0.4^2 = 0.16 = 16%

b The population is small and selection is taking place.

18.4

1 a stabilising
  b stabilising
  c directional
  d disruptive
  e disruptive
  f directional
  g stabilising

2 The light coloured (non-melanic) form because pollution control means buildings are no longer black. The melanic form is therefore more conspicuous than the light form and so preferentially eaten by predators. The light form is more likely to survive and reproduce to give more light-coloured offspring There is a selection pressure favouring the light form that has led to it outnumbering the melanic form.

18.5

1 A species is a group of individuals that share similar genes and are capable of breeding with one another to produce fertile offspring. In other words they belong to the same gene pool.

2 Speciation is the evolution of new species from existing species.

3 Geographical isolation occurs when a physical barrier, such as mountains or oceans, prevents two populations from breeding with one another.

4 Geographically isolated populations may experience different environmental conditions. In each

population, phenotypes that are best suited to the particular environmental conditions are selected. The composition of the alleles in each gene pool therefore changes as they pass to subsequent generations. The composition of the gene pool of each population becomes increasingly different over time. Being geographically isolated, individuals of each population cannot breed with one another and so the two gene pools remain separate and different.

5 Allopatric is speciation as a result of two populations becoming reproductively isolated because they are geographically separated and so unable to interbreed. Sympatric is speciation as a result of populations that live together being reproductively isolated for other reasons e.g. they have different breeding seasons which do not overlap.

19.1

1 ecology
2 biotic
3 abiotic
4 community
5 population
6 habitat
7 the carrying capacity

19.2

1 Certain factors limit growth, e.g. availability of food, accumulation of waste, disease.

2 Biotic factors involve the activities of living organisms.

Abiotic factors involve the non-living part of the environment.

3 a low light intensity
  b lack of water
  c low temperature

4 a Using a standard scale, most of the points plotted for the population of the world would be so close together as to be indistinguishable from each other on the graph. A logarithmic scale separates these points.

b

Time/years before present (BP)	Estimated human population/ billions	Log human population
12 000	1	0.00
10 000	5	0.70
8000	10	1.00
6000	20	1.35
4000	35	1.54
2000	200	2.30
0	600	2.88

c As the time scale is back in time from the present day, the values can be treated as minus values and so the scale plotted from -12 000 years to 0 years (present). The points are joined by a series of straight lines rather than a line of best fit because we cannot be certain that the intervening values would fall on the curve plotted. This is because human populations can fluctuate over relatively short periods e.g. due to diseases, famine etc.

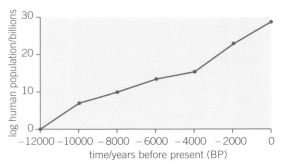

The influence of abiotic factors on plant population

1 a   1   c   2
   b   3   d   3

2 The pH is too high for species X and the temperature is too low for species Y.

The growth and size of human populations

1 a   Number of births = $\dfrac{25 \times 1\,000\,000}{1000} = 25\,000$

   b   Number of deaths = $\dfrac{20 \times 1\,000\,000}{1000} = 20\,000$

   c   percentage population growth =
       $\dfrac{\text{population change during 2007}}{\text{population at the start of 2007}} \times 100$

       $= \dfrac{(25\,000 - 20\,000)}{1\,000\,000} \times 100$

       $= \dfrac{5000 \times 100}{1\,000\,000} = 0.5\%$.

2 a   Stage 3   d   Stage 1
   b   Stage 1   e   Stage 4
   c   Stage 2

3. Pyramid A represents stage 4 because there is a low birth rate (narrow base to the pyramid) and a low death rate (sides are fairly vertical and many people live beyond 65 years).

    Pyramid B represents stage 2 because there is a high birth rate (wide base to the pyramid) and a falling death rate (sides slope upwards and some, but not many, people live beyond 65 years).

19.3

1 Intraspecific competition occurs when individuals of the same species compete with one another for resources.

Interspecific competition occurs when individuals of different species compete for resources.

2 Any 2 from: food / water / breeding sites (or any other relevant factor, e.g. light, minerals).

The effects of interspecific competition on population size

1 After 1985 the rise in the grey squirrels population is mirrored by a fall in the red squirrel population.

2 Lack of food / adverse weather, e.g. cold winters / increase in number of squirrel predators / new disease.

3 Grey squirrels have more chance of finding fruits / nuts / seeds that have fallen to the ground as well as those that are still on the trees / bushes.

4 The sea presents a barrier to the grey squirrels reaching islands. The red squirrels already present on the islands have little or no competition from grey squirrels and so flourish.

Competing to the death

1 Population increases slowly at first and then at an accelerating / exponential / logarithmic rate to around 8 days. The growth rate then slows, reaching a maximum at around 12 days which is maintained at a constant level up to 20 days.

2 Population growth is faster initially. Maximum size is reached earlier. Maximum size is reduced to less than half. Size is not maintained at a constant level (it falls to zero).

3 *P. caudatum* is unsuccessful in competing with *P. aurelia* for yeast / food. Most available food is taken by *P. aurelia* and *P. caudatum* starves, leading to a population crash.

4 Some of the yeast / food is taken by *P. caudatum*, leaving less for the population growth of *P. aurelia*.

5 After 20 days all *P. caudatum* have died. *P. aurelia* has no competition for food and so it reaches its previous maximum. *P. aurelia* is in effect 'alone' again.

Effects of abiotic and biotic factors on population size

1 mark-release-recapture technique

2 Increase in population is 1320 (in 1995) minus 260 (in 1993) = 1060.

   Time period = 2 years.

   Mean annual growth in population is therefore: $= \dfrac{1060}{2} = 530$.

3 The more acorns produced in the autumn, the larger the deer mice population the following spring. The fewer acorns produced, the smaller the deer mice population.

4 1992

5 The population of deer mice would fall as more oak leaves are eaten by gypsy moth caterpillars so there will be less food (acorns) to support the deer mice population / some deer mice will starve.

6 a warm spring → more acorn seed is set → more acorns produced in autumn → more food is available over winter → more deer mice survive and breed → deer mice population increases → more predation of gypsy moth pupae by deer mice → smaller gypsy moth population

 b more owls → more predation on deer mice → smaller deer mice population → fewer gypsy moth pupae are eaten → larger gypsy moth population → more oak leaves eaten → less energy available from photosynthesis for the production of acorns → fewer acorns.

19.4

1 The range and variety of laboratory habitats is much smaller than in natural ones. This means that in nature there is a greater range of hiding places and so the prey has more space and places to escape the predator and survive.

2 With fewer predators, fewer prey are taken as food. The death rate of prey is reduced. Assuming the birth rate remains unchanged the population size increases.

3 Graph showing population fluctuations (peaks and troughs) of A. Species B mirrors these changes after a time lag. The population size of B is, for the most part, smaller than A. B eats A → population of A falls → fewer A for B to eat → population of B falls → fewer B means fewer A are eaten → population of A rises → more A means more food for B → population of B rises.

The Canadian lynx and the snowshoe hare

1 The assumption is made that the relative numbers of each type of fur traded represents the relative size of each animal's population at the time.

2 The population size of the snowshoe hare fluctuated in a series of peaks and troughs. Each peak and trough was repeated about every 10 years. The population size of the Canadian lynx also fluctuated in a 10-year cycle of peaks and troughs. The relative pattern of peaks and troughs is similar for the lynx and the snowshoe hare. The rise in the population size of the lynx often (but not always) followed that of the snowshoe hare.

3 The snowshoe hare population increases due to the low numbers of Canadian lynx that feed on them → more hares mean more food for the lynx, whose population therefore increases as fewer starve / more are able to raise young → more lynx means

there is more predation of hares, whose population therefore decreases → fewer hares means less food for the lynx, many of which starve and so their population decreases.

4 4 times

5 Addition of food – because the population increased more in every year that data were collected.

6 Both food supply and predation influence hare population size. Food supply has a greater influence than predation but a combination of both factors has an even greater influence than either of the other two separately.

19.5

1 $100 \times \dfrac{80}{5} = 1600$

2 a Population over-estimated (appears larger) as there will be proportionally fewer marked individuals in the second sample.

 b Population over-estimated / appears larger as there will be proportionally fewer marked individuals in the second sample because all the 'new' individuals will be unmarked.

 c No difference because the proportion of marked and unmarked individuals killed should be the same.

3 $(120 \times 120) \div 960 = 15$

19.6

1 pioneer species

2 primary colonisers (pioneer species) photosynthesise and fix nitrogen → these die and form a soil with nutrients → further colonisers can survive in this soil → environment is a little less hostile → more habitats and food sources available → other species are able to survive → increased biodiversity

3 climax community

Warming to succession

1 Biomass increases very slowly and so the line curves gently upwards at first (up to 60 years) because there is little nitrogen in the soil and therefore growth and hence net production of the pioneer species (*Dryas*) is small.

Biomass increases at a greater but constant rate and so the curve becomes a straight line with an upward gradient, from 60 to 120 years as soil nitrogen levels rise. Increased levels of soil nitrogen remove this limit on growth (net primary productivity) therefore large species, such as alder, and later spruce, establish themselves and hence biomass increases more rapidly.

Biomass increase slows and finally stops and so the curve flattens out after 150 years because soil nitrogen levels fall as plants take it into their biomass – nitrogen again limits plant growth (net primary productivity).

2   a   Nitrogen from the atmosphere is fixed into compounds, e.g. proteins and amino acids by the nitrogen-fixing species (lichens, *Dryas* and alder). When these die or shed their leaves this nitrogen is released when decomposers break them down into ammonium compounds (ammonification) which are then broken down by nitrifying bacteria into nitrites and nitrates.

   b   More nitrogen is being absorbed by the increased biomass of the plants. The nitrogen-fixing lichens, *Dryas* and alder have been replaced by spruce that does not fix nitrogen therefore less nitrogen is being added to the soil.

3   a   (Pioneer) species are taking advantage of new habitats and lack of competition to rapidly colonise the empty land.

   b   Spruce is becoming dominant and out-competing the other species, such as lichen, *Dryas* and alder, for light, nutrients, etc. These other species are eliminated from the community.

4   Transects are better because there is a gradient of environmental factors that produce a series of changes over a long distance. Transects also ensure that every community is sampled, which may not be the case with random sampling.

19.7

1   The species within the habitat possess unique genes that at some point in the future may be useful. Conserving habitats maintains biodiversity. The greater the variety of habitats, the greater their potential to enrich our lives and provide enjoyment.

2   Cut back reeds to prevent them becoming dominant. Remove dead vegetation to prevent build-up and thus stop fens drying out. Pump water into fens to keep them waterlogged. Cut back grasses and shrubs to prevent succession.

Conflicting interests

1   96 (32% of 300)

2   It might increase the population of grouse as harriers would have alternative sources of food and therefore eat fewer grouse chicks. Alternatively it might lead to a large increase in harriers that then prey on grouse (especially once the supply of voles and meadow pipits has been exhausted). This would lead to a decrease in the grouse population.

3   The moorland would undergo secondary succession, finally reaching its climax community of deciduous (oak) woodland.

4   A selection from each of the following arguments:

For	Against
• The harrier is a very rare bird – there are only 750 pairs in the UK.	• The harrier is a major predator of grouse and so could threaten the already declining grouse population.
• Previous persecution led to its extinction on the UK mainland and this could happen again.	• If the grouse population is reduced / eliminated and / or the harrier population is not controlled, this could adversely affect the populations of alternative harrier prey, such as voles and meadow pipits.
• Harriers are part of our natural heritage and their population should not be controlled other than by natural means.	• Reduction / elimination of grouse population could make grouse shooting uneconomic and, unless money is found from elsewhere, the moorland habitat might be lost along with the species that live there, and so reduce biodiversity.

5   a   A long time is needed to allow population changes in both species as each only breeds once a year.

   b   The conflicting interests of conservationists and grouse managers mean that agreement on issues such as the population ceiling for hen harriers is unlikely without independent arbitration. An independent body can ensure that the experiment is carried out properly and that the results are interpreted without bias from parties with a vested interest.

   c   This ensures that hen harrier populations rise within as short a time as possible so that results can be analysed and decisions on future policy made. If this takes too long, the harrier may already be eliminated from some regions.

   d   This ensures a wide range of different biotic and abiotic conditions as well as a range of different individuals. Some areas may not be typical and some individuals may not be totally cooperative and this may skew the results. A number of varied sites / individuals will minimise the effects of any such anomalies.

   e   They fear it might further reduce the currently dangerously low harrier population. They fear it might set a precedent for other species and other experiments.

6 Where views conflict, evidence is essential to support or discount any claims made. Scientists can produce this evidence in carefully devised, controlled and unbiased experiments. The scientific evidence helps decisions to be made that are more likely to have the desired effect.

20.1

1 deletion and addition because the bases are deleted from one chromosome and added to a different one.

2 a deletion

b inversion

c substitution

d addition

3 a It will cause a frame shift causing triplets (codons) in a sequence to be read differently because each has been shifted to the right by one base. If the additional base is inserted early in the sequence most codons will be changed, so will the amino acids they code for. The resultant polypeptide will be very different from normal.

b If the additional base is inserted at the end of the sequence few, if any codons will be changed. Few, if any, amino acids they code for will differ and the resultant polypeptide will be normal or near normal.

4 Where the duplicated bases are consecutive, the frame shift is three bases long and so the subsequent codons are not affected. The polypeptide will have an additional amino acid but otherwise be unchanged. If the bases are separate, the frame shift will initially be one base long, becoming two bases long after the second duplicate base is added. Codons after both the duplications will be changed and the polypeptide will have many different amino acids (but not necessarily all – degenerate code). After the third duplicate base the codons will be unchanged.

5 Some codons will be changed to ones that code for the same amino acid (degenerate code). The frame shift might not alter some codons because the replacement bases are the same as the originals. (e.g. GCT TTT CGA – a single base frame shift to the right does not alter the TTT codon).

Mutagenic agents

1 The codons in mRNA will be CAU AAA UAA (Note: In mRNA, guanine is coded for by cytosine in DNA, adenine by thymine and uracil by adenine as usual, but after the change, cytosine becomes uracil in DNA and this codes for adenine in mRNA.)

2 substitution gene mutation

3 The active site of DNA polymerase can no longer fit the DNA molecule because the shapes of some DNA bases have been altered by X-rays.

4 The replication of DNA requires DNA polymerase and so the process cannot continue.

5 Public opinion, special interest groups such as the owners of shops selling or using sunbeds, manufacturers, consumers, professional bodies (e.g. members of the medical profession), the media and other scientists.

20.2

1 Totipotent cells are cells with the ability to develop into any other cell of the organism.

2 Totipotent – can differentiate into any type of cell in the body and comprise the first few cells that form from the zygote. Pluripotent – can differentiate into almost any type of cell (but not all) and are found in the embryo and young fetus. Multipotent – can differentiate into a limited number of cells and found in the umbilical cord and some adult tissues e.g. bone marrow. Unipotent – can only differentiate into a single type of cell and are found in adult tissue such as the skin.

3 In skin cells, the gene that codes for keratin is expressed, but not the gene for myosin. The genetic code for keratin is translated into the protein keratin, which the cell therefore produces, but the genetic code for myosin is not translated. In muscle cells, the gene for myosin is expressed but not the gene for keratin. In the same way, the genetic code for myosin rather than keratin is translated and so only myosin is produced.

4 Skin cells, being on the outside of the body, are subjected to much wear and tear and so need replacing frequently. Many other organs are less prone to damage and need little cell replacement.

Human embryonic stem cells and the treatment of disease

1 Any properly structured and evaluated accounts that make scientifically accurate points in a reasoned fashion are acceptable, for example:

For: Huge potential to cure debilitating diseases; wrong to allow suffering when can be relieved; embryos created for other purposes (IVF) so why not stem cells; embryos of less than 14 days not recognisably human and so do not command same respect as adults or fetuses; no risk of research escalating or including fetuses because current legislation prevents this; adult stem cells not as suitable as embryonic stem cells and it may be many years before they are, in meantime many people suffer unnecessarily.

Against: wrong to use humans, including potential humans as a means to an end; embryos are human, they have human genes, and deserve the same respect and treatment as adult humans; is a 'slippery slope' to the use of older embryos and fetuses for research; could lead to research and development of human cloning and, although banned in UK the information could be used elsewhere; undermines respect for life; adult stem cells are an available alternative and energies should be directed towards developing these.

Growth of plant tissue cultures

1 differentiation
2 'in vitro'
3 a clone b mitosis
4 IAA and 2,4-D
5 In test tube 1 the low concentration of IAA produces moderate shoot development but when a high concentration of cytokinin is added (test tube 3) the presence of cytokinin influences the effects of the IAA by reducing shoot development to a 'little'.

20.3

1 Transcriptional factors stimulate transcription of a gene.

2 Oestrogen diffuses through the phospholipid portion of a cell-surface membrane into the cytoplasm of a cell, where it binds with a site on a receptor portion of the transcriptional factor. Oestrogen changes the shape of the DNA binding site on the transcription factor so it can now bind with DNA. The transcriptional factor now enters the nucleus through a nuclear pore and binds with DNA, stimulating transcription of the gene that makes up that portion of DNA, i.e. it stimulates gene expression.

Gene expression in haemoglobin

1 It allows the fetus to load its haemoglobin with oxygen from the mother's haemoglobin where the two blood supplies come close to each other (at the placenta).

2 alpha = 50% beta = 20% gamma = 30%

3 The gene for gamma-globulin is expressed less while the gene for beta-globulin is expressed more.

4 Expression of the gene for gamma-globulin is progressively reduced as a result of either preventing transcription, and hence preventing the production of mRNA, or by the breakdown of mRNA before its genetic code can be translated.

5 A possible therapy would be to express (switch on) the gene for gamma-globulin and prevent the expression of (switch off) the gene for beta-globulin. This would result in haemoglobin being of the fetal rather than the adult type.

20.4

1 Epigenetics is the process by which environmental factors can cause heritable changes in gene function without changing the base sequence of DNA.

2 decreased histone acetylation and increased DNA methylation.

3 The other strand would have complementary bases (i.e. GCUA instead of CGAU respectively). It is unlikely that these opposite base pairings would complement a sequence on the mRNA. The siRNA, with enzyme attached, would therefore not bind to the mRNA and so would be unaffected.

4 a Some siRNA that blocks a particular gene could be added to cells. By observing the effects (or lack of them) we could determine what the role of the blocked gene is.

   b siRNA could be used to prevent the disease by blocking the gene that causes it.

5 a Chromatin would be more condensed (tightly packed).

   b Transcription would cease.

Nature versus nurture

1 Environmental factors

2 If the influence was totally genetic the plants that were genetically identical (those in the same column) would have the same phenotype regardless of where they were grown. The greater the environmental influence the greater the differences between the genetically identical plants. There are major differences so the main influence is environmental.

3 Environmental conditions at high altitude are more extreme than those at low altitude and less suited for photosynthesis (colder, windier, less soil). Plants from high altitude have evolved to survive in these extremes. The conditions at medium and low altitude therefore present few problems and they thrive. Plants that have evolved at low altitude, by contrast, find harsher conditions at medium and high altitude and struggle to grow.

Prader-Willi syndrome

1 Epigenetic tags are usually erased in sperm and eggs or during early fetal development and so do not silence the genes.

2 The genes that are deleted might code for polypeptides/proteins/enzymes/hormones that are essential to the reproductive process. For example, they might code for enzymes/proteins needed to make functioning ovaries/eggs or testes/sperm or for the synthesis of hormones needed for the development of these organs/gametes.

20.5

1 Fat cells of the breasts tend to produce more oestrogens after the menopause. These locally produced oestrogens release an inhibitor molecule that prevents transcription causing proto-oncogenes of breast tissue to develop into oncogenes. These oncogenes increase the rate of cell division leading to the development of a tumour (breast cancer).

2 Proto-oncogenes **increase** the rate of cell division and so their activation produces a mass of cells (tumour) but tumour suppressor genes **decrease** the rate of cell division and so their deactivation produces tumour.

3 Cells of a benign tumour produce adhesion molecules that make them stick together and are surrounded by a capsule of dense tissue. The tumour therefore remains as a compact structure and so surgical removal is likely to remove **all** tumour cells.

4 Malignant tumours spread to other regions of the body and so even though surgery can remove the more obvious larger ones, tiny ones will require other therapies to prevent these re-growing into new tumours.

5 HADC removes acetyl groups from histones, inhibiting transcription and switching off the gene. Some cancers are the result of genes (e.g. tumour suppressor genes) that normally help repair DNA (and so prevent cancers) being switched off. By inhibiting HADC, phenylbutyric acid could prevent the removal of acetyl groups from histones and switch the 'protective' gene back on.

Risk factors and cancer

1 air pollution / inhaled substances (carcinogens), e.g. asbestos at work

2 a positive correlation between the number of cases of lung cancer in men and the number of cigarettes smoked

3 It is unlikely that a coincidence would have occurred many times over.

4 No. While the data clearly point to the likelihood that smoking causes lung cancer, they do not provide experimental evidence that specifically links smoking to lung cancer.

5 The experiment that showed that the derivative of benzopyrene caused changes to DNA at precisely the same three points as the mutations of the gene in a cancer cell.

6 This is a single case. The link between early death and lung cancer is about probabilities not certainties. Statistically it is unlikely, but not impossible, for smokers to live to be very old.

20.6

1 A genome is all the genetic material in an organism. A proteome is all the proteins produced by the genome.

2 Simple organisms generally have just one, circular piece of DNA that is not associated with histones and there is little, if any, non-coding DNA. Complex organisms have considerably more DNA and the majority of this does not code for proteins.

3 It allows identification of those proteins that act as antigens on the surfaces of the pathogens. These antigens can then be used to produce effective vaccines against the disease.

21.1

1 recombinant

2 reverse transcriptase

3 complementary (cDNA)

4 DNA polymerase

5 restriction endonucleases

6 blunt

7 sticky

8 CTTAAG

21.2

1 A vector transfers genes (DNA) from one organism into another.

2 To show which cells (bacteria) have taken up the plasmid (gene).

3 Results can be obtained more easily and more quickly – because, with antibiotic-resistance markers, the bacterial cells with the required gene are killed, so replica plating is necessary to obtain the cells with the gene. With fluorescent gene markers, the bacterial cells are not killed and so there is no need to carry out replica plating.

4 a B, C, D, J, K and L – because those that did not take up the plasmid will not have taken up the gene for ampicillin resistance and so will be the ones that are killed on the ampicillin plate, i.e. the colonies that have disappeared.

   b E, F and I – because those with the plasmid containing gene X will have a non-functional gene for tetracycline resistance and therefore the colonies will have been killed on the tetracycline plate, i.e. the colonies will have disappeared.

21.3

1 Primers are short pieces of DNA that have a set of bases complementary to those at the end of the DNA fragment to be copied.

2 Primers attach to the end of a DNA strand that is to be copied and provide the starting sequences for DNA polymerase to begin DNA cloning. DNA polymerase can only attach nucleotides to the end

of an existing chain. They also prevent the two separate strands from rejoining.

3 Because the sequences at the opposite ends of the two strands of DNA are different.

4 Hydrogen bonds

5 Biological contaminants may contain DNA and this DNA would also be copied.

Evaluation of DNA technology

Arguments should be reasoned, logical and based on sound science, and should use specific examples rather than vague references.

1 Whichever aspects are chosen the beneficial aspects to humans should be clear, e.g. genetically modified crops that can be grown in extreme conditions – greater productivity; more food; less poverty and hunger in some human populations.

2 Arguments must relate directly to the aspects chosen in question 1 and should oppose the use of the technology, for example, genetically modified crops that can be grown in extreme conditions – risk of damaging the ecological balance; risk of the gene passing to other organisms; dangers from unforeseen by-products of the plants' metabolism; dangers from possible mutations of the genes; the economic consequences for developing countries.

Treatment of severe combined immunodeficiency using gene therapy

1 Using restriction endonucleases to cut the gene from the length of DNA that carries it / using reverse transcriptase to make cDNA from mRNA produced by the gene / manufacturing it in a 'gene machine'.

2 A virus with RNA as its nucleic acid and which can make a copy of DNA from this RNA using the enzyme reverse transcriptase.

3 Defective gene – ineffective or no enzyme ADA produced – toxins not destroyed by enzyme ADA – T cells destroyed by toxins – no/little immunity to infection – an infection causes death.

4 T cells only live for 6-12 months and those that replace them do not possess the gene.

5 Bone marrow stem cells divide to produce T cells and so there is a constant supply of the ADA gene and hence the enzyme ADA.

21.4

1 A DNA probe is a short, single-stranded section of DNA that has some label attached that makes it easily identifiable.

2 Determine the order of nucleotides on the mutated gene by DNA sequencing – produce a fragment of DNA that has complementary bases to the mutated portion of the gene – label the fragments to form

a DNA probe – make multiple copies of the DNA probe using PCR techniques – add the probe to DNA fragments from the individual being tested. If the donor has a mutant allele, the probe will bind to the complementary bases on the donor DNA. These fragments will now be labelled and can be distinguished from the rest of the DNA.

3 a Tumour suppressor genes inhibit cell division.

b He/she might change their lifestyle to reduce the risk of cancer, e.g. by giving up smoking, losing weight, eating more healthily and avoiding mutagens as far as possible; checking more regularly for early symptoms of cancer; choosing to undergo treatment.

21.5

1 PCR is used to increase the quantity of DNA because the quantity available, e.g. at a crime scene, is often very small.

2 a Suspect B – because the bands on this suspect's genetic fingerprint match those of the genetic fingerprint of blood found at the crime scene.

b To eliminate the victim as the source of the blood sample found at the scene.

3 The chemicals may inhibit some of the restriction endonucleases, which would then fail to cut some sections of DNA. There would therefore be a greater number of longer DNA fragments than normal and the fingerprint would be different.

4 In a person with the allele for Huntington's disease, some of the DNA fragments will be larger than those in a person without the allele because of the extra repeating units on the gene. These will travel a shorter distance in the electrophoresis gel and so there will be more thick bands nearest the start of the fingerprint (where the initial sample was located).

5 Genetic fingerprints can determine how closely any two individuals are related. The closer the match between their fingerprints, the closer they are related. Therefore, to avoid the problems caused by inbreeding, it is advisable to mate animals whose fingerprints differ the most.

Locating DNA fragments

1 restriction endonuclease

2 A = 2 B = 4 C = 5 D = 1 E = 6 F = 3

Explanation – the shorter the fragments (those with fewer base pairs) the further they travel and the longer the fragments the less distance they travel.

3 5 times – because 5 cuts produce 6 fragments.

Gel electrophoresis and DNA sequencing

1 two

2 cytosine

3 CACTGTTCAT

Practical questions – answers

Practical 1

1 (a) Sketch a graph of your expected results.
       Remember to label your axes.

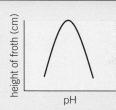

   (b) List all variables that need to be controlled and how you would control them.
        Temperature – Keep celery extract and H_2O_2 in a thermostatically controlled water bath at 30°C
        Enzyme concentration – use the same source of celery extract which has been mixed evenly
        Substrate concentration – use same volume and concentration of H_2O_2

2 (a) Repeat each pH at least twice and calculate a mean.
   (b) This method is very subjective to decide on the highest point of the froth.
        Change method to using a gas syringe to collect the O_2 gas released.
        Celery extract may contain varying concentrations of enzyme.
        Change method to use a pure source of a specific concentration of enzyme.

Practical 2

1 (a) $\dfrac{12}{150} \times 100 = 8\%$

   (b) No stain used / not root tip / cells not dividing in this small sample / more than one layer of cells as not squashed firmly enough.

2 (a) As distance increases from the root tip, the mitotic index decreases.
        Above 1 mm an increase in distance from root tip has little effect on the mitotic index / plateaus.
        Correctly quote paired set of data.

   (b) Meristem tissue only nearest the tip has the ability to divide and there is less meristem tissue as the distance increases from the tip.
        Nearest the tip gets more damage, therefore needs to do more cell division to repair the tissue.

Practical 3

1

Concentration of sucrose (mol dm^{-3})	Volume of distilled water (cm^3)	Volume of 0.5 mol dm^{-3} stock solution sucrose (cm^3)
0.0	20	0
0.1	16	4
0.2	12	8
0.3	8	12
0.4	4	16
0.5	0	20

2 (a) Calculate the % change in mass of potato tissue.

Concentration of sucrose (mol dm^{-3})	Mass before submerging in solution (g)	Mass after submerging in solution (g)	Mass change (g)	Percentage change in mass of potato tissue (%)
0.0	4.5	5.0	+0.5	**+11.1**
0.1	3.9	4.3	+0.4	**+10.3**
0.2	4.3	4.5	+0.2	**+4.7**
0.3	4.1	4.2	−0.1	**−2.4**
0.4	4.4	3.7	−0.7	**−15.9**
0.5	4.4	3.6	−0.8	**−18.2**

(b) There are different starting and finishing masses.
The mass change is very small; therefore a % change is easier to compare real differences.

3 (a) Correctly labelled axes with units;
uniform axes;
plots taking up over $\frac{1}{2}$ space of graph;
accurate plots;
smooth line of best fit.

(b) Use this graph to find the concentration of sucrose (where curve crosses x-axis). Between 0.25 and 0.3 mol dm^{-3}

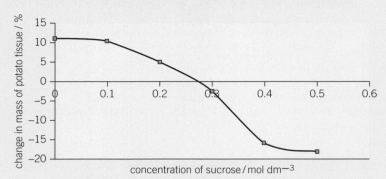

(c) Use a data resource with listed sucrose concentrations and water potentials to find the water potential for the sucrose solution read off the graph.

Practical 4

1 (a) The red pigment is water soluble and held in the vacuole;
The cell-surface membrane is selectively permeable and some pigment diffuses out.

(b) As temperature increases from 20 to 40°C, there is a small increase in absorbance reflecting a small increase in the permeability of the cell-surface membrane.
Above 50°C there is a steep increase in the permeability of the cell-surface membrane.

(c) The proteins embedded in the cell-surface membrane become denatured.
The structure of the cell-surface membrane has been permanently disrupted so is now fully permeable and most of the pigment diffuses out.

Practical 5

1 (a) Spiracles.

(b) Control water vapour loss by closing spiracles if need to conserve water.

(c) High temperature environment causes more water to evaporate;
hairs trap water <u>vapour</u> and this reduces water potential gradient and therefore water vapour loss.

2 (a) Penetrate deep into muscle tissues;
delivers more air / oxygen to muscles.

(b) Chitin

(c) Smaller diameters are more permeable to gases and get closer to body cells for gaseous exchange by diffusion.

Practical 6

1 (a) To sterilise the equipment/ to kill any microbes on the equipment.

(b) 1   Washing hands / cleaning work surface with disinfectant
2   Flame sterilising the inoculating loop
3   Flaming the neck of the culture tube containing the bacteria
4   Streaking the plate with the inoculating loop **quickly**
5   Only lifting the lid of the petri dish a small amount

(c) To kill any harmful / pathogenic bacteria so they don't harm anyone.

Practical 7

(a) Ensure pigment spot is above solvent/ensure atmosphere in container was saturated with solvent before running.

(b) Solutes/pigments dissolve in solvent; solvent moves up paper; distance moved by solutes/pigments depends on their relative solubility/molecular size.

(c) (i) Relative flow (Rf) is a physical constant; for a specific solute in a specific solvent; it is the distance moved by the solute divided by the distance moved by the solvent (front);

Rf = distance moved by the compound ÷ distance moved by the solvent.

(ii) $B = \dfrac{35}{93} = 0.38$; (0.376) $C = \dfrac{36}{93} = 0.39$; (0.387)

(iii) 2-way chromatography/run with a different solvent

Practical 8

1 Colour change and inferences that can made from the results:

Tube 1 (leaf extract + DCPIP) colour changes until it is the same colour as tube 4 (leaf extract + distilled water).

Tube 2 (isolation medium + DCPIP) no colour change. This shows that the DCPIP does not decolourise when exposed to light.

Tube 3 (leaf extract + DCPIP in the dark) no colour change. It can therefore be inferred that the loss of colour in tube 1 is due to the effect of light on the extract.

Tube 4 (leaf extract + distilled water) no colour change. This shows that the extract does not change colour in the light. It acts as a colour standard for the extract without DCPIP.

Tube 5 (supernatant + DCPIP) no colour change if the supernatant is clear; if it is slightly green there may be some decolouring.

The results should indicate that the light-dependent reactions of photosynthesis are restricted to the chloroplasts that have been extracted.

2 Carbon dioxide will have no effect, because it is not involved in the light-dependent reactions.

3 Students should describe a procedure in which light intensity is varied but temperature is controlled.

Practical 9

1 **a**

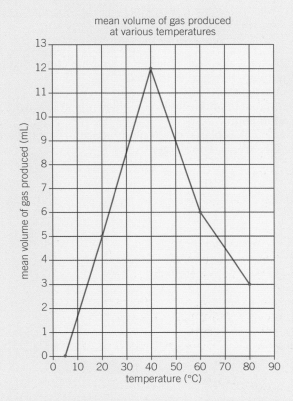

mean volume of gas produced at various temperatures

    b)   40°C

    c)   the smallest quantity of glucose – answer 2

Practical 10

1   **(a)**   Darkness/shelter/humidity/moisture/protection from predators

    (b)   Cover one side to put into darkness/ make one side more moist – place equal numbers of woodlice in each side – leave for a set time and then compare the numbers in each side and note the preference.

    (c)   The woodlice would prefer the moist / dark side

    (d)   Protection from predators / prevention of water loss

Practical 11

    (a)   Use the same volume or concentration of Benedict's solution, use the same volume of glucose solution, calibrate the colorimeter, boil for the same length of time

    (b)   6.5

    (c)   Boil/heat with Benedict's then test the filtrate. Treat the filtrate with sucrose/ invertase/acid

Practical 12

    (a)   Set up coordinate grid/use tapes along 2 sides; generate random numbers for co-ordinates; co-ordinate indicates centre of sampling quadrat.

    (b)   Advantage: unbiased/allows statistical testing; Disadvantage: coverage may be uneven/unrepresentative/large areas may be missed.

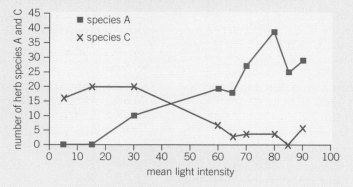

    (i)   Correct axes (light intensity on x-axis) / suitable scale / accurate plotting / joining points with a ruled / straight line / key/curves labelled.

    (ii)   Number increases as average light intensity decreases/converse.

    (iii)   Shade tolerant/loving plant/able to photosynthesise efficiently at low light intensities/low compensation point.

    (d)   Thinner epidermis / large or many chloroplasts / chloroplasts concentrated towards upper surface / high chlorophyll concentration / larger leaves.

Glossary

A

abiotic an ecological factor that makes up part of the non-biological environment of an organism, e.g. temperature, pH, rainfall and humidity. See also *biotic*.

acetylcholine one of a group of chemicals, called *neurotransmitters*, released by *neurones*. It diffuses across the gap (*synapse*) between adjacent neurones and so passes an impulse from one neurone to the next.

action potential change that occurs in the electrical charge across the membrane of an *axon* when it is stimulated and a nerve impulse passes.

activation energy energy required to bring about a reaction. The activation energy is lowered by the presence of *enzymes*.

active site a group of amino acids that makes up the region of an *enzyme* into which the *substrate* fits in order to catalyse a reaction.

active transport movement of a substance from a region where it is in a low concentration to a region where it is in a high concentration. The process requires the expenditure of *metabolic* energy.

adenosine triphosphate (ATP) an activated nucleotide found in all living cells that acts as an energy carrier. The *hydrolysis* of ATP leads to the formation of adenosine diphosphate (ADP) and inorganic phosphate, with the release of energy.

adrenaline a hormone produced by the adrenal glands in times of stress that prepares the body for an emergency.

aerobic connected with the presence of free oxygen. Aerobic respiration requires free oxygen to release energy from glucose. See also *anaerobic*.

allele one of a number of alternative forms of a *gene*. For example, the gene for the shape of pea seeds has two alleles: one for 'round' and one for 'wrinkled'.

anaerobic connected with the absence of oxygen. Anaerobic respiration releases energy from glucose or other foods without the presence of oxygen. See also *aerobic*.

antibiotic resistance the development in microorganisms of mechanisms that prevent *antibiotics* from killing them.

antibody a protein produced by *lymphocytes* in response to the presence of the appropriate antigen.

antidiuretic hormone (ADH) a hormone produced by the *hypothalamus* that passes to the posterior *pituitary gland* from where it is secreted. ADH reduces the volume of water in urine by increasing water reabsorption in the kidneys.

antigen a molecule that triggers an immune response by *lymphocytes*. ATP see adenosine triphosphate.

ATP (adenosine triphosphate) *nucleotide* found in all living organisms, which is produced during respiration and is important in the transfer of energy.

autonomic nervous system part of the nervous system, controlling the muscles and glands, that is not under voluntary control.

autosome a chromosome which is not a sex chromosome.

axon a process extending from a *neurone* that conducts *action potentials* away from the cell body.

B

biodiversity the range and variety of genes, species and habitats within a particular region.

biomass the total mass of living material, normally measured in a specific area over a given period of time.

biotic an ecological factor that makes up part of the living environment of an organism. Examples include food availability, competition and predation. See also *abiotic*.

biosensor a device that uses biological molecules to measure the level of certain chemicals.

C

carcinogen a chemical, a form of radiation, or other agent that causes *cancer*.

cardiac output the total volume of blood that the heart can pump each minute. It is calculated as the volume of blood pumped at each beat (*stroke volume*) multiplied by the number of heart beats per minute (heart rate).

carrier molecule (carrier protein) a protein on the surface of a cell that helps to transport molecules and ions across plasma membranes.

climax community the organisms that make up the final stage of ecological succession.

clone a group of genetically identical cells or organisms formed from a single parent as the result of asexual reproduction or by artificial means.

codon a sequence of three adjacent *nucleotides* in mRNA that codes for one amino acid.

community all the living organisms present in an *ecosystem* at a given time.

consumer any organism that obtains energy by 'eating' another. Organisms feeding on plants are known as primary consumers and organisms feeding on primary consumers are known as secondary consumers. See also *producer*.

cuticle exposed non-cellular outer layer of certain animals and the leaves of plants. It is waxy and impermeable to water. It therefore helps to reduce water loss.

D

deciduous term applied to plants that shed all their leaves together at one season.

denaturation permanent changes due to the unravelling of the three-dimensional structure of a protein as a result of factors such as changes in temperature or pH.

depolarisation temporary reversal of charges on the cell-surface membrane of a *neurone* that takes place when a nerve impulse is transmitted.

diffusion the movement of molecules or ions from a region where they are in high concentration to one where their concentration is lower.

diploid a term applied to cells in which the nucleus contains two sets of *chromosomes*. See also *haploid*.

dominant allele a term applied to an allele that is always expressed in the phenotype of an organism. See also recessive allele.

E

ecological niche describes how an organism fits into its environment. It describes what a species is like, where it occurs, how it behaves, its interactions with other species and how it responds to its environment.

ecosystem all the living and nonliving components of a particular area.

effector an organ that responds to stimulation by a nerve impulse resulting in a change or response.

electron negatively charged subatomic particle that orbits the positively charged nucleus of all atoms.

enzyme a protein or RNA that acts as a catalyst and so alters the speed of a biochemical reaction.

eukaryotic cell a cell that has a membrane-bound nucleus and *chromosomes*. The cell also possesses a variety of other membranous organelles, such as mitochondria and endoplasmic reticulum. See also *prokaryotic cell*.

F

facilitated diffusion diffusion involving the presence of protein *carrier molecules* to allow the passive movement of substances across plasma membranes.

G

gamete reproductive (sex) cell that fuses with another gamete during fertilisation.

gene section of DNA on a *chromosome* coding for one or more polypeptides.

gene pool the total number of *alleles* in a particular population at a specific time.

gene marker a section of DNA that is used to indicate the location of a *gene* or other section of DNA.

gene mutation a change to one or more *nucleotide* bases in DNA resulting in a change in *genotype* which may be inherited.

generator potential *depolarisation* of the membrane of a receptor cell as a result of a stimulus.

glucagon a hormone produced by α cells of the *islets of Langerhans* in the pancreas that increases blood glucose levels by initiating the breakdown of glycogen to glucose.

glycolysis first part of cellular respiration in which glucose is broken down anaerobically in the cytoplasm to two molecules of pyruvate.

glycoprotein substance made up of a carbohydrate molecule and a protein molecule. Parts of cell surface membrane and certain hormones are glycoproteins.

H

habitat the place where an organism normally lives and which is characterised by physical conditions and the types of other organisms present.

haploid term referring to cells that contain only a single copy of each *chromosome*, e.g. the sex cells (*gametes*).

heterozygous condition in which the *alleles* of a particular gene are different.

homeostasis the maintenance of a more or less constant internal environment.

homologous chromosomes a pair of *chromosomes*, one maternal and one paternal, that have the same gene *loci* and therefore determine the same features. They are not necessarily identical, however, as individual *alleles* of the same *gene* may vary, e.g. one chromosome may carry the allele for blue eyes, the other the allele for brown eyes. Homologous chromosomes are capable of pairing during *meiosis*.

human genome project international scientific project to map the entire sequence of all the base pairs of the genes in a single human cell.

hydrogen bond chemical bond formed between the positive charge on a hydrogen atom and the negative charge on another atom of an adjacent molecule, e.g. between the hydrogen atom of one water molecule and the oxygen atom of an adjacent water molecule.

hydrolysis the breaking down of large molecules into smaller ones by the addition of water molecules. See also *condensation*.

hyperthermia a condition that results from the core body temperature rising above normal.

hypothalamus region of the brain adjoining the pituitary gland that acts as the control centre for the *autonomic nervous system* and regulates body temperature and fluid balance.

hypothermia a condition that results from the core body temperature falling below normal.

I

insulin a hormone, produced by the α cells of the *islets of Langerhans* in the pancreas, which decreases blood glucose levels by, amongst other things, increasing the rate of conversion of glucose to glycogen.

intraspecific competition competition between organisms of the same species.

ion an atom or group of atoms that has lost or gained one or more *electrons*. Ions therefore have either a positive or negative charge.

isotope variations of a chemical element that have the same number of protons and *electrons* but different numbers of neutrons. While their chemical properties are similar they differ in mass. One example is carbon which has a relative atomic mass of 12 and an isotope with a relative atomic mass of 14.

K

Krebs cycle series of aerobic biochemical reactions in the matrix of the mitochondria of most *eukaryotic cells* by which energy is obtained through the oxidation of acetylcoenzyme A produced from the breakdown of glucose.

L

ligament a tough, fibrous connective tissue, rich in collagen, that joins bone to bone. See also *tendon*.

limiting factor a variable that limits the rate of a chemical reaction.

loop of Henle the portion of the *nephron* that forms a hairpin loop that extends into the medulla of the kidney. It has a role in the reabsorption of water.

M

meiosis the type of nuclear division in which the number of *chromosomes* is halved. See also *mitosis*.

mesophyll tissue found between the two layers of epidermis in a plant leaf comprising an upper layer of *palisade cells* and a lower layer of spongy cells.

mitosis the type of nuclear division in which the daughter cells have the same number of *chromosomes* as the parent cell. See also *meiosis*.

mutagen any agent that induces a *mutation*.

mutation a sudden change in the amount or the arrangement of the genetic material in the cell.

mutualism a nutritional relationship between two species in which both gain some advantage.

N

NADP (nicotinamide adenine dinucleotide phosphate) a molecule that carries *electrons* produced in the *light-dependent reaction* of photosynthesis.

nephron basic functional unit of the mammalian kidney responsible for the formation of urine.

neurone a nerve cell, comprising a cell body, *axon* and *dendrites*, which is adapted to conduct *action potentials*.

neurotransmitter one of a number of chemicals that are involved in communication between adjacent neurones or between nerve cells and muscles. Two important examples are *acetylcholine* and *noradrenaline*.

niche see *ecological niche*.

nucleotides complex chemicals made up of an organic base, a sugar and a phosphate. They are the basic units of which the nucleic acids DNA and RNA are made.

O

osmosis the passage of water from a region of high *water potential* to a region where its *water potential* is lower, through a selectively permeable membrane.

oxidation chemical reaction involving the loss of *electrons*.

oxidation-reduction a chemical reaction in which electrons are transferred from one substance to another substance. The substance losing electrons is oxidised and the substance gaining the electrons is reduced.

P

pathogen any microorganism that causes disease.

phagocytosis mechanism by which cells engulf particles to form a vesicle or a vacuole.

phenotype the characteristics of an organism, often visible, resulting from both its *genotype* and the effects of the environment.

phospholipid triglyceride in which one of the three fatty acid molecules is replaced by a phosphate molecule. Phospholipids are important in the structure and functioning of plasma membranes.

Pituitary gland gland of the endocrine (hormone) system situated at the base of the brain. It has two parts, anterior and posterior.

plasmid a small circular piece of DNA found in bacterial cells.

population a group of individuals of the same species that occupy the same *habitat* at the same time.

producer an organism that synthesises organic molecules from simple inorganic ones such as carbon dioxide and water. Most producers are photosynthetic and form the first trophic level in a food chain. See also *consumer*.

prokaryotic cell a cell of an organism belonging to the kingdom Prokaryotae that is characterised by lacking a nucleus and membrane-bound organelles. Examples include bacteria. See also *eukaryotic cell*.

proton positively charged sub-atomic particle found in the nucleus of an atom. See also *electron*.

R

receptor a cell adapted to detect changes in the environment.

recessive allele the condition in which the effect of an allele is apparent in the *phenotype* of a *diploid* organism **only** in the presence of another identical allele. See also *dominant allele*.

reduction chemical process involving the gain of *electrons*.

renal capsule the cup shaped portion of the start of the nephron that encloses the glomerulus.

repolarisation return to the *resting potential* in the axon of a neurone after an *action potential*.

restriction endonucleases a group of enzymes that cut DNA molecules at a specific sequence of bases called a recognition sequence.

S

saltatory conduction propagation of a nerve impulse along a *myelinated dendron* or *axon* in which the action *potential* jumps from one *node of Ranvier* to another.

saprobiontic microorganism also known as a saprophyte, this is an organism that obtains its food from the dead or decaying remains of other organisms.

sarcomere a section of myofibril between two Z-lines that forms the basic structural unit of *skeletal muscle*.

selection process that results in the best-adapted individuals in a *population* surviving to breed and so pass their favourable *alleles* to the next generation.

selection pressure the environmental force altering the frequency of alleles in a *population*.

sinoatrial node (SAN) an area of heart muscle in the right atrium that controls and coordinates the contraction of the heart. Also known as the pacemaker.

smooth muscle also known as involuntary or unstriated muscle, smooth muscle is found in the alimentary canal and the walls of blood vessels. Its contraction is not under conscious control. See also *skeletal muscle*.

sodium–potassium pump protein channels across cell-surface membranes that use ATP to move sodium *ions* out of the cell in exchange for potassium ions that move in.

species a group of similar organisms that can breed together to produce fertile offspring.

stoma (plural stomata) a pore, mostly found in the lower epidermis of a leaf, through which gases diffuse in and out of the leaf.

stroma matrix of a chloroplast where the *light-independent reaction* of photosynthesis takes place.

substrate-level phosphorylation the formation of ATP by the direct transfer of a phosphate group from a reactive intermediate to ADP.

synapse a junction between *neurones* in which they do not touch but have a narrow gap, the synaptic cleft, across which a *neurotransmitter* can pass.

T

tendon tough, flexible, but inelastic, connective tissue that joins muscle to bone. See also *ligament*.

thylakoid series of flattened membranous sacs in a chloroplast that contain chlorophyll and the associated molecules needed for the *light-dependent* reaction of photosynthesis.

tissue fluid fluid that surrounds the cells of the body. Its composition is similar to that of blood plasma except that it lacks proteins. It supplies nutrients to the cells and removes waste products.

transcription formation of messenger RNA molecules from the DNA that makes up a particular *gene*. It is the first stage of protein synthesis.

transducer cells cells that convert a non-electrical signal, such as light or sound, into an electrical (nervous) signal and vice versa.

transpiration evaporation of water from a plant.

tumour a swelling in an organism that is made up of cells that continue to divide in an abnormal way.

tumour suppressor gene a gene that maintains normal rates of cell division and so prevents the development of tumours.

V

vasoconstriction narrowing of the internal diameter of blood vessels. See also *vasodilation*.

vasodilation widening of the internal diameter of blood vessels. See also *vasoconstriction*.

vector a carrier. The term may refer to something such as a *plasmid*, which carries DNA into a cell, or to an organism that carries a *parasite* to its host.

voltage-gated channel protein channel across a cell-surface membrane that opens and closes according to changes in the electrical potential across the membrane.

W

water potential the pressure created by water molecules. It is the measure of the extent to which a solution gives out water. The greater the number of water molecules present, the higher (less negative) the water potential. Pure water has a water potential of zero.

X

xylem vessels dead, hollow, elongated tubes, with lignified side walls and no end walls, that transport water in most plants.

Index

Acknowledgements

The authors wish to thank Graham Read for his invaluable help with the manuscript, Mitch Fitton for her meticulous editing, James Penny for the mathematical aspects, Louise Garcia for her work on the practice questions, Ellena Bale and Simon Ditchfield with their help on the practical elements and, of course, Alison Schrecker and Amy Johnson from OUP for their support, hard work and encouragement.

Cover: Blend Images Photography/Veer; **p2-3**: Maks Narodenko/Shutterstock; **p5**: Dr Jeremy Burgess/Science Photo Library; **p6**: Biophoto Associates/Science Photo Library; **p9**: Dr Kenneth R Miller/Science Photo Library; **p14**: Martyn F Chillmaid/Science Photo Library; **p25**: ISM/Science Photo Library; **p29**: Eye of Science/Science Photo Library; **p30**: David Levenson/Alamy; **p35**: Suzanne L & Joseph T Collins/Science Photo Library; **p38**: Stuart Wilson/Science Photo Library; **p41**: Martin Dohrn/Science Photo Library; **p44**: Hugh Spencer/Science Photo Library; **p45**: Chris Gomersall/Alamy; **p47**: Nigel Cattlin/Alamy; **p49(T)**: Vladimir Sazonov/Shutterstock; **p49(B)**: Djgis/Shutterstock; **p50**: Geogphotos/Alamy; **p55**: Maks Narodenko/Shutterstock; **p60-61**: Martynowi. Cz/Shutterstock; **p62**: Stefan Sollfors/Alamy; **p63(B)**: Jerome Wexler/Science Photo Library; **p63(T)**: Chris Martin Bahr/Science Photo Library; **p65**: Martin Shields/Science Photo Library; **p71**: Martyn F Chillmaid/Science Photo Library; **p73**: Anatomical Travelogue/Science Photo Library; **p75**: Omikron/Getty Images; **p77**: Medi-Mation/Getty Images; **p84**: Jean-Claude Revy, ISM/Science Photo Library; **p85**: Steve Gschmeissner/Science Photo Library; **p90**: CNRI/Science Photo Library; **p93**: Steve Gschmeissner/Science Photo Library; **p97**: CNRI/Science Photo Library; **p101**: Steve Percival/Science Photo Library; **p105**: Astrid & Hanns-Frieder Michler/Science Photo Library; **p106**: Biology Media/Science Photo Library; **p111**: Ictor/iStockphoto; **p115(T)**: Digital Vision/Getty Images; **p115(B)**: Bigroloimages/Shutterstock; **p116**: Morganlefaye/iStockphoto; **p117(B)**: Stanislav Fosenbauer/Shutterstock; **p117(T)**: Plinney/iStockphoto; **p121**: Christopher Leggett/Alamy; **p123**: Astrid & Hanns-Frieder Michler/Science Photo Library; **p125**: JC Revy, ISM/Science Photo Library; **p127**: Image Point Fr/Shutterstock; **p132(T)**: Astrid & Hanns-Frieder Michler/Science Photo Library; **p132(C)**: Science Vu, Visuals Unlimited/Science Photo Library; **p132(B)**: Prof P Motta/Dept of Anatomy/University "La Sapienza", Rome/Science Photo Library; **p137**: Thomas Deerinck, NCMIR/Science Photo Library; **p147**: Martynowi.Cz/Shutterstock; **p152-153**: Calvin Chan/Shutterstock; **p156**: Mark Burnett/Science Photo Library; **p165**: Cecil36/iStockphoto; **p168**: Andrew Roland/Shutterstock; **p169**: Biophoto Associates/Science Photo Library; **p176(L)**: Faslooff/iStockphoto; **p176(R)**: Eric Isselee/Shutterstock; **p181(T)**: Guy J Sagi/Shutterstock; **p181(C)**: Schankz/Shutterstock; **p181(B)**: Isarescheewin/Shutterstock; **p186**: Michael W Tweedie/Science Photo Library; **p188**: Jeanne White/Science Photo Library; **p195**: Claude Nuridsany & Marie Perennou/Science Photo Library; **p202**: Tsz01/iStockphoto; **p203**: Nrt/Shutterstock; **p204(T)**: Soopysue/iStockphoto; **p204(B)**: Juniors Bildarchiv Gmbh/Alamy; **p205**: Chris Gomersall/Alamy; **p206**: Mikenorton/Shutterstock; **p208**: Adisa/iStockphoto; **p212(T)**: Rickochet/iStockphoto; **p212(B)**: Whiteway/iStockphoto; **p215**: Tom Brakefield/Getty Images; **p218**: Alamy; **p221(T)**: Raywoo/Shutterstock; **p221(B)**: Erni/Shutterstock; **p222**: Raywoo/Shutterstock; **p223(T)**: Simon Fraser/Science Photo Library; **p223(C)**: Simon Fraser/Science Photo Library; **p223(B)**: Simon Fraser/Science Photo Library; **p224**: Daniel J Rao/Shutterstock; **p225(L)**: Peter Schwarz/Shutterstock; **p225(R)**: Duncan Shaw/Science Photo Library; **p229**: Calvin Chan/Shutterstock; **p234-235**: Science Photo/Shutterstock; **p237**: Cs333/Shutterstock; **p239**: Alamy; **p242**: Dr Yorgos Nikas/Science Photo Library; **p244**: Rosenfeld Images Ltd/Science Photo Library; **p249**: Yuri Arcurs/Shutterstock; **p260**: Gcpics/Shutterstock; **p266**: Dr Gopal Murti/Science Photo Library; **p272**: Dr Gopal Murti/Science Photo Library; **p275**: F1Online Digitale Bildagentur Gmbh/Alamy; **p276**: Victorio Castellani/Alamy; **p284**: Alamy; **p285**: Janine Wiedel Photolibrary/Alamy; **p286**: Stockfolio/Alamy; **p289**: David Parker/Science Photo Library; **p295**: Science Photo/Shutterstock; **p340**: Eye of Science/Science Photo Library;

Artwork by Q2A Media